Poly(vinylpyridine)

Poly(vinylpyridine) is one of the most important polymers, with wide-ranging applications in organic chemistry, polymer chemistry, medicine, and industry. This book contains more than 1,650 references about "poly(vinylpyridine)", which have been reviewed and analyzed, and it includes complete details of reactions, syntheses, and applications related to this polymer. Throughout this book, reactions and their conditions are explained in detail using reaction schemes, providing readers with immediate comprehension of each reaction's purpose. On the whole, this book is a very simple and complete resource on poly(vinylpyridine), which allows audiences to access information on this polymer quickly and easily. Students, academicians, and researchers in the fields of chemistry and polymers, as well as researchers active in the polymer industry, will find this book a valuable reference.

Kaveh Parvanak Boroujeni earned his B.Sc. degree in Applied Chemistry from Babolsar University (1998) and his M.Sc. (2000) and Ph.D. degrees (2004) in Polymer Chemistry from Shiraz University, Iran, under the direction of Professor Bahman Tamami. He currently serves as an Associate Professor of Polymer Chemistry in the Department of Chemistry at Shahrekord University in Iran, working on the synthesis and applications of functional polymers, nanocatalysts, metal–organic frameworks, heat-resistant polymers, photocatalysts, and carbon nitride- and boron nitride-based materials.

Poly(vinylpyridine)

Structure, Properties, Synthesis, and Applications

Kaveh Parvanak Boroujeni

CRC Press
Taylor & Francis Group
Boca Raton London New York

CRC Press is an imprint of the
Taylor & Francis Group, an **informa** business

Designed cover image: Kaveh Parvanak Boroujeni

First edition published 2026
by CRC Press
2385 NW Executive Center Drive, Suite 320, Boca Raton FL 33431

and by CRC Press
4 Park Square, Milton Park, Abingdon, Oxon, OX14 4RN

CRC Press is an imprint of Taylor & Francis Group, LLC

ISBN: 978-1-041-10638-8 (hbk)
ISBN: 978-1-041-10639-5 (pbk)
ISBN: 978-1-003-65610-4 (ebk)

DOI: 10.1201/9781003656104

Typeset in Times
by codeMantra

Contents

Preface

Poly(vinylpyridine) (PVPy) has garnered increasing interest across diverse research disciplines, including academic investigations, materials science and technology, and medical sciences. Annually, numerous scientific sites report an ever-growing tally of studies on PVPy—evidence that its potential continues to expand in areas, from catalysis and separation membranes to drug-delivery vehicles and smart nanostructures. However, the extant literature on PVPy is primarily confined to journal articles, patents, and encyclopedia entries that lack comprehensive coverage. To the best of our knowledge, no book on PVPy currently exists, highlighting a clear gap in both library collections and research resources. The publication of this PVPy book would therefore be timely, attractive, and invaluable.

This book is organized into four chapters. In Chapter 1, I have consolidated all known properties, characteristics, and applications of PVPy through a systematic survey of approximately 1200 scientific articles. Each section begins with a brief historical overview and traces the development of the topic to the present, thereby granting readers rapid and easy access to concepts and obviating the need for difficult and time-consuming searches among the literature. Chapter 2 presents a brief overview of functional polymers—including their importance, classification, synthesis methods, and advantages over non-polymeric counterparts—and establishes the conceptual basis necessary to better understand the following chapters. Chapters 3 and 4 introduce PVPy as a support for catalysts and reagents, citing about 450 references. Mostly, polymer reactants or catalysts are detailed using schematic representations of reactions along with corresponding reaction conditions and synthetic routes. This allows readers to understand each transformation quickly, making the book stand out.

Given the above, I anticipate that this book—representing the first up-to-date comprehensive and critical volume on PVPy, covering more than 70 years of published scientific articles—will serve both as a widely cited reference for academic and industrial researchers and as an educational asset for students across all levels.

Kaveh Parvanak Boroujeni, 2025

Acknowledgments

I dedicate this book to my beloved wife, *Faranak*, whose unwavering support made this book possible, and to my parents for their constant encouragement.

I thank all the faculty members of Shiraz and Babolsar Universities—especially my thesis supervisor, Professor Bahman Tamami, and my friend, Dr. Abdul Hamid Fadavi—for their valuable input. My appreciation also goes to the editorial and production teams at CRC Press and Taylor & Francis for their excellent preparation, editing, and professionalism, as well as to the Research Council of Shahrekord University for its support. Finally, I am grateful to all readers and warmly welcome any criticisms or suggestions regarding this volume.

List of Abbreviations

Abbreviation	Full Term
AAc	Acrylic acid
AAM	Acrylamide
AN	Acrylonitrile
ATRP	Atom transfer radical polymerization
AIBN	Azobis(isobutyronitrile)
HBS	Benzenesulfonic acid
BPO	Benzoylperoxide
BDTA	Benzyl dithiobenzoate
BCP	Block copolymer
BET	Brunauer, Emmett, and Teller
CS	Chitosan
CSA	Camphorsulfonic acid
CTA	Chain transfer agent
CT	Charge-transfer
CRP	Controlled radical polymerization
cmc	Critical micelle concentration
CDB	Cumyl dithiobenzoate
DP	Degree of polymerization
DCE	Dichloroethane
DMSO	Dimethylsulfoxide
DPE	Diphenylethylene
Boc	Di-*tert*-butyl dicarbonate
DVB	Divinylbenzene
DTG	Derivative thermogravimetry
DSC	Differential scanning calorimetry
DBB	1,4-Dibromobutane
DSSCs	Dye-sensitized solar cells
Eg	Energy gap
EEGE	Ethoxy ethyl glycidyl ether
EGDMA	Ethylene glycol dimethacrylate
EG	Ethylene glycol
EO	Ethylene oxide
χ**AB**	Flory–Huggins interaction parameter
T_g	Glass transition temperature
GCE	Glassy carbon electrode
GOX	Glucose oxidase
GO	Graphene oxide
GTP	Group transfer polymerizations
HSAB	Hard-soft acid-base
HDPE	High-density polyethylene
ICD	Induced circular dichroism

Abbreviation	Full Term
ILs	Iionic liquids
IEP	Isoelectric point
IUPAC	International Union of Pure and Applied Chemistry
LbL	Layer by layer
LEDs	Light-emitting diodes
LOD	Limit of detection
LCST	Lower critical solution temperature
MAA	Methacrylic acid
MSA	Methane sulfonic acid
MA	Methyl acrylate
MMA	Methyl methacrylate
2-M-5-V-Py	2-Methyl-5-vinylpyridine
2-M-6-V-Py	2-Methyl-6-vinylpyridine
MWCNTs	Multiwalled carbon nanotubes
NASI	*N*-(Acryloyloxy)succinimide
DMF	*N*,*N*-Dimethylformamide
MBAAM	*N*,*N*′-Methylenebisacrylamide
NaMnt	Na-Montmorillonite
NPs	Nanoparticles
EPHP	*N*-ethylpiperidine hypophosphite
HEP	*n*-Heptane
NMRP	Nitroxide-mediated radical polymerization
NMP	*N*-methylpyrrolidone
NLO	Nonlinear optics
NVP	*N*-Vinyl pyrrolidone
NVCz	*N*-Vinylcarbazole
OLEDs	Organic light-emitting diodes
PDP	Pentadecylphenol
PEO	Poly(ethylene oxide)
PMAA	Poly(methacrylic acid)
PVP	Poly(*N*-vinylpyrolidone)
P_2VPyNO	Poly(2-vinylpyridine *N*-oxide)
P_2VPy	Poly(2-vinylpyridine)
P_3VPy	Poly(3-vinylpyridine)
$P_4VPy\text{-}BuSO_3$	Poly(4-vinylpyridine-butanesultone)
P_4VPyNO	Poly(4-vinylpyridine *N*-oxide)
P_4VPy	Poly(4-vinylpyridine)
PBI	Poly(benzimidazole)
PBzA	Poly(benzyl acrylate)
PBuMA	Poly(butyl methacrylate)
PDVB	Poly(divinylbenzene)
PDA	Polydopamine
PEG	Poly(ethylene glycol)
PET	Poly(ethylene terephthalate)
PCL	Poly(ε-caprolactone)
PILs	Poly(ionic liquids)

Abbreviation	Full Term
PI	Poly(isoprene)
PMMA	Poly(methyl methacrylate)
PDMAAM	Poly-(*N*,*N*-dimethylacrylamide)
PNIPAM	Poly(*N*-isopropylacrylamide)
PtBA	Poly(*tert*-butyl acrylate)
PVA	Poly(vinyl alcohol)
PVC	Poly(vinyl chloride)
PVDF	Poly(vinylidene fluoride)
PVPh	Poly(vinyl phenol)
P(VDF)	Poly(vinylidene fluoride)
PVP	Poly(*N*-vinylpyrrolidone)
PAA	Polyacrylic acid
PAAM	Polyacrylamide
PAN	Polyacrylonitrile
PDI	Polydispersity index
PEL	Polyelectrolytes
PEI	Polyethyleneimine
PISA	Polymerization-induced self-assembly
PS	Polystyrene
PTFE	Polytetrafluoroethylene
PU	Polyurethane
PVIM	Polyvinylimidazoles
PZs	Polyzwitterions
PVA	Poly(vinyl alcohol)
PCC	Pyridinium chlorochromate
PCM	Phase change material
PTCs	Phase transfer catalysts
PL	Photoluminescence
PDC	Pyridinium dichromate
QDs	Quantum dots
DQ	Degree of quaternization
REM-GTP	Rare earth metal-mediated group transfer polymerizations
ROS	Reactive oxygen species
RAFT	Reversible addition-fragmentation chain transfer
RDRP	Reversible deactivation radical polymerization
SWCNTs	Single-walled carbon nanotubes
SDS	Sodium dodecylsulfate
SFEP	Surfactant-free emulsion polymerization
SVA	Solvent vapor annealing
sPS	Syndiotactic polystyrene
TCNQ	Tetracyanoquinodimethane
THF	Tetrahydrofuran
TGA,	Thermogravimetric analysis
TPU	Thermoplastic polyurethane
TOL	Toluene
TMS	Trimethylsilyl

Abbreviation	Full Term
TTMSS	Tris(trimethylsilyl)silane
TEMPO	2,2,6,6-Tetramethylpiperidine-*N*-oxyl
VIM	1-Vinylimidazole)
2-VPy	2-Vinylpyridine
3-VPy	3-Vinylpyridine
bipy	4,4′-Bipyridine
4-VPy	4-Vinylpyridine
SAXS	Small-angle X-ray scattering measurements

1 Poly(vinylpyridine)
Structure, Properties, Synthesis, and Applications

1.1 STRUCTURE AND PHYSICAL PROPERTIES

Pyridine (Py) is a basic heterocyclic organic compound and is derived from two components: "pyr", which signifies "fire" and "idine", which denotes "aromatic base". It possesses an estimated dipole moment of 3.61 D and is fully soluble in water. Its derivatives exhibit a wide range of biomedical properties such as antitumor, analgesic, anti-inflammatory, antiviral, antibacterial, antifungal, antihistaminic, and anti-diabetic effects, along with applications in bio-imaging [1]. Also, they are present in the structure of numerous products, such as vitamins, pharmaceuticals, insecticides, food flavorings, paints, adhesives, water-resistant textiles, and rubber goods [2]. The pyridine derivatives are illustrated in Scheme 1.1.

Vinylpyridine monomers (2-VPy, 3-VPy, 4-VPy, 2-M-5-V-Py, and 2-M-6-V-Py) are one of the most important derivatives of pyridine. Due to their strong propensity to polymerize, stabilization is necessary for their storage and transportation, with common stabilizers including phenols, amines, thiols, and polysulfides. 2-VPy, 4-VPy, 2-M-5-V-Py, and 2-M-6-V-Py are the most prevalent available compounds, but the 3-VPy isomer, compared with 4-Vpy and 2-VPy, is not easily accessible. Vinylpyridines cause significant skin irritation, and exposure to either the vapor or liquid typically results in inflammation or blistering. Additionally, they exhibit considerable toxicity when inhaled. Among vinylpyridines, 2-VPy and 4-VPy have been investigated the most. 2-VPy and 4-VPy are colorless liquids, although samples are

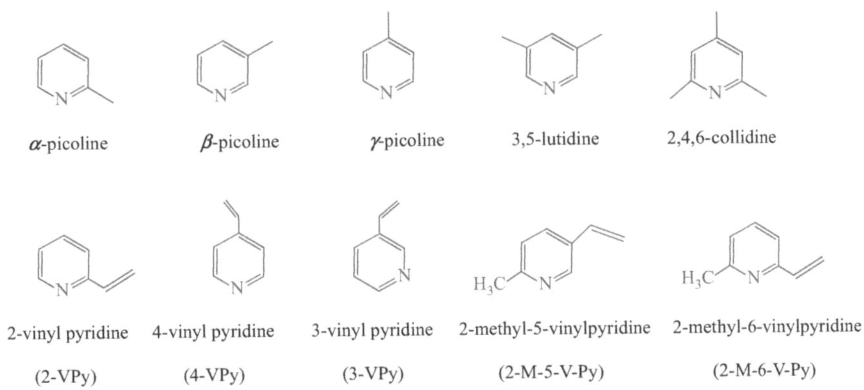

SCHEME 1.1 The molecular structure of some pyridine derivatives.

DOI: 10.1201/9781003656104-1

often brown. Some physical properties of these monomers are listed in Table 1.1 and the rest are available in handbooks [3–7]. pK_a values of 2-VPy, 3-VPy, and 4-VPy were found to be approximately 4.98, 5.63, and 5.5, respectively, which are comparable to pK_a of pyridine (5.22) [7,8].

Industrially, the synthetic method for 2-VPy and 4-VPy is the reaction of 2- or 4-methylpyridine with aqueous solution of formaldehyde, followed by dehydration of the obtained alcohol (Scheme 1.2) [7,9]. Also, the reaction of 2- or 4-methylpyridine with formaldehyde results in the formation of pyridylethanols, which upon dehydration yield 2-VPy and 4-VPy [9]. Dehydration of 3-hydroxyethyl pyridine through treatment with thionyl chloride, followed by the reaction with KOH, has been reported for the preparation of 3-VPy [10]. Dehydrohalogenation of the corresponding halogenoethyl derivatives using a base catalyst, is another basic and popular synthetic approach to the libraries in the preparation of vinylpyridine isomers [9,11].

Vinylpyridines have the ability to undergo polymerization via double bond to afford the corresponding polymers. 2-VPy, 3-VPy, and 4-VPy are polymerized into poly(2-vinylpyridine) (or poly(2-pyridylethylene)), poly(3-vinylpyridine) (or poly(3-pyridylethylene)), and poly(4-vinylpyridine) (or poly(4-pyridylethylene)), respectively, (Scheme 1.3a). Poly(2-vinylpyridine), poly(3-vinylpyridine), and poly(4-vinylpyridine) were abbreviated hereinafter to P_2VPy, P_3VPy, and P_4VPy, respectively. Usually, (PVPy)s exhibit significant resistance to electrophilic aromatic substitution owing to the electron-withdrawing effect of the ring nitrogen atom. They have structural amphilicity and chemical reactivity compared with common surfactants and can be protonated by acids, or coordinated with metal ions through nitrogen atom in the pyridine ring. They are genuine industrial polymers, in contrast to polystyrene (denoted as PS hereafter), and their unit cost is significantly higher than that of PS. (PVPy)s have gained a progressively increasing interest in many fields due to its intriguing properties, including nonhygroscopic nature, ease of functionalization,

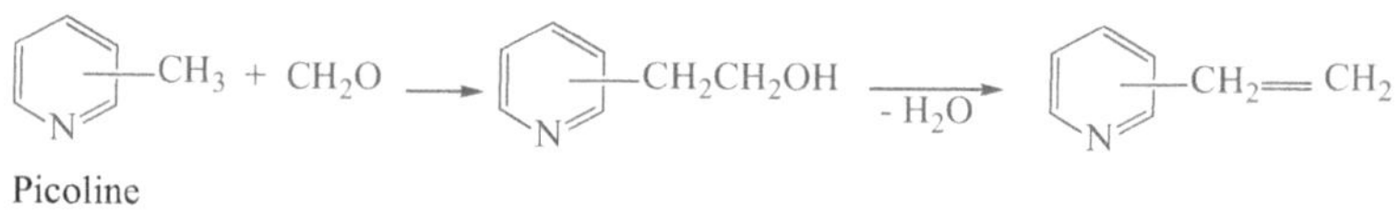

SCHEME 1.2 The industrial synthetic pathway for 2-VPy and 4-VPy isomers [7].

TABLE 1.1
Some Properties of Vinylpyridine Monomers [7]

	Value	
Property	**2-VPy**	**4-VPy**
Density at 20°C (g/cm^3)	0.977	0.988
Boiling point	70°C/4 kPa	65°C/2 kPa
	110°C/20 kPa	120°C/4 kPa
Solubility in water (g/L 20°C)	27.5	29
pK_a	4.98	5.5

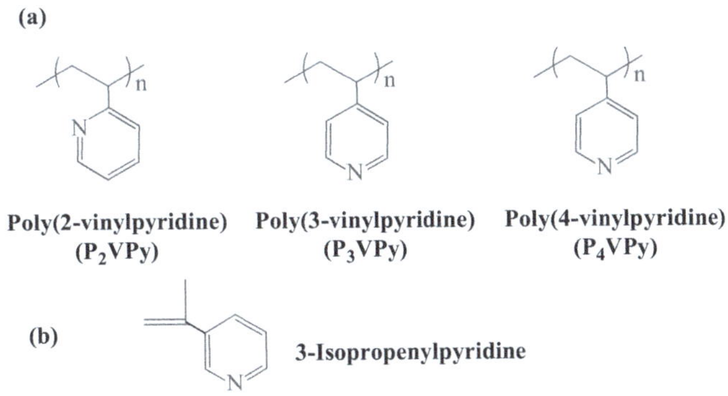

SCHEME 1.3 Molecular structures of (a) P_2VPy, P_3VPy, and P_4VPy; and (b) 3-isopropenyl pyridine.

high availability of functional groups, effective filtering capabilities, and ability to swell in a variety of organic solvents.

Numerous techniques have been established for the polymerization of vinylpyridines. Among these, free radical polymerization is one of the most prevalent methods [6]. This technique is favored for its straightforward operation; however, it often produces polymers characterized by a wide molecular weight distribution and restricted control over their microstructure. Over time, researchers have enhanced free radical methods by incorporating controlled radical polymerization (CRP) techniques to achieve polymers with more precise definitions, which are detailed in the following sections. Also, due to the presence of electron-withdrawing pyridine rings in vinylpyridines, their vinyl groups possess a degree of electrophilicity, rendering them appropriate for anionic polymerization when conducted under carefully chosen conditions [12]. Cationic polymerization of vinylpyridines have rarely been studied well, since the nitrogen atom within the pyridine ring possesses lone-pair electrons that function as a Lewis base, potentially disrupting the initiation process by deactivating cationic species. It is worth noting that while the literature extensively covers the polymerization of 2-VPy and 4-VPy, the reported studies on the polymerization of 3-VPy have very rarely been explored, probably owing to the high cost of this monomer [13]. However, 3-VPy is gradually polymerized at room temperature, and the polymerization rate increases with increasing temperature [9]. Polymerization and copolymerization of 3-isopropenylpyridine, which has a similar structure to vinylpyridine monomers (Scheme 1.3b), have also been investigated, and the results showed that it does not homopolymerize, but does copolymerize with methyl methacrylate, butadiene, and styrene (ST) [14]. A summary of some physical characteristics of P_2VPy and P_4VPy, collected from multiple literature sources, is presented in Table 1.2.

The positioning of the pyridinyl nitrogen of P_4VPy and P_2VPy causes these polymers exhibit different chemical properties. In P_4VPy, the nitrogen atom is located at the para position in relation to the attachment point of the vinyl group. This arrangement results in the nitrogen being situated further away from the polymer backbone, thereby enhancing the accessibility of its lone pair for interactions such as protonation or coordination with metal ions. Conversely, in P_2VPy, the nitrogen is found at

TABLE 1.2
Some Reported Values of Physical Properties of P_2VPy and P_4VPy and Some of Their Derivatives [7,15]

Property	Value	
	P_2VPy	**P_4VPy**
Specific gravity	0.96 g/cm^3	1.15 g/cm^3
T_g	84°C	142°C
	90°C [16][a]	–
	76°C [16][b]	–
	–	160°C [17][c]
TM_{ax}[d]	~415°C, a heating rate of 10°C/min N_2 [18]	~415°C a heating rate of 10°C/min N_2 [18]
Dielectric constant	~4 [16]	~3.5–4.5
pK_a	3.50, at 50% ionization in H_2O [19]	3.95, at 50% ionization in H_2O [19]
	–	3.25, in 45% ethanol/55% water [15]
	4.5 [20]	5.8 [21]
χ[e]	0.499 in EtOH [22]	2.1 ± 0.3 in EtOH [23]
χ_{AB}[f]	0.26 (P_2VPy-*b*-P_4VPy) [24]	
	0.11 (P_2VPy-*b*-PS) [25]	0.33 (P_4VPy-*b*-PS) [26]
Flash point	46.7°C	11°C
Refractive index	1.549 at 20°C	–
HSAB class[g]	Borderline	Borderline [27]
Electrical d.c. conductivity	–	$<10^{-14}$ S/cm [28]
	–	2.86×10^{-7} to 3.46×10^{-6} S/cm[h] [28]
Solubility	Soluble in $CHCl_3$, THF, and benzene; insoluble in water	Soluble in acetic acid and lower alcohols (3% solution or greater) and DMF; insoluble in water
	Soluble in water, trifluoroethanol, methanol, isopropanol, and mixed solvents of acetone:water and *p*-dioxane:water[i] [29]	Soluble in water, trifluoroethanol, and mixed solvents of acetone:water and *p*-dioxane:water[j] [29]

[a] Atactic polymer; [b]isotactic polymer; [c]P_4VPy (2% DVB); [d]maximum degradation temperature; [e]polymer–solvent interaction constant; [f]Flory–Huggins interaction parameter; [g]hard-soft acid-base principle; [h]polymers containing pyridine *N*-oxide units at room temperature; [i]poly(2-vinylpyridine-*N*-oxide); [j]poly(4-vinylpyridine-*N*-oxide).

the ortho position, in close proximity to the polymer backbone. This closeness can lead to steric hindrance, potentially limiting the nitrogen's capacity to engage freely with external reagents or protons. Consequently, P_2VPy typically demonstrates diminished reactivity in reactions that depend on the nucleophilic properties of the pyridinyl nitrogen. For example, only the P_2VPy can be made with different tacticities which exhibit unique physical and chemical properties which will be discussed in some detail in the following sections (Scheme 1.4) [30–32]. In contrast to P_4VPy, the complexation of P_2VPy possessing different tacticities with chiral acids leads to

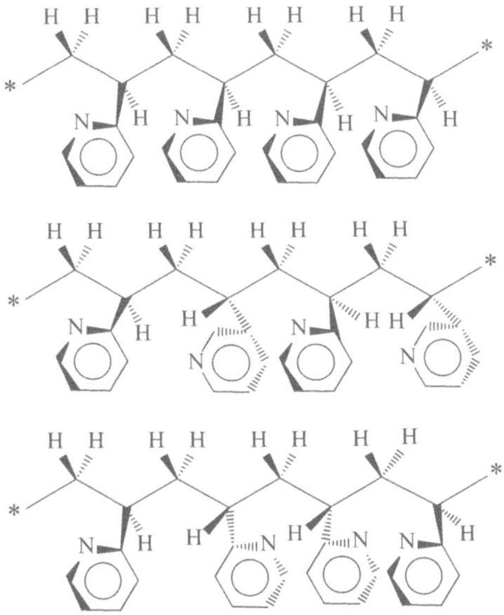

SCHEME 1.4 Different tacticities of P_2VPy.

the formation of helical chain conformation and eventually results the induced circular dichroism behavior of P_2VPy [33]. Also, it has been reported that the conductivity of solutions of the living P_2VPy in tetrahydrofuran (THF) is much less than those of living polymers of 4-VPy, maybe due to the possibility of interaction between the sodium counterion and the ring nitrogen atom for the case of 2-VPy [34]. Structural properties of P_2VPy by molecular dynamics simulations have been studied [35–37].

P_4VPy is more readily accessible to binding of host sterically demanding guests and nanoparticles (NPs), which also results in improved binding stability [38–40]. Scheme 1.5 shows several possible associations between the pyridinyl nitrogen with different attached species including hydrogen halides (as smallest objects) and NPs (as largest objects). In this latter, the association of larger guests to P_2VPy would be more difficult and complicated with the steric restrictions of the P_2VPy repeating unit. For instance, it has been reported that the gold NPs (Au NPs) are preferentially immobilized onto P_4VPy domains of P_4VPy-*b*-P_2VPy [38,41]. Thus, for immobilization of large molecular guests and NPs onto block copolymer (BCP) systems bearing P_4VPy or P_2VPy blocks, careful considerations are made about choosing which polymer to use depending upon the loading level and the binding ability desired [42]. In a recent study, based on the investigation of the effect of steric hindrance of a series of catalysts, bearing the similar pyridinic ring but having the different degree of steric crowding at their nitrogen active sites, on cycloaddition of CO_2 to epichlorohydrin, the order shown in Scheme 1.6 has been proposed [8] (details are discussed in Chapter 3).

Copolymers incorporating 2-VPy or 4-VPy with other vinylic monomers such as butadiene, styrene, esters of acrylic acid, and acrylonitrile are a flexible category of well-defined polymer architectures with tailorable properties, obtained by tuning the

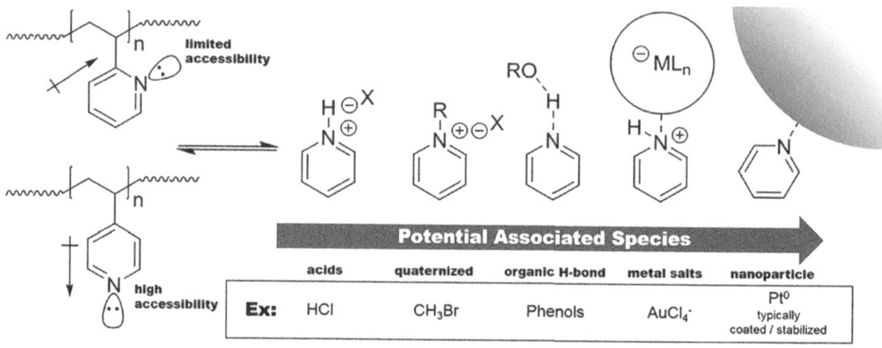

SCHEME 1.5 Configurations of P_2VPy and P_4VPy repeating units emphasize the heightened steric hindrance of the nitrogen lone pair in P_2VPy. A selection of potential associated species is anticipated to exhibit a stronger affinity for P_4VPy (compared to P_2VPy) as the size of the species increases [38]. (Reproduced with permission from Ref. [38]; Copyright 2019 American Chemical Society).

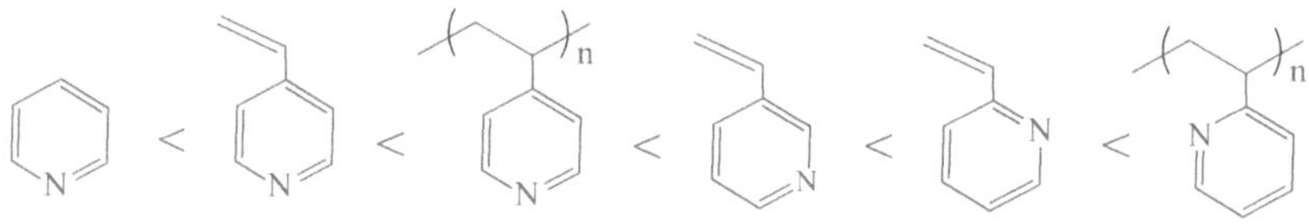

SCHEME 1.6 The order of steric hindrance at the pyridinic nitrogen atom of pyridine, vinylpyridine monomers, and PVPy polymers [8].

individual segments. The inclusion of pyridine groups introduces basic sites along the polymer chain that can be further modified or serve as coordination sites with different species, making these copolymers attractive in materials science and biological applications [43]. Reactivity ratios (*r*) for copolymerization of 2-VPy and 4-VPy with some monomers were given in Ref. [4], and some of them are summarized in Table 1.3. Among these, BCPs of vinylpyridine monomers with styrene, especially PS-*b*-P_2VPy and PS-*b*-P_4VPy, have been widely investigated, due to unique properties and applications, and will be introduced in the following discussions. At 50°C in benzene solvent using azobis(isobutyronitrile) (AIBN) as initiator, the r_1 and r_2 values for the copolymerization of 2-VPy with styrene are 1.1 and 0.55, respectively, and for 4-VPy with styrene are 0.75 and 0.55, respectively, where r_1 and r_2 represent the reactivity ratios of pyridine monomers and styrene, respectively [4]. Also, copolymerization of VPy monomers with alkynes such as phenylacetylene has been investigated [44].

In addition to linear PVPy, cross-linked PVPy is also available. It is a network type of polymer formed by the bonding of individual polymer chains. This interconnected architecture provides numerous benefits over linear polymers, including increased mechanical strength, superior solvent resistance, and enhanced thermal stability. Hydrogels are three-dimensional network structures composed of hydrophilic homo- or copolymers, which have attracted significant interest for their applications such as in contact lenses, soft tissue implants, and drug delivery systems [45,46].

TABLE 1.3
Reactivity Ratios for Copolymerization of 2-VPy and 4-VPy with Some Monomers at Different Temperatures [4,6,9]

Monomer 1	Monomer 2	r_1	r_2	Temperature (°C)
Styrene	2-VPy	1.1	0.55	50
	4-VPy	0.75	0.55	50
Methacrylic acid (MAA)	2-VPy	0.58	1.55	70
Methyl acrylate (MA)	2-VPy	0.2	2.03	60
	4-VPy	0.22	1.7	60
Acrylic acid butyl ester	2-VPy	2.5	0.097	60
	4-VPy	5.15 ± 0.09	0.46 ± 0.09	60
Methacrylic acid methyl ester	2-VPy	1.13	0.3	60
	4-VPy	1.77	0.18	60
Acrylic acid (AAc)	2-VPy	1.38 ± 0.04	0.44 ± 0.02	–
Ethyl acrylate	2-VPy	0.23 ± 0.05	0.19 ± 0.06	75
Acrylonitrile (AN)	2-VPy	4	0.083	
	4-VPy	0.41 ± 0.09	0.11 ± 0.005	60
Butadiene	2-VPy	0.90	0.94	50
n-Vinyl caprolactam	4-VPy	4.43	0.096	
Vinyl imidazole	4-VPy	1.07	0.21	60
Isoprene	2-VPy	0.47 ± 0.07	0.59 ± 0.05	60
Vinyl acetate	2-VPy	10	0.3	65

A common strategy for synthesizing cross-linked PVPy entails the copolymerization of vinylpyridine monomers with a bifunctional or multifunctional monomer (known as cross-linking agent) possessing two or more reactive vinyl groups. The most common cross-linking agents for PVPy are divinylbenzene (DVB), ethylene glycol dimethacrylate (EGDMA), *N,N'*-methylenebisacrylamide (MBAAM), and 2,6-divinylpyridine. Scheme 1.7 shows synthetic pathway cross-linked P_4VPy using DVB as a cross-linker. An alternative method for the preparation of cross-linked PVPy involves initially synthesizing a linear PVPy, followed by a cross-linking reaction with dihalides or epoxide compounds as cross-linkers. For example, by combination of two successive parallel reactions (polymerization and alkylation) on vinylpyridine monomers using a bis-haloalkane (e.g., *p*-xylylene dichloride), a rigid openwork structure of cross-linked networks, denoted as hypercross-linked polymer, has been synthesized (Scheme 1.8) [47]. They have an extensive internal surface and swell in both nonpolar and polar solvents, which make them good candidates for application in ion exchangers or catalysis. Cross-linked polymers can be produced not only by utilizing monomers with a functionality greater than 2, but also through the application of irradiation, vulcanization, or other chemical reactions. For example, a three-dimensional network structure of P_2VPy has been synthesized by application of gamma irradiation [22].

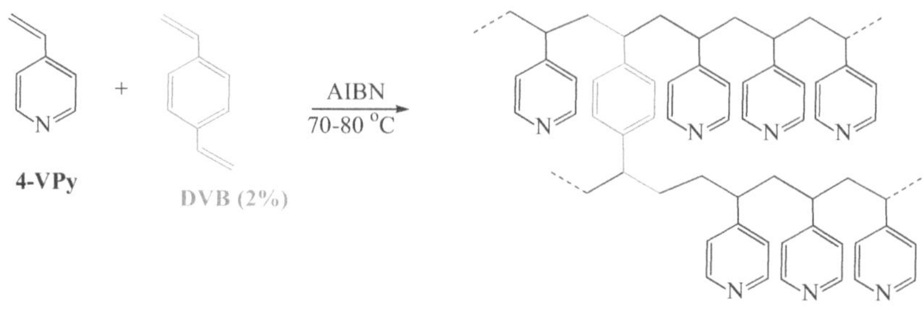

Poly(4-vinylpyridine-*co*-divinylbenzene) (P_4VPy (2% DVB))

SCHEME 1.7 The copolymerization of 4-VPy monomers with DVB [48,49].

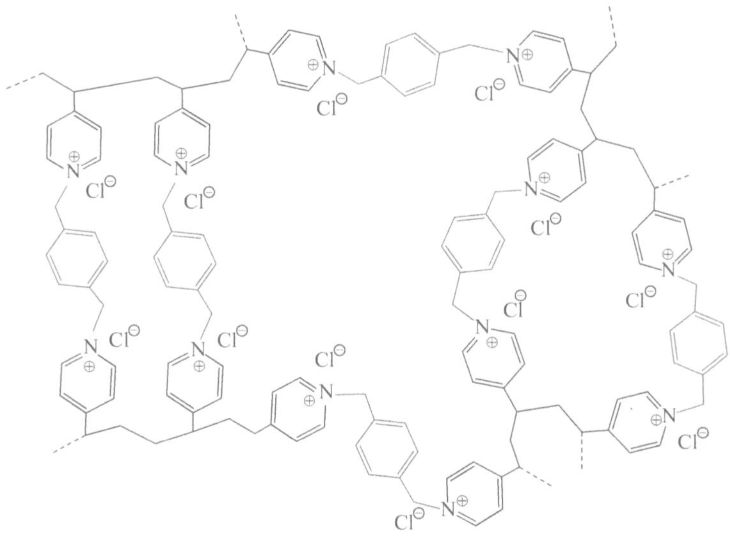

SCHEME 1.8 A rigid openwork structure of hypercross-linked P_4VPy [47].

Usually, a minimum level of cross-linking agent (at least 0.5%) causes the formation of insoluble polymer in organic solvents. When PVPy was used as a support for catalysts and reagents, it is about 2%–10% cross-linked which needs to an appropriate polymer swelling solvent wherein the reactants diffuse into the porous structure and access the active sites [48,49]. It is worth mentioning that cross-linking serves as an effective and robust approach for enhancing the stability of BCP nanoassemblies with well-defined structure and several stimuli-responsive characters [50]. Cross-linking of internal P_4VPy domains of poly(α-methylstryrene-*b*-4-VPy) (PMS-*b*-P_4VPy) using 1,4-dibromobutane (DBB) [51,52], cross-linking of internal P_2VPy domains of PMS-*b*-P_2VPy [53] using 1,4-diiodobutane, and in situ cross-linking of poly(methyl methacrylate-*b*-4-VPy) (PMMA-*b*-P_4VPy) particles using DVB [54] have been reported.

PVPy polymers exhibit geometrical parameters and conformational characteristics that closely resemble those of PS; thus, it can be expected that they possess nearly identical spectral characteristics and fluorescence properties [32,55]. Nevertheless, the two polymers demonstrate some differing behaviors in these aspects, which can be attributed to the heterocyclic nature of PVPy polymers, influenced by solvent–polymer interactions and other factors such as pH [56].

P_4VPy and P_2VPy appear to exhibit different pK_a values in their excited state compared to their ground state, similar to what was observed for pyridines [57]. In their photo-excited states, they readily undergo protonation. Investigations indicate that upon ultraviolet (UV) irradiation of (PVPy)s, they exhibit the emission characteristics which are highly dependent on the excitation wavelength (λ_{ex}). Several factors such as quaternization and hydrogen bonding also affect their emission properties [58]. A series of studies on the emission properties of PVPy solutions revealed that through UV irradiation, the wavelengths of the generated emission bands (λ_{em}) were influenced not only by the excitation wavelength but also by the pH of the medium. Alcoholic or neutral solutions [56,59,60], acidic solutions [61,62], and gel systems [63,64] of (PVPy)s exhibited UV, UV and near-visible, and visible emissions, respectively. Also, the thin films and bulk solids of (PVPy)s, excited at wavelengths greater than 300 nm, showed UV and visible emissions [57,65,66]. For example, the influence of pH on the emission of P_2VPy in the solvent mixture of ethanol:water (1:1) has been investigated [59]. The findings, as illustrated in Scheme 1.9, indicated that (1) in solutions devoid of acid or with low acid concentrations, only an emission peak at 350 nm, due to hydrogen bonding between solvents and pyridine moieties, is detected; (2) an increase in acid concentration leads to the emergence of a new emission peak around 390 nm, attributed to the emission from a protonated form of the heterocycles; (3) at elevated acid concentrations, the 390-nm emission is diminished, likely due to the formation of charge-transfer complexes (or interactions) between the counterion and the protonated rings. It is noteworthy that, under the same conditions, this behavior is not observed in 2-ethylpyridine (as a model compound), indicating that this is specific for the polymer structures. At room temperature, the fluorescence

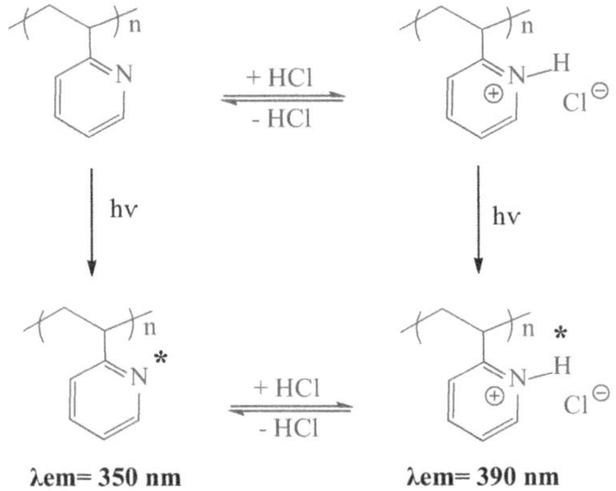

SCHEME 1.9 Ground and excited-state equilibria in P_2VPy acid-base system at the excitation wavelength of 260 nm [59].

spectrum of P_4VPy exhibits a singular band, where its position is different in protic and aprotic solvents. At $\lambda_{ex}=280\,nm$, λ_{em} for P_4VPy in methanol and H_2O/HCl was found to be 310–330 and 370 nm, respectively [56].

Also, photoluminescence (PL) characteristics of a system comprising P_4VPy in pyridine solvent under UV irradiation at 380 nm have been surveyed [63]. Through this process, solution undergoes a phase transition and transforms into a gel with a quasi-crystal structural order, resulting in the change of the PL maximum from 440 to 480 nm and in the optical properties of P_4VPy. By excitation at 480 nm, the PL emissions can reach 570 nm and the area that was irradiated glows with yellow light. It is proposed that this phenomenon is attributed to short-range interactions between the solvent and polymer side chains, interchain interactions, and multiple hydrogen bonds.

The fluorescence properties of P_4VPy and P_2VPy in aqueous solutions of H_2SO_4 were studied, revealing that the fluorescence intensities of both polymers increased with higher acid concentrations, with P_2VPy exhibiting an intensity six times greater than that of P_4VPy [67]. These differences can be explained by the distinct microenvironment factors of the pyridinium groups in each polymer. In this study [67], P_2VPy-Tb^{3+} and P_4VPy-Tb^{3+} complexes were also examined and the findings indicated an increased fluorescence intensity in both complexes, maybe due to the ligand–metal energy transfer effects. Recent investigations have revealed that altering the stereoregularity of P_2VPy significantly influences its phosphorescent characteristics [32]. A system comprising (−)-camphorsulfonic acid ((−)-CSA) and P_2VPy demonstrates that changing the stereoregularity polymeric chains from atactic to isotactic increases the fluorescence emission intensity at wavelengths exceeding 500 nm, prolongs the phosphorescence lifetime to as much as 364 ms, and broadens the fluorescence emission spectrum to encompass the entire visible range.

Owing to the unique emission properties exhibited by (PVPy)s and their derivatives, they have potential applications in fluorescence-based imaging, UV-sensitive coatings and sensors, optoelectronic devices such as light-emitting diodes (LEDs) or UV detectors, and in photolithography. Some of these properties of (PVPy)s will be discussed in the following section.

1.2 PROPERTIES OF PVPY POLYMERS

1.2.1 QUATERNIZATION OF PVPY

Quaternization of PVPy involves the chemical modification of its pyridine rings to produce pyridinium salts. They can be considered as members of polyelectrolytes, which utilize the advantages offered by both polymers and electrolytes [68]. Also, they exhibit increased sensitivity to external environmental factors, such as solvent polarity and pH, due to the presence of existing charges. Furthermore, quaternization facilitates the incorporation of the positive charges into the polymer backbone, leading to electron delocalization effects that are related to the presence of electron-rich pyridines and electron-deficient pyridiniums [69]. The nitrogen atom of the pyridine ring in PVPy can undergo quaternization with alkyl halides or aryl halides (Scheme 1.10, route 1) [58,70–83]. Another strategy to make quaternized PVPy is the

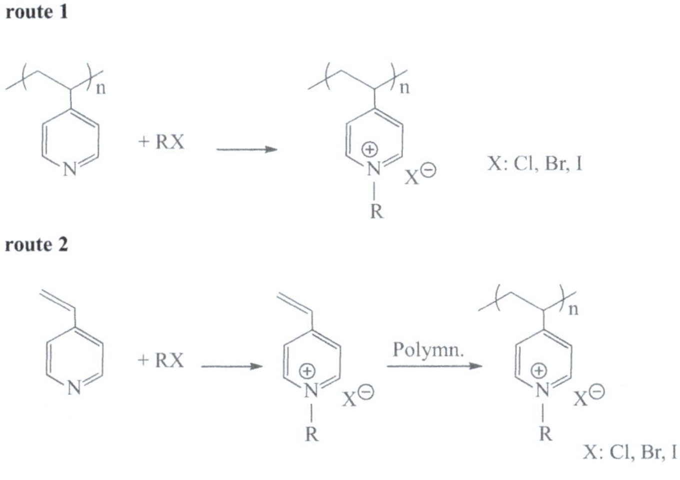

SCHEME 1.10 Synthetic routes of quaternized P_4VPy.

spontaneous polymerization of VPy monomers upon quaternization using various alkylating agents (Scheme 1.10, route 2) [84–89]. Majority of quaternary ammonium monomers are hydrolytically unstable, and hence, route 2 becomes less common. The extent of quaternization can vary based on the reaction conditions, such as reaction time, temperature, and concentration of the alkyl halide. Partial quaternization results in some pyridine moieties remaining unmodified, while complete quaternization transforms almost all pyridine units into pyridinium groups. In most cases, the degrees of quaternization (DQs) are about 70% and the obtained polymers are classified as polyelectrolytes (PEL). While both P_4VPy and P_2VPy can undergo quaternization, P_2VPy is less efficient and slower probably because of the steric hindrance of the polymer backbone [90,91]. A macroporous anion-exchange resin (namely Reillex™ HPQ), which is a quaternized form of P_4VPy (Scheme 1.11) with the quaternization extent of 70%, is commercially available [92,93]. The components of the resin include a copolymer backbone of 1-methyl-4-vinylpyridine (70%, by weight) and DVB (30%). DVB is used as 80% DVB and 20% ethylvinylbenzene.

As mentioned before, if quaternizing agent becomes bifunctional, e.g. a dihalide, PVPy is cross-linked [94]. The quaternization can then be driven to completion using a monofunctional quaternizing agent, e.g., an alkyl halide, giving an anion-exchange resin of known structure. However, this process is diffusion-limited and practically

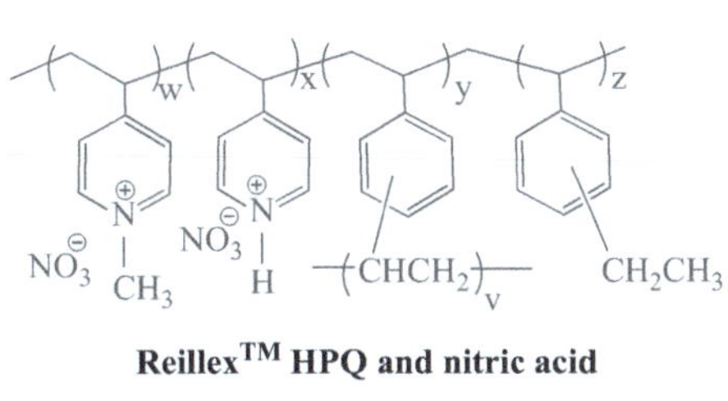

SCHEME 1.11 Structure of commercially available Reillex™ HPQ [92].

impossible. On the other hand, if PVPy is first partially quaternized with an alkyl halide, then the bifunctional agent is added; most probably the "dangling" bonds are created due to reaction of the bifunctional agent at one end only. To overcome this dilemma, it is preferable to use the simultaneous addition of monofunctional and bifunctional quaternizing agents (Scheme 1.12). For example, P_4VPy was quaternized and cross-linked with a mixture of α,α′-dibromo-1,10-ditolyldecanedione-1,10 (LCX) and *n*-butyl bromide (Scheme 1.13) [95].

Through the quaternization of P_4VPy, the flexible P_4VPy matrix is converted to a significantly more rigid one. Mechanism and kinetics of this process and change in the polymer conformations have been investigated [89,91,93]. The incorporation of the positive charges along the polymer chain transforms the pristine hydrophobic or moderately hydrophilic P_4VPy into a more soluble material in water, making the polymer usable in hydrogels, membranes, viscosity and surface modifiers, adsorbents, and NP dispersions [58,96–99]. On the other hand, the fixed positive charges through the quaternization process can change ionic conductivity of the resulting polymer, which are of paramount importance for applications in fuel cells technology, nonlinear optical (NLO) devices, and sensors [100–102]. Also, by regulating the extent of quaternization, scientists can precisely adjust the equilibrium between the hydrophobic backbone and the hydrophilic ionic segments, which can

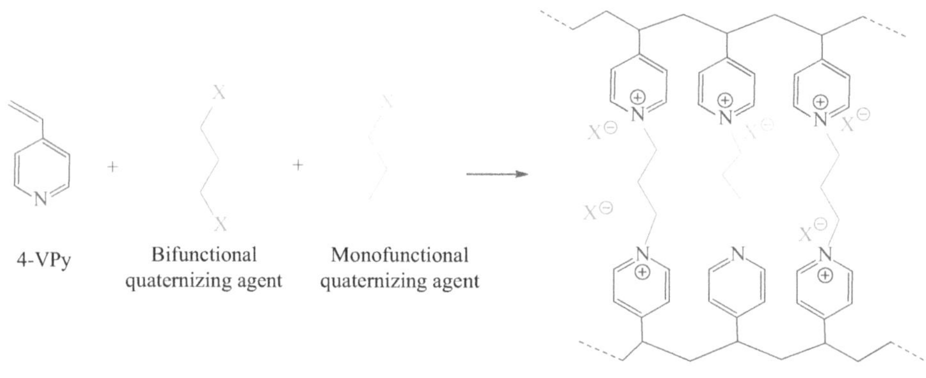

SCHEME 1.12 Cross-linking of P_4VPy through quaternization process.

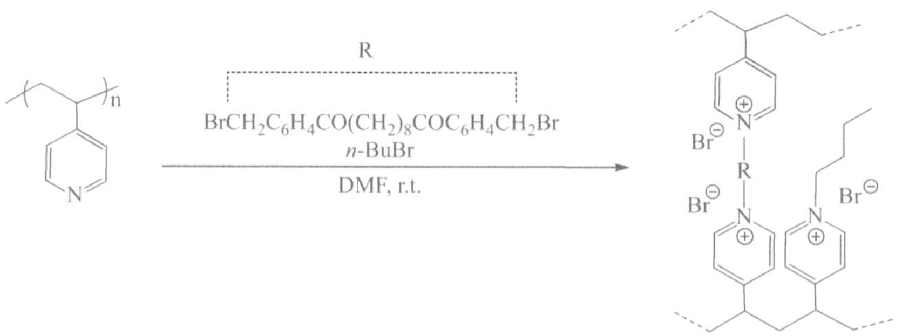

SCHEME 1.13 Preparation of quaternized and cross-linked P_4VPy using a long-chain dibenzyl bromide (i.e., LCX) and *n*-butyl bromide [95].

be beneficial in the development of multicomponent polymeric systems for uses such as drug delivery. Quaternization has the potential to influence dipole–dipole or ionic interactions within the polymer, which may lead to elevated glass transition temperatures and alter the mechanical properties of the material. In this respect, the overall impact is contingent upon the DQ and the characteristics of the introduced alkyl group. The thermal stability [103], swelling and solubility behavior [58], and ionic conductivity [101] of P_4VPy get affected upon quaternization, which will be discussed later. For example, it has been reported that UV–Vis absorption peaks of all *N*-methyl quaternized derivatives of P_4VPy, with DQ of 35%–100%, exhibited a red shift of the λ_{max} from 206 to 217 nm through increasing DQ [58]. The optical energy gap (E_g) of P_4VPy and its quaternized derivatives with 35%, 46%, 85%, and 100% of DQ were measured to be 4.1, 2.83, 2.74, 2.75, and 2.85 eV, respectively. As it can be seen through the quaternization process, the E_g values have reduced and are on a satisfactory level for semiconducting and optoelectronic applications of these materials. In the above study, the solvation properties of both P_4VPy and its quaternized forms were also evaluated in binary solvent mixtures of dimethylsulfoxide (DMSO) and water by a solvatochromic model dye, confirming the formation of complexes of the type PVPy···HOH···dye. The thermodynamic behavior associated with the potentiometric titration of quaternization of P_4VPy by octyl bromide has also been studied [104]. Thermodynamic and potentiometric titration analysis of poly(*N*-octyl-4-vinylpyridinium bromide) has shown the hydrophobic-hydrophobic and hydrophobic-hydrophilic effects in polymer. Quaternized forms of P_4VPy with random distributions of positive and negative charge have been applied as polyampholytes and polyelectrolytes [105–108]. Also, they have been used as polysoaps and phase transfer catalysts and will be discussed in detail in the following chapters. Quaternized PVPy has been used as alternative charge control agents in order to impart a positive charge to the toner particles instead of their low-molecular weight counterparts such as quaternary ammonium compounds which have some drawbacks such as incompatibility with the thermoplastic toner resin and leaching problems [108]. In a recent study, a series of poly(ionic liquids) (PILs) were synthesized from *N*-alkyl quaternized P_4VPy with varying alkyl chain lengths, bearing different counterions (Scheme 1.14) [97]. These PILs were evaluated for their effectiveness in removing chromium(VI) from aqueous solutions, with $P_4VPy\text{-}C_nBF_4^-$ demonstrating superior performance. $P_4VPy\text{-}C_n(CF_3SO_2)_2N^-$ and $P_4VPy\text{-}C_nPF_6^-$ were found to be excessively viscous and unstable solid, respectively. The findings indicated that while $P_4VPy\text{-}CnBr^-$ with $n=2$, 4, and 5 is soluble in both water and methanol, the synthesized PILs are insoluble in water but soluble in acetonitrile, dimethylformamide, and DMSO, suggesting that PILs prefer polar aprotic solvents with higher dielectric constants more. Additionally, the structure of the anion was shown to affect thermal stability, ranked as $CF_3SO_3^- > (CF_3SO_2)_2N^- > BF_4^- > Br^-$.

1.2.2 Basicity Properties

The basicity observed in PVPy primarily originates from the pyridine ring integrated into the polymer backbone. The nitrogen atom in the pyridine ring possesses a lone pair that can readily accept a proton, thereby contributing to the polymer's basic

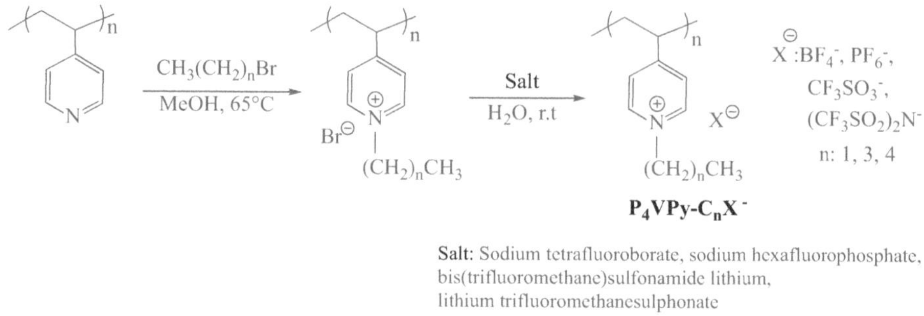

SCHEME 1.14 Synthetic pathways for a series of PILs derived from quaternized P_4VPy [97].

properties. In contrast to the isolated pyridine molecule, in PVPy the effective pK_a is influenced not only by the pyridine functional group but also by various factors, including solvent and the microenvironment effects, the hydrogen bonding in solution of polymer, steric accessibility, degree of polymerization (DP) and distribution of pyridine groups, temperature, and ionic strength. Among these, the steric hindrance at the basic site plays an important role. In P_4VPy, the nitrogen is located further from the backbone compared to P_2VPy, which typically makes it more accessible, and thus this distinction has affected the pK_a values of polymers. Based on the factors mentioned above, the reported pK_a values for P_4VPy and P_2VPy are sparse and inconsistent in literature but often falling between 4 and 6 (unless otherwise specified), with the pK_a of P_4VPy being higher than that of P_2VPy. pK_a values of 4.5 and 5.8 have been introduced for P_2VPy [20] and P_4VPy [21], respectively. Recently, the following order for pK_a values of Py, vinylpyridines, and PVPy polymers has been reported: P_4VPy ($pK_a = 5.98$) > P_2VPy ($pK_a = 5.96$) > 4-VPy ($pK_a = 5.5$) > Py ($pK_a = 5.22$) > 2-VPy ($pK_a = 4.98$) [8].

Several studies have been conducted in the past regarding the influence of protonation levels on various solution characteristics of PVPy polymers [109–111]. Dependence of pK_a of P_2VPy and P_4VPy on the degree of protonation (α) at different temperatures during potentiometric titration of these polymers with benzenesulfonic acid (HBS) and HCl has been investigated [112]. It was found that pK_a values of P_2VPy titrated with HCl become lower than those of P_4VPy with the increase in α (Figure 1.1). Hydrogen bond formation of Cl^- counterion with the pyridinium proton (formation of a charge-transfer complex between them) is suggested to be responsible for the higher pK_a of P_4VPy, which is likely due to the longer spacing of the charges on P_4VPy than on P_2VPy. Based on the results of titration of P_2VPy with HBS, pK_a values are higher at temperatures above 25°C and 40°C and lower at temperatures below 10°C than the values obtained from titration of this polymer with HCl at the corresponding temperatures (Figures 1.1 and 1.2). However, there is no significant difference between HCl and HBS titrations for P_4VPy. Once more, these findings were ascribed to the collaboration of hydrophobic and electrostatic interactions between the polycations and BS^- counterions. Indeed, even a minor variation in the location of the charged sites in P_4VPy and P_2VPy can considerably influence the properties of counterion–polymer interactions [112].

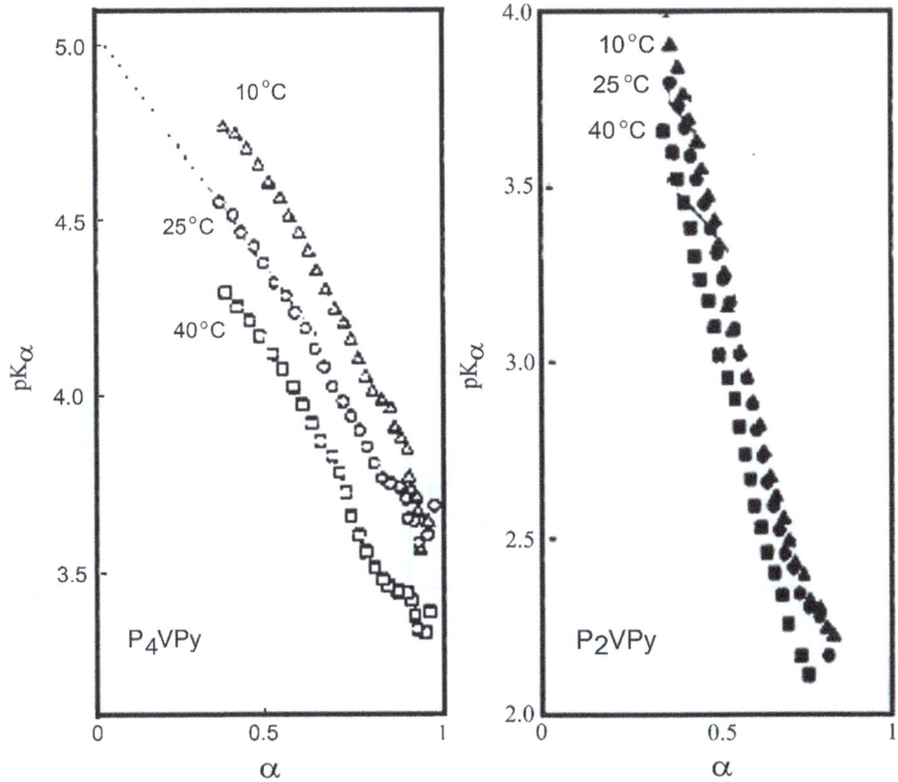

FIGURE 1.1 Dependence of pK_a value of PVPy polymers on α for HCl titration in aqueous solutions [112]. (Reproduced with permission from Ref. [112]; Copyright 1989 American Chemical Society).

The degree to which counterions bind to polyelectrolytes influences various properties. Potentiometric titrations of P_2VPy, P_3VPy, and P_4VPy were conducted utilizing HBr, HCl, HNO_3, and $HClO_4$ [113]. In this study, based on the calculated thermochemical radius and dehydration energy parameters, it is revealed that for all three polymers, Cl^- exhibited the lowest affinity for the charged pyridinium groups, while the affinities for Br^- and NO_3^- were almost identical to one another. The extent of binding for ClO_4^- varied among the polymers examined, with P_4VPy showing the highest binding affinity. The results of the titration suggest that the pK_a values of the pyridinium moieties typically rise with an increased level of protonation. The pK_a values at $\alpha = 50\%$ for the titrations of P_2VPy, P_3VPy, and P_4VPy using HBr at 25°C were determined to be 2.13, 3.03, and 3.34, respectively. Titrations conducted with $HClO_4$ yielded pK_a values of 2, 3.15, and 3.62 for P_2VPy, P_3VPy, and P_4VPy, respectively.

1.2.3 Solubility and Swelling Properties

PVPy polymers demonstrates notable solubility and swelling characteristics attributed to their pyridine functional groups, which engage with solvents via hydrogen

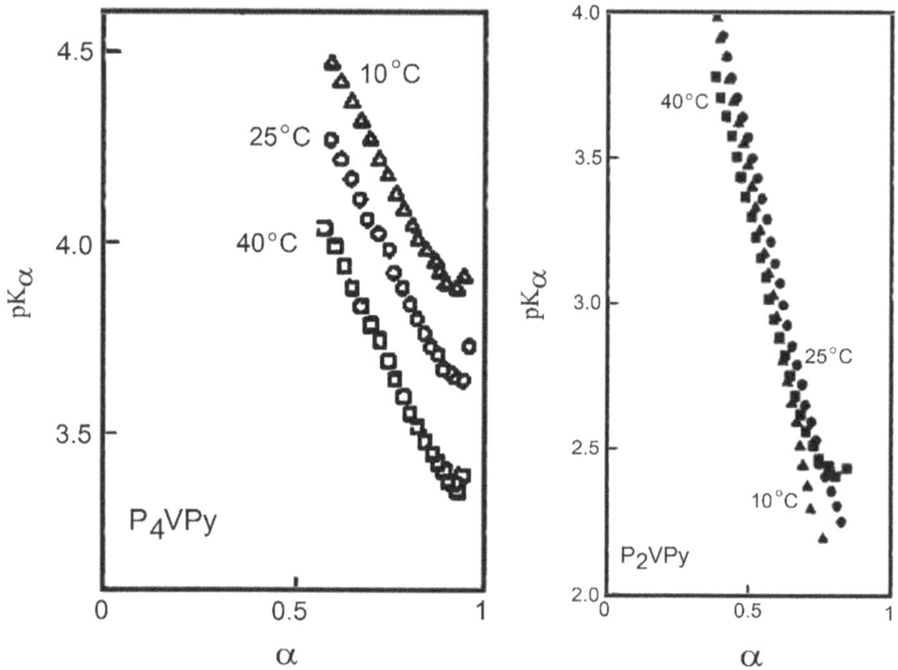

FIGURE 1.2 Dependence of pK_a value of PVPy polymers on α for HBS titration in aqueous solutions [112]. (Reproduced with permission from Ref. [112]; Copyright 1989 American Chemical Society).

bonding and dipole interactions. While P_2VPy is soluble in $CHCl_3$, THF, and benzene, P_4VPy requires addition of polar cosolvents (e.g., DMSO, *N,N*-dimethylformamide (DMF), and *N*-methylpyrrolidone (NMP)) [38,114]. It was observed that the solubility of both isomers increases by addition of common alcohols to their mixtures, due to favorable hydrogen bonding interactions [38]. The main reason for observing the solubility differences in P_4VPy compared with P_2VPy isomers results from the fact that the nitrogen atom of P_2VPy is protected from chemical reactions due to the steric hindrance of the polymer chain [4].

Even though the group contribution method suggests that P_2VPy and P_4VPy possess comparable solubility parameters, as they are merely isomeric forms [115,116], it seems that P_4VPy is likely to exhibit a greater Flory parameter χ (indicating the interaction between the polymer and solvent) compared to P_2VPy, owing to its enhanced miscibility in solvents with higher polarity [38]. In fact, P_2VPy presents lower values of the solubility parameter due to its low hydrogen bonding strength [38]. The Flory parameter for P_4VPy [23,117] and P_2VPy [22,118] has been reported. χ values of P_4VPy and P_2VPy in ethanol were found to be 2.1 ± 0.3 and 0.499, respectively.

A clear example demonstrating the different solubility behavior of two polymers is the observation that poly(styrene-*b*-vinylphenol) (PS-*b*-PVPh) displays macrophase separation with addition of 90 wt.% P_2VPy, while under the same conditions, addition of 90 wt.% P_4VPy remains the blend system to sphere structures without macrophase separation. This was to be expected, since the lower fraction of hydrogen-bonded pyridine group of P_2VPy (in comparison with that of P_4VPy) induces macrophase separation [115]. In other words, P_4VPy exhibits more miscibility in the copolymer structure due to its more capability in forming hydrogen bonds with the phenolic hydroxyl groups present in the copolymer structure.

An additional example of the different effect of PVPy polymers on the solubility properties of systems that include these polymers is the influence of light irradiation on the solubility of P_4VPy-*co*-FISS and P_2VPy-*co*-FISS copolymers in different solvents (Scheme 1.15). While irradiation with light (UV) improves dissolution of copolymer of 2-VPy and 9-fluorenylideneimino *p*-styrenesulfonate (FISS), the copolymer from 4-VPy becomes insoluble in the solvents on irradiation [119]. Upon irradiation, the cleavage of –O–N– bond of FISS occurs, and the subsequently formed radical absorbs a hydrogen atom from the reaction solvent to produce sulfonic acid (SA) pendent group which acts as a polar promoter for solubilizing copolymer (Scheme 1.16). The findings indicate that for P_2VPy copolymers, the dissolution of the irradiated polymer increased as the irradiation time was extended, whereas for P_4VPy copolymers, a reduction in the dissolved fraction of the polymer was noted. The irradiated P_4VPy copolymers remained insoluble even after addition of a salt such as NaCl. This behavior can be attributed to formation of tight ion pair complexes between 4-VPy and SA moieties, through hydrogen bonding interactions. The formation of such complexes is impossible for the irradiated P_2VPy copolymers due to steric hindrance.

PVPy polymers are neutral in the absence of protons (H^+) at $pH \geq 7$ and are insoluble in water. When the aqueous acid for PVPy polymers is used as a solvent, the polymers are protonated and soluble in aqueous solution. P_2VPy and P_4VPy are insoluble in water until more than ca. 35% of the pyridine groups is charged, e.g., by protonation [4]. The hydrophilicity of PVPy polymers will increase upon lowering pH; this presents opportunistic pH-responsive behavior for PVPy

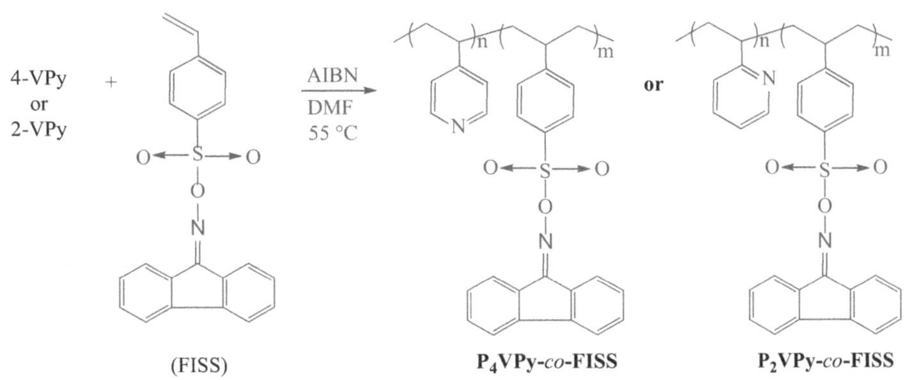

SCHEME 1.15 Synthesis of copolymers of 2-VPy or 4-VPy and FISS [119].

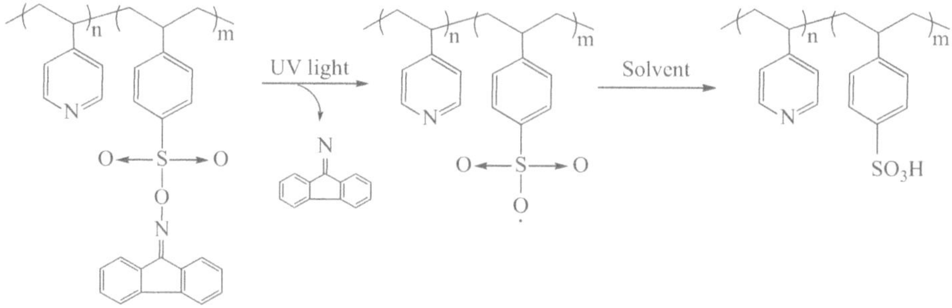

SCHEME 1.16 The cleavage of –O–N– bond of FISS in P_4VPy-*co*-FISS using light and generation of sulfonic acid pendent groups [119].

segments, which will be discussed latter. In fact, when the nitrogen atom of PVPy is protonated or quaternized, the polymer can be considered as a polyelectrolyte. Generally, in low ionic strength, chain polyelectrolyte is expanded due to an electrostatic repulsion. On the other hand, when the ionic strength is increased, chain polyelectrolytes exhibit a coiled conformation. The protonation of PVPy segments within their carrier systems can improve the solubility of these systems. For example, it has been reported that protonation of pyridine groups within CNTs-P_4VPy or CNTs-P_2VPy systems (CNTs: carbon nanotubes) contributes to the solubility and dispersibility of CNTs in both organic solvents and aqueous solutions at about pH 2 [120–122].

The influence of the kind and concentration of acid on the solution properties of PVPy polymers has been investigated [111,123–126]. In all cases, the solubility of the polymer increased when it was immersed in an acidic solution. However, discrepancies were noted regarding the behavior of the polymers in H_2SO_4 compared to their behavior in HCl solution. For example, the solution properties of P_4VPy in aq. H_2SO_4 revealed that a phase separation appears along with a decrease in the H_2SO_4 concentration, and when Na_2SO_4 was added to the solution, the phase separation disappeared [123]. The intrinsic viscosity of this solution increased with an increase in the concentration of H_2SO_4. It is likely that phase separation takes place as a result of the creation of SO_4^{2-} bridges, and when the concentration of SO_4^{2-} rises (with addition of Na_2SO_4), these bridges disintegrate, leading to a clear solution. However, such phase separation behavior was not observed at any concentration of H_2SO_4 in the case of P_2VPy. In copolymers that include P_4VPy and P_2VPy segments, it has been noted that both polymer blocks start to exhibit insolubility (leading to phase separation) at a pH level above approximately 5, while remaining soluble at pH levels below 5 [127–129]. Also, the swelling and deswelling characteristics of hydrogels, which consist of a cross-linked network structure and incorporate PVPy polymers, are affected by variations in solution pH as a result of the protonation and deprotonation of their PVPy segments [130,131].

Besides the protonation of PVPy polymers, their quaternization may also improve their solubility. In quaternized forms of PVPy, every other carbon atom contains an ionic, alkyl, or arylpyridinium group, which are constrained to remain near each other, regardless of the average concentration of the solution [132–134]. The generated charges within the polymer structure through the quaternization process, facilitate the polymer–solvents interactions. For example, while P_4VPy is soluble only in very polar solvents [23,123], the quaternized forms of P_4VPy exhibited an enhanced solubility in both aqueous and nonaqueous solvents [134–136].

Cross-linked polymers possess a three-dimensional network structure that inhibits the dissolution of polymer chains in a solvent; however, when exposed to a solvent, the polymer can absorb it and swell. Swelling of a cross-linked polymer network causes the polymer chains to stretch. The swelling of polymers was significantly affected by various parameters such as the external pH, ionic strength, temperature, and so on [137]. It is reasonable to assume that as the cross-linking degrees of polymer increases, the swelling of polymer network decreases. Cross-linked PVPy polymers, similar to other network structures, are subjected to swelling when exposed to an appropriate solvent. For example, as reported previously (Scheme 1.13), an anion-exchange resin from the simultaneous quaternization and cross-linking of P_4VPy with a mixture of LCX and *n*-butyl bromide has been prepared, and its swelling property in different solvents was surveyed [95]. The results showed that with increasing degree of cross-linking, the swelling ratios (q_0, corresponding to the equilibrium concentration of electrolyte) decreased. For instance, at 2.5%, 5%, and 10% LCX, the q_0 values of the obtained P_4VPy in water were found to be 136, 51, and 22, respectively. The q_0 values for the examined solvents for all percentages of LCX were in the following order: $H_2O > MeOH > DMF$. The results showed that the Flory parameter χ decreased with increasing percentage of cross-linking of the polymer in both the aqueous solutions and methanol.

Swelling properties and polymer–solvent interaction in cross-linked P_2VPy obtained by application of gamma irradiation have been surveyed [22]. The results showed that swelling ratio, solvent content, and volume of adsorbed solvent decrease with an increase in the applied irradiation dose, due to increase in the cross-link density. The calculated χ value (in ethanol) was found to be 0.499 for all cases (gamma irradiations from 182 to 315 kGy), which lies in the range of other polymer-good solvent systems, such as poly(vinyl alcohol) (PVA)–water, poly(*N*-vinylpyrrolidone) (PVP)–water, and poly(methacrylic acid) (PMAA)–water with χ values in the range 0.40–0.59 [22]. Swelling behavior of cross-linked copolymers of 2-VPy and 4-VPy with styrene versus pH and ionic strength has been investigated [19,131]. As shown in Figures 1.3a and b, the extent of swelling for PS-*co*-P_2VPy and PS-*co*-P_4VPy (1% DVB) above pH 4.5 and 5, respectively, is very low. With decreasing pH by 2, the swelling increased and at pH 2, water content (gram of water/gram of dry beads, wt.%) reached ca. 80% for both copolymers. The increasing of water content at low pH arises undoubtedly from the ionization of copolymer gels. At high pH values, unprotonated PVPy blocks dominate the overall swelling behavior. It is apparent that the change of the cross-linking degrees has negligible effect on the water content of both PS-*co*-P_2VPy and PS-*co*-P_4VPy. The results also revealed that with increasing ratio of ST/2-VPy or ST/4-VPy, the water content of copolymers declined over the whole pH range studied, due to an increase in the hydrophobicity of copolymers. Further, the effect of addition of dichloroethane (DCE), which is a good solvating agent for both PS and PVPy, was surveyed. An increase in the water content was observed with increasing proportion of DCE to monomer over the whole pH range, due to an increase in the mobility of the polymer chains. As it can be seen from the figure, the swelling transition happens at lower pH (4.5) for PS-*co*-P_2VPy (Figure 1.3b) in comparison with that for PS-*co*-P_4VPy (pH 5, Figure 1.3a). In addition, in the case of PS-*co*-P_2VPy, a rather sharp decrease in the water content is observed. The first difference can be due to more basicity of P_4VPy than P_2VPy based on their pK_a values which are 3.95 and 3.50, respectively, at 50% ionization. Therefore, with decreasing pH, the protonation of P_4VPy is firstly initiated, resulting in an earlier swelling transition. The second difference can be due to the more hydrophobicity of P_4VPy than P_2VPy. Thus, PS-*co*-P_4VPy is less sensitive to pH changes (showing a less steep swelling transition). The influence of the ionic strength on swelling of copolymers was also investigated. Figure 1.3c shows the water content changes versus ionic strength for PS-*co*-P_4VPy. In all copolymers, with increasing ionic strength, the water content first increased until to a maximum and then gradually decreased. The position of maxima is shifted to higher ionic strength values with increasing pH. For copolymers in an ionized state, after the maximum swelling, with further increasing in ionic strength, the ionic swelling pressure is lowered, resulting in a decrease of swelling (Donnan theory) [131,138]. The increase of the water content of copolymers with increasing ionic strengths below the maximum swelling

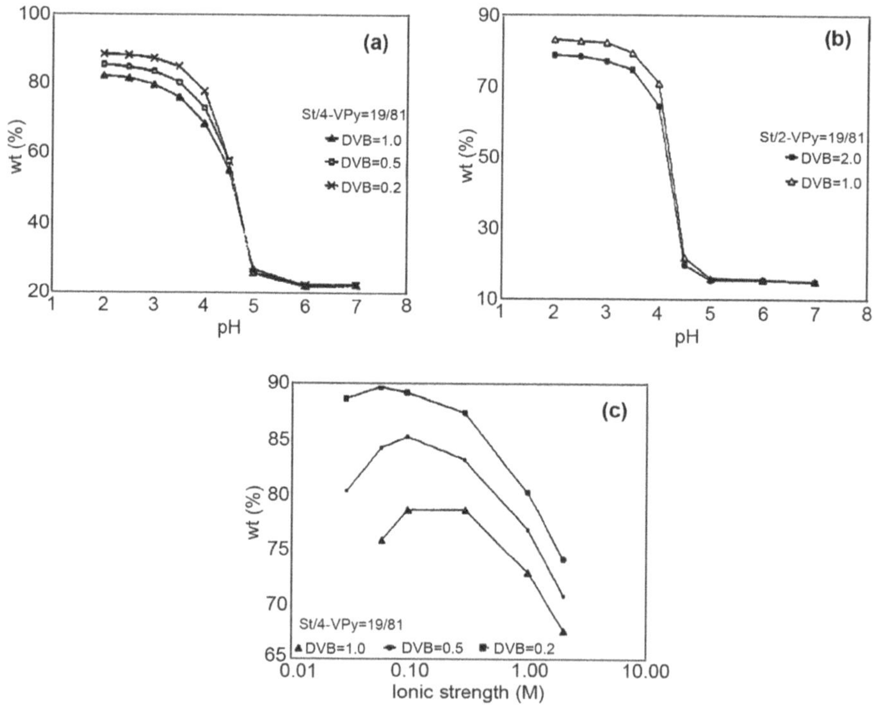

FIGURE 1.3 The influence of the pH changes of citrate buffer (pH 3, ionic strength = 0.3 M, C = 0.05 M) on water content (wt(%)) of (a) PS-*co*-P_4VPy and (b) PS-*co*-P_2VPy (19:81 molar ratio of ST:VPy) with different DVB contents, and (c) the influence of ionic strength on water content (wt(%)) of PS-*co*-P_4VPy with different DVB contents in citrate buffer [19]. (Reproduced with permission from Ref. [19]; Copyright 1993 American Chemical Society).

value is difficult to explain but may be due to "salting-in" effect [139]. The results revealed that the water content of copolymers with lower DVB levels is more affected by ionic strength of buffer solution. However, at higher ionic strengths, differences between the water content of copolymers with different DVB content are less.

Gas permeability, diffusivity, and solubility of P_4VPy films casted on a liquid mercury surface have been investigated [140]. The findings indicated that permeability and solubility are inversely correlated with pressure, whereas diffusivity exhibits a direct correlation with pressure. Gases with greater condensability and stronger interaction exhibited higher solubility in polymers. Gas permeability is in the following order: $He > H_2 > CO_2 > O_2 > CH_4 > N_2$ at 619 cm Hg and 35°C. Also, other reports have shown that P_4VPy has good permselectivity for gas separation [141]; this is described in the section on applications of PVPy polymers in membrane technology.

1.2.4 Thermal Stability

Both P_4VPy and P_2VPy are thermally stable polymers, without significant differences between them. Thermal stability and degradation behavior of these polymers [18,142–147] have been investigated. A mass loss of about 4% was observed for both polymers under 300°C under N_2 atmosphere. Both polymers (with almost the same molecular weights, M_n: ~60,000) show the maximum degradation temperatures (TM_{ax}) of ca. 415°C along with complete decomposition [18]. However, some investigations revealed different degradation temperatures for these polymers [148]; for instance, it was observed that the thermal degradation processes begin at ~350°C and ~380°C for P_4VPy and P_2VPy, respectively [147]. Several reports indicate that P_2VPy exhibited higher degradation temperatures than P_4VPy, demonstrating the significant resistance of this polymer [148,149].

The pyrolysis mass spectrometry findings for P_2VPy revealed that this polymer is decomposed through several paths including sequential H-transfer reactions leading to mostly pyridine, protonated oligomers, and 2-VPy monomer (Scheme 1.17) [18]. However, decomposition of P_4VPy involves one degradation step and the main degradation product is the monomer, with traces of pyridine and 4-methylpyridine [18,144,150]. Also, the theoretical investigations (the atom superposition and electron delocalization molecular orbital (ASED-MO) method) showed that 4-VPy is the only observed species on thermodegradation of P_4VPy [151].

The glass transition temperature (T_g) of a polymer refers to the temperature at which it transitions from a rigid, glassy state to a soft, rubbery state. This happens at different temperatures depending on polymer structure, molecular weight, DP, and measurement conditions (test atmosphere and heating rate). T_g values of P_2VPy and P_4VPy were found to be 84°C and 142°C, respectively [15]. Significant differences between T_g values of these polymers show that P_4VPy segments will require higher temperatures to thermally process. Usually, more rigid backbones (like aromatic rings), higher molecular weight, and cross-linked structures, which restrict chain movements, raise T_g of the polymer. In addition, plasticizers and additives lower T_g by increasing free volume and chain mobility. For example, reduction of T_g of P_2VPy and P_4VPy by the addition of carbon dioxide, which is highly relevant for foam extrusion, has recently been studied [152]. The T_g of P_2VPy and P_4VPy were noted to

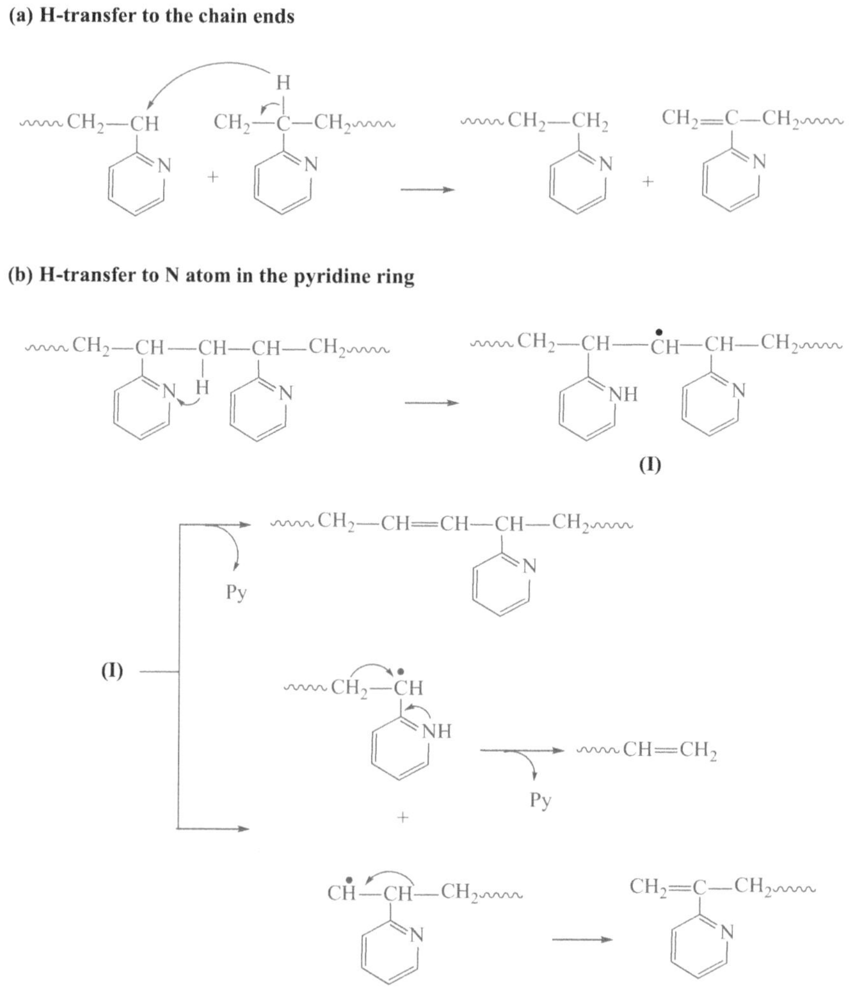

SCHEME 1.17 The degradation mechanism of P_2VPy through H-transfer to the chain ends (a) and H-transfer to N atom in the pyridine ring (b) [18].

decline from 100°C to 91°C and from 150°C to 142°C, respectively, when subjected to a carbon dioxide pressure of 100 bar.

As mentioned above, one of the factors affecting thermal stability of polymers is the degree of cross-linking. One of the most important types of cross-linked PVPy polymers that has been used repeatedly as a support for various species such as oxidizing and reducing agents is P_4VPy cross-linked by 2% DVB (Reillex® 402 ion-exchange resin). High-resolution thermogravimetry (TG) of P_4VPy (2% DVB) under air and nitrogen atmosphere has been investigated [149]. Figure 1.4a shows thermogravimetric analysis (TGA), derivative thermogravimetry (DTG), and heating rate curves of this polymer. The DTG of samples indicated that the first major degradation process includes three steps (small and medium shoulders besides a strong peak). The thermodegradation steps are probably due to the breakage of head-to-head linkages and cross-linked bonds as well as the degradation of the polymer chains. For this copolymer, the degradation temperature (T_d) and the temperatures at the maximum weight-loss rate (T_{dm1}, T_{dm2}, T_{dm3}) under nitrogen atmosphere were found to be 322°C, 336°C, 358°C, 597°C, respectively. The degradation temperatures were higher in nitrogen than in air atmosphere. The results showed that the P_4VPy (2% DVB) has lower heat resistance, larger carbonaceous residues (greater char-forming ability), and slower rate of weight loss than P_4VPy. These outcomes partially align with the thermal degradation properties of poly(styrene-*co*-divinylbenzene) [153]. In this study, the activation energy (E) of P_4VPy (2% DVB) obtained from kinetic studies on degradation of this copolymer was also found to be 78 kJ/mol under nitrogen atmosphere which is much lower than the bond dissociation energies of P_4VPy [149,154], revealing that in comparison with P_4VPy, thermal degradation of P_4VPy (2% DVB) initiates at lower temperatures, maybe due to the presence of some weak thermolabile bonds in its structure. Figure 1.4b shows bond dissociation energies in P_4VPy [149]. The T_g value of P_4VPy (2% DVB) was found to be 160°C in N_2 [17].

Another influential factor noted in the thermal stability of PVPy polymers is the quantitative quaternization of the nitrogen on pyridine rings. Quaternization of PVPy causes a decrease in thermal stability compared with pristine polymer and the decrease being a function of the extent of quaternization [155]. For example, it has been reported that while P_4VPy exhibits a slight weight loss by TGA until 385°C, quaternization with bromomalonamide lowers its thermal stability to about 300°C [155]. The degradation in the polymeric 1-alkyl or 1-arylpyridinium salts generally involves two stages: first loss of the alkyl or aryl groups and then scission of the polymer backbone. Thermal characteristics of a series of *N*-methyl quaternized derivatives of P_4VPy with different DQs were studied [58]. TGA measurements of P_4VPy

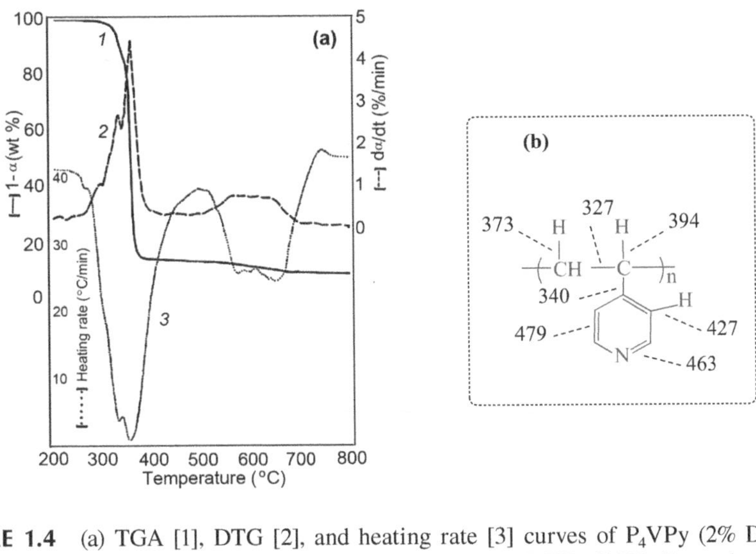

FIGURE 1.4 (a) TGA [1], DTG [2], and heating rate [3] curves of P_4VPy (2% DVB) in nitrogen flow and (b) bond dissociation energies (kJ/mol) in P_4VPy [149]. (Reproduced with permission from Ref. [149]; Copyright 1999 Elsevier).

and their quaternized forms under N_2 atmosphere and under a flow of air are depicted in Figure 1.5a and b. The results showed that the thermal stability decreases with quaternization, due to Hoffman elimination reactions of the polymeric quaternary salts, although the quaternized polymers still have satisfactory thermal resistance up to 300°C. Under an N_2 atmosphere, the major mass loss, which relates to the thermal degradation of the polymers, occurs at about 300°C–400°C and 400°C–450°C for the quaternized polymers and P_4VPy, respectively. Under an air atmosphere, the main degradation event occurs at approximately 250°C–380°C and 300°C–400°C for the quaternized polymers and P_4VPy, respectively. For the quaternized polymers, thermal degradation of the formed products under O_2 (oxides) creates an additional degradation step around 500°C–600°C. The quaternization leads to elevation of the T_g, due to the conversion of the flexible polymer matrix to a significantly more rigid one (Figure 1.5c) [58].

PVPy polymers upon exposure to protic acids convert to polymeric pyridinium salts; this process can affect their thermal stability. The thermal stability of P_2VPy-HCl has been investigated [156]. TGA and DTG curves of pristine P_2VPy and P_2VPy-HCl are shown in Figure 1.6. In TGA curve of P_2VPy-HCl, 26% weight loss started at 255°C is mainly assigned to the removal of HCl molecule, and a loss of 70% at 400°C is due to decomposition of P_2VPy as it is seen for the pristine P_2VPy. As it can be seen from the figure, P_2VPy-HCl has lower onset temperature of thermal

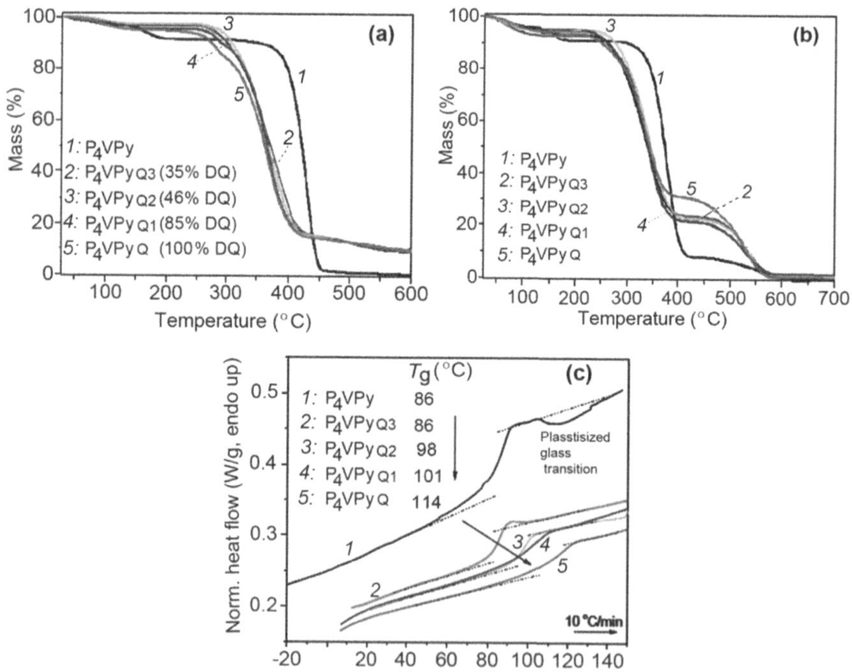

FIGURE 1.5 TGA (under (a) N_2 and (b) air) and (c) heat flow curves of P_4VPy and the quaternized derivatives [58]. DQ: degree of quaternization). (Reproduced with permission from Ref. [58]; Copyright 2022 MDPI).

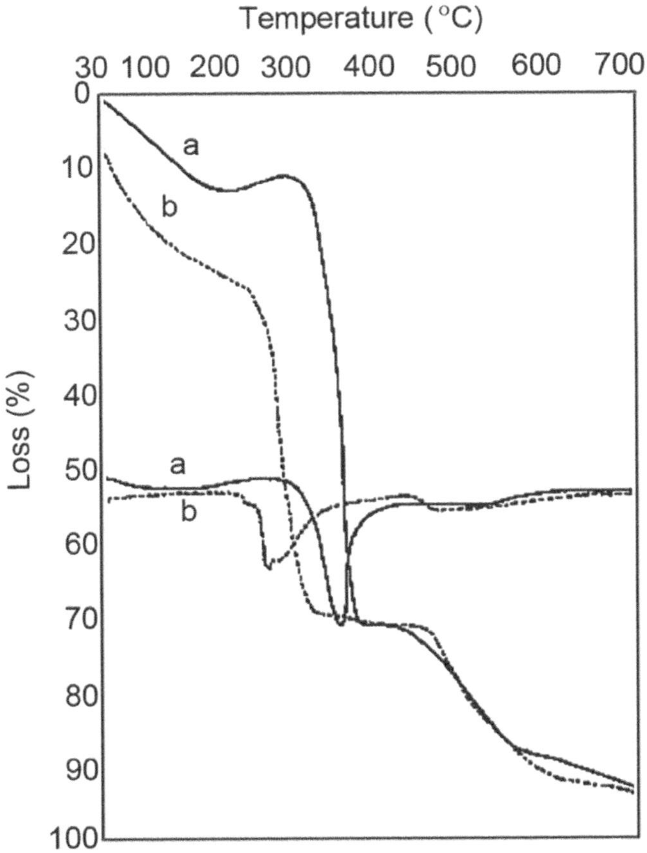

FIGURE 1.6 TGA and DTG curves of (a) P_2VPy and (b) P_2VPy-HCl [156]. (Reproduced with permission from Ref. [156]. Copyright 2001 Elsevier).

degradation than P_2VPy. Also, a detailed study of thermal behavior of P_4VPy doped with a weak acid, *p*-toluene sulfonic acid (PTSA), revealed that the T_g values gradually decrease with increasing amount of PTSA, due to the plasticizing effect. The T_g values of P_4VPy and $P_4VPy + 3\%$ PTSA were found to be 153°C and 148°C, respectively [157]. Plasticizing effect has also affected the thermal stability of binary systems formed by P_4VPy and dihydric phenols [158], i.e., a decreasing of the T_g values. Plasticizing effect increases the free volume between the polymer chains; hence, less energy is needed to produce motion on chain segments, resulting in decreasing of T_g.

The thermal behaviors of poly ion complexes produced from mixing of anionic lignosulfonic acid (LSA, bearing hydroxyl, methoxy, and sulfonate groups) solution with cationic forms of P_4VPy or P_2VPy in DMSO or dilute HCl were studied [159]. The LSA/P_2VPy complexes exhibited higher T_g values than that of the pristine P_2VPy, implying that there is a high ionic interaction between LSA and P_2VPy. It is interesting to note that the LSA/P_4VPy complexes showed no glass transition, probably due to the existence of much higher degree of interactions in LSA/P_4VPy

complexes than that in LSA/P_2VPy complexes, resulting in stronger suppression of molecular mobility in the complexes of P_4VPy.

Incorporation of metal ions into PVPy structure can exert the variable effects on the heat resistance of the resultant polymer-bound metal complex [27,103,160,161]. Metal salts exhibit distinct behavior compared to low-molecular weight plasticizers, as they usually elevate the T_g of the polymer by creating rigid coordination complexes. For example, it has been reported that the ruthenium and osmium metal ions increased the T_g of P_4VPy [162]. Also, the thermal analysis of the P_4VPy (2% DVB)-Cu(II) complex prepared from the reaction of P_4VPy (2% DVB) with $CuSO_4{\cdot}5H_2O$ indicated that the T_g value of a complex with the Cu^{2+} content of 0.793 mmol per gram resin was 187°C, which is higher than that of the P_4VPy-DVB copolymer (160°C) [17]. The T_g of the P_4VPy (2% DVB)-Cu(II) increased with increasing Cu^{2+} ions content, probably because of increasing the rigidity of the copolymer complex. Figure 1.7 shows TGA and DTG thermograms of this polymer complex and the P_4VPy (2% DVB). It is worth noting that the addition of metal ion to P_4VPy solution in methanol decreases the reduced viscosity of the solution, because the coiling up of P_4VPy chains due to interchain chelation [163].

Likely, a series of P_4VPy-Cu(II) complexes from the reaction of $CuSO_4$, $CuCl_2$, and $Cu(SCN)_2$ with the linear P_4VPy were prepared, and the anion influence on thermal properties of the obtained polymer complexes was surveyed [164]. Figure 1.8a shows typical differential scanning calorimetry (DSC) curves for these complexes with their T_g values. The results revealed that in these polymer complexes *N*-ligand chelation can improve T_g of the resultant polymers through increasing the polymer reticulation degree. As shown in Figure 1.8a, the polymer complexes formed from $CuCl_2$ exhibited two T_g values which can be due to the higher rigidity of the polymer complex and different Cu(II) coordination numbers for the polymeric chains. As it can be seen from TGA curve in Figure 1.8b, the polymer complexes have higher

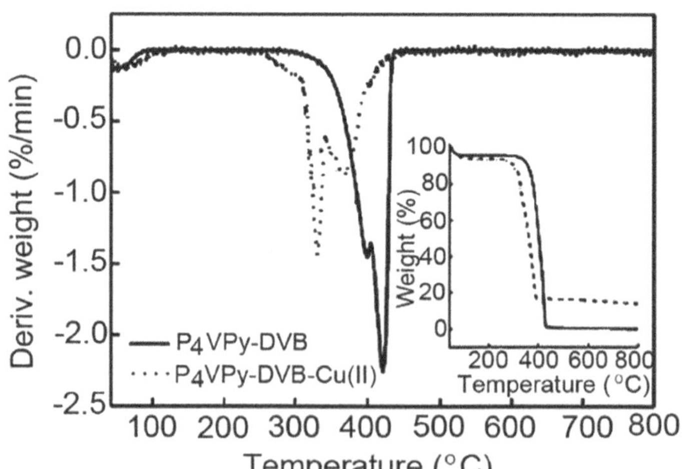

FIGURE 1.7 DTG and TGA thermograms of P_4VPy (2% DVB)-Cu(II) complex and P_4VPy (2% DVB) under N_2 at a heating rate of 10°C/min [17]. (Reproduced with permission from Ref. [17]. Copyright 2003 Elsevier).

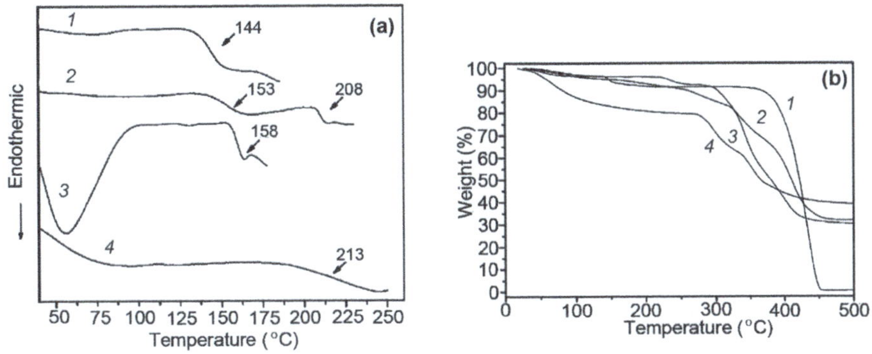

FIGURE 1.8 (a) DSC and TGA curves of P_4VPy [1], $P_4VPy/CuCl_2$ [2], $P_4VPy/Cu(SCN)_2$ [3], and $P_4VPy/CuSO_4$ [4] complexes [164]. [Reproduced with permission from Ref. [164]; Copyright 2004 Elsevier).

thermal stability than pure P_4VPy, because the electronic density of the pyridine rings is affected by copper ions which eventually lead to weakening of all chemical bonds in the polymer complex. Furthermore, from the infrared (IR) and Raman spectra, it was clear that in $P_4VPy/CuSO_4$, sulfate anion is monocoordinated and thiocyanate anion in $P_4VPy/Cu(SCN)_2$ forms an *N*-bonded complex (Scheme 1.18). Like $P_4VPy/Cu(SCN)_2$ and $P_4VPy/CuSO_4$ complexes mentioned above, T_g values of linear P_2VPy-$LiClO_4$ increased through increasing the molar percentage of $LiClO_4$ due to the formation of transient physical cross-links (Scheme 1.19) [165]. While T_g of P_2VPy is 107°C, P_2VPy-$LiClO_4$ containing 10 mol.% $LiClO_4$ has a T_g of 142°C. The thermal behavior of linear P_2VPy-$Zn(ClO_4)_2$ and P_4VPy-$Zn(ClO_4)_2$ is similar to that of P_2VPy-$LiClO_4$ [166]. Copper iodide (CuI) and copper chloride (CuCl) complexes of PVPy polymers have been prepared, and their T_g values were studied as a function of metal concentration [167,168]. For example in the P_4VPy-CuI and P_2VPy-CuI complexes, it was observed that T_g values increase with increasing CuI concentrations [167]. In the case of P_4VPy-CuI complex the maximum enhancement in T_g was achieved with 15 mol.% of CuI. In P_4VPy-CuI complex, each transition metal can undergo interaction with more than one pyridine ring, resulting in an enhancement of thermal properties. However, at higher concentrations of CuI (>15 mol.%), T_g of P_4VPy-CuI decreases, maybe due to steric effects. With 50 mol.% of CuI, T_g values of P_4VPy-CuI and P_2VPy-CuI complexes are almost same (ca. 180°C). A similar behavior of P_4VPy-CuI complex has also been observed by P_4VPy-zinc salts in which the T_g values decrease at high amounts of zinc salts in polymer complex [166]. Recently, P_4VPy complexes of $CeCl_3$, $LaCl_3$, $SaCl_3$, and $TiCl_3$ were prepared and their thermal stability was investigated [169]. The TG data showed that the initial degradation temperatures of all polymer complexes (which ranged from 286°C to 301°C) were less than that of pure P_4VPy (331°C), while the final degradation temperatures (ranged from 536°C to 651°C) were higher than that of P_4VPy (402°C).

Although a lot of researches indicated that incorporating metals into the polymer matrix of PVPy raises the T_g value, a few reports also note that this incorporation can

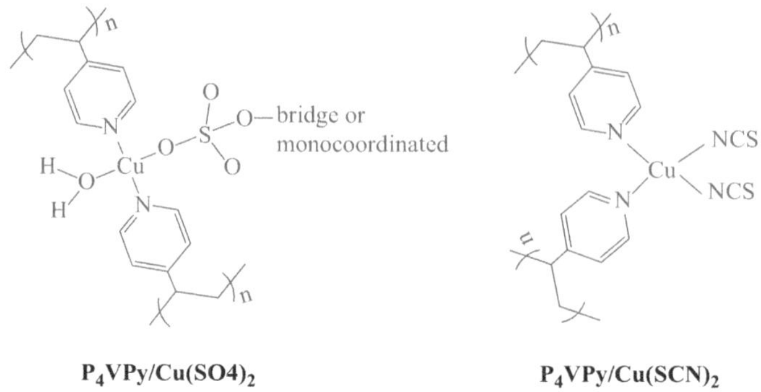

SCHEME 1.18 The chemical bonds in the metallic center of $P_4VPy/Cu(SCN)_2$ and $P_4VPy/CuSO_4$ complexes [164].

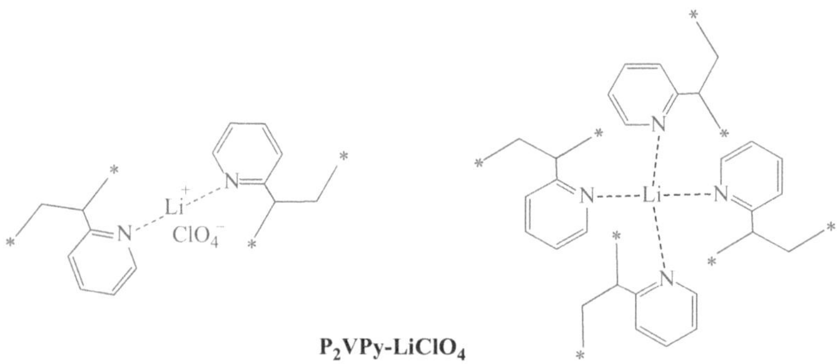

SCHEME 1.19 The possible coordination bonding in P_2VPy-$LiClO_4$ which causes the cross-linking of polymer [165].

lead to a reduction in the T_g. For example, as shown in Scheme 1.20, reaction of P_4VPy with Cu(II), Co(II), and Ni(II) chlorides produced chelate complexes in which the ratio of metal to ligand was 2:1 (based on elemental analyses and IR) [142]. All the resultant polymer complexes showed lower thermal stability than P_4VPy. The reported values of T_g for P_4VPy, P_4VPy-$CuCl_2$, P_4VPy-$NiCl_2$, and P_4VPy-$CoCl_2$ were 99, 76, 75, and 74°C, respectively. All polymer complexes exhibited two degradation steps, and weight remaining after 500°C could be considered to be related to metal oxides.

In polymer blends based on PVPy (discussed above) when strong interactions can take place between the blend components, T_g values are higher than that of individual polymers. For example, while T_g values of P_2VPy and poly(vinyl phenol) (PVPh) are 84°C and 150°C, respectively, the blend P_2VPy/PVPh (4:1) shows T_g of 170°C [170]. Also, the blend P_4VPy/PVPh (50:50) exhibits T_g of 210°C, much higher than that observed for P_4VPy (142°C) and PVPh (138°C), resulting from complete miscibility between the two polymers due to the hydrogen bonding interactions [171]. The blends PAMPS/P_2VPy and PAMPS/P_4VPy, constructed from poly(2-acrylamido-2-methyl-propane sulfonic acid) (PAMPS) and PVPy with the PAMPS mole fractions of 0.65,

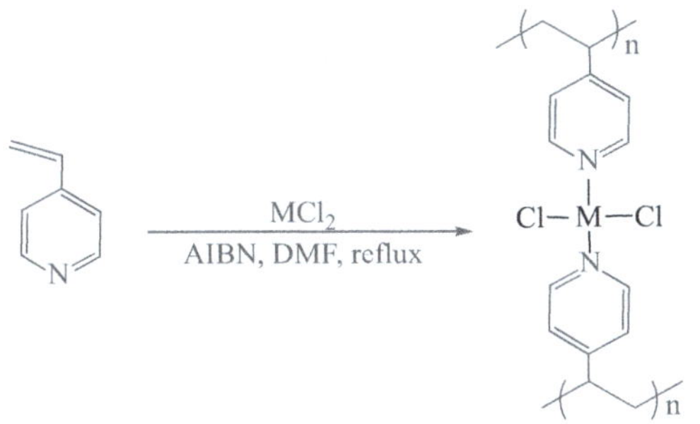

SCHEME 1.20 Synthesis of polymer complexes of 4-VPy with transition metal chlorides [142].

have T_g values of 132°C and 166°C, respectively, which are higher than those found for P_2VPy (84°C), P_4VPy (142°C), and PAMPS (75°C) [172]. The PS/PVPh (50:50) and PS/P_4VPy (50:50) blends have T_g values of 125°C and 118°C, respectively, which are intermediate between those of the individual polymers, because each of these blends contain two incompatible components. The T_g value of PS alone is 88°C.

Finally, the effect of nanoscale restrictions, geometrical confinement in the nanometer range, on the dynamic glass transition of P_2VPy has been investigated [173–183]. For example, the glass transition behavior of P_2VPy confined within alumina nanopores (down to 20 nm size) exhibited that an out-of-equilibrium behavior announced for other glass-forming systems confined at the nanoscale, is not evident [173].

Complexation between metal complexes and PVPy structure has not only an influence on the thermal stability of the polymer but also on the conductivity. For example, the following order for the d.c. conductivity (S/cm) of a series of metal complexes of P_2VPy and $CuCl_2$, $CoCl_2$, $PdCl_2$, and H_2PtCl_6 has been reported: Pt-P_2VPy (3.32×10^{-4}) > Pd-P_2VPy (5.85×10^{-5}) > Co-P_2VPy (1.19×10^{-9}) > Cu-P_2VPy (8.1×10^{-13}) > P_2VPy (5.85×10^{-5}) [161]. The results showed that the order of conductivity (S/cm) of these materials is Pt-P_2VPy (3.32×10^{-4}) > Pd-P_2VPy (5.85×10^{-5}) > Co-P_2VPy (1.19×10^{-9}) > Cu-P_2VPy (8.1×10^{-13}) > P_2VPy (5.85×10^{-5}). The improved conductivity of complexes is proposed to be ascribed to intrachain coupling through d–π orbital overlap of the metal and pyridine ring, respectively, as well as the formation of a metal–chlorine–metal bridge structure. Conducting materials, based on well-defined BCPs of PVPy polymers, have gained a progressively increasing interest due to their self-assembly nature which will be discussed in the following section.

1.2.5 Self-Assembly Behavior

BCPs are constructed from blocks of chemically distinct repeat units [184–197]. Since the first report on BCPs in the 1950s [198], their potential to assemble into different morphologies has captivated the research community. In BCPs, if the blocks are chemically or thermally incompatible, they undergo phase separation. Pleasingly, in the linear BCPs in which the blocks are connected covalently, the phase segregation of the blocks only occurs on a nanoscale (5–100 nm), because these bonds restrict segregation of phases to a length scale and lead to microphase separation into domains [199,200]. Composites of such BCPs and inorganic materials can form different ordered structures with unique function. In these compounds, the convenient size and shape tunability of microdomains can be obtained by adjusting their chemical compositions [201].

Among BCPs that can form self-assembled nanostructures, those containing at least one segment of PVPy polymers have gained tremendous attention in the past decade [38,192,197,202–229]. Notably, only P_2VPy and P_4VPy were used in manufacturing of BCPs and no BCPs with P_3VPy as a segment were discovered in literature, probably due to the high monomer cost. BCP of VPy and ethylene oxide (EO) monomers (PVPy-*b*-PEO) is the best example to explain assembly within an aqueous environment. While PVPy segments are miscible with PEO in the bulk state, these segments will separate into microphases in aqueous solution. This can be explained on the fact that PVPy polymers are insoluble in nonacidic water solutions while PEO is miscible. When PVPy-*b*-PEO is added to an aqueous media, the resultant $PEO+H_2O$ has enough incompatibility with PVPy, which leads the system to form micelle structures in which PVPy is located in the core [38].

Through the microphase separation of PVP-containing BCPs and generation of different morphologies, the PVPy domains retain their responsive properties. For example, pH-responsive behavior can be selectively achieved in the microphase separation of PVPy-containing BCPs. In these systems, adjusting the solution pH and thus controlling swelling/deswelling of PVPy units allows tunable pore structure, size, and water flux [228]. The pH-responsiveness of these BCPs can be fruitfully exploited for applications such as drug delivery, sensors, and membranes [38]. Also, the basicity and coordinative ability of PVPy domains provide their convenient chemical modification in the nanoscale BCP domains. In fact, the chelating ability of pyridine moiety of PVPy domains in binding organic/inorganic guests allows a wide range of applications of the corresponding BCPs in electronic, magnetic, catalytic, conductive, optical, and cargo-releasing materials [38,228].

The most investigated PVPy-containing systems with self-assembly behavior are PS-*b*-P_2VPy and PS-*b*-P_4VPy [38,200,230–235]. The self-assembly of PS-*b*-P_2VPy alone has been discussed in a recent review [234]. A great advantage of choosing PS-*b*-P_2VPy in above investigations is that PS and P_2VPy exhibit nearly identical T_g and as a result, the segmental mobility is similar for the copolymer samples having different styrene/2-VPy compositions [236]. Owing to their unique physical and chemical properties, they have gained much interest for many applications in chemistry and materials science, including (nano)lithography [237–239], biomedical materials [240,241], membrane filtration [242,243], optical devices [244], and nanoporous structure [245]. Besides copolymers of PVPy polymers and PS, copolymers of PVPy polymers and other polymers have been also used for the design of other nanostructured assemblies with unique properties [38,246–248]. For example, self-assemblies of a polyurethane (PU) and P_4VPy (PU-*b*-P_4VPy) [249], P_4VP-*b*-poly(*N*-acryloylpiperidine) [226], poly(3-hexylthiophene)-*b*-P_2VPy [250,251], poly(3-hexylthiophene)-*b*-P_4VPy [252], poly(3-dodecylthiophene)-*b*-P_2VPy [253], poly(diethylhexyloxy-*p*-phenylenevinylene)-*b*-P_4VPy [254], poly(isoprene-*b*-styrene-*b*-4-VPy) (PI-*b*-PS-*b*-P_4VPy) [255], P_2VPy-*b*-poly-(*N*,*N*-dimethylacrylamide)

(P_4VPy-*b*-PDMAAM) [256], PS-*b*-P_2VPy-*b*-PEO [257], and diblock copolymers of PVPy and polyacrylates or polymethacrylates [258–263] have also been investigated.

The most driving force in the self-assembly of AB-type BCPs involves the minimization of free energy between the two segments in solution and between each block and solvent. The first is expressed by Flory–Huggins interaction parameter, χ_{AB}, and the second by polymer–solvent interaction parameters, χ_{AS} and χ_{BS}. However, the phase separation and self-assembly behavior of BCPs can be manipulated by the following factors: (1) the total DP, N (the length of the polymer chains); (2) the copolymer composition, f (or the volume fractions of two blocks), and (3) χ_{AB}, showing the incompatibility between two segments. χ_{AB} is mostly positive, indicating that two blacks are immiscible. These parameters can be modified to facilitate AB-type BCPs with different morphology. The microphase separation tendencies between different blocks is improved by increasing the repulsion between them, and therefore, a higher χ parameter can cause the ordered morphologies [264]. Generally, the product of χ and N, $\chi_{AB}N$ (the segregation strength), determines whether they are miscible or not [199,265,266]. It has been shown that in symmetric diblock copolymers ($f_A = f_B = 0.5$), the critical segregation strength must exceed 10.5 in order to attain microphase separation [265]. Also, to achieve self-assembled structures with domain sizes smaller than 10 nm, BCPs should have a high χ and a low N [29]. In this context, a strategy is the integration of blocks with significant differences in polarity, like in poly(3-hydroxystyrene-*b*-dimethylsiloxane [267].

Past investigations well documented that the P_2VPy and especially P_4VPy show "high-χ" BCPs when segmented with low-polarity counterparts such as PS [264]. It is due to the fact that there is a correlation between χ and solubility parameters; a high χ is observed in the case of a high difference in solubility parameters between the segments. $\chi_{PS\text{-}P2VPy}$ and $\chi_{PS\text{-}P4VPy}$ at $T_g > 150°C$ and a segment averaged $V_{ref} \approx 99.5$ Å^3 were found to be approximately 0.11 and 0.33, respectively [25,26]. V_{ref} is reference volume, a parameter that includes a defined volume of the desired BCP. It is usually taken as the average molar volume of the two BCP blocks. Because PS is considered as a hydrophobic segment, the threefold increase in $\chi_{PS\text{-}P4VPy}$ over $\chi_{PS\text{-}P2VPy}$ indicates the greater hydrophilicity of the P_4VPy domains than that of P_2VPy [26]. P_2VPy and P_4VPy can construct a lamellar P_4VPy-*b*-P_2VPy system [41]. The Flory's interaction parameter of P_4VPy-*b*-P_2VPy is between PS-*b*-P_4VPy and PS-*b*-P_2VPy; $\chi_{P2VPy\text{-}P4VPy}$ at 150°C and $V_{ref} = 118$ Å^3 was determined to be ≈0.26 [24]. Other PVPy-containing BCPs with relatively high χ value are poly(4-*tert*-butylstyrene-*b*-2-VPy) ($\chi \approx 0.11$) [38] and poly(4-*tert*-butoxystyrene-*b*-4-VPy) ($\chi \approx 0.39$) [268,269]. Zwitterionic BCPs of 4-VPy-propane-1-sulfonate (PVPS) with PS, PEO, or L-lactide consisting of a high χ and a low N have been introduced as ultrahigh-χ systems (about 0.8) with sub-5 nm microdomains [270]. Based on the recent theoretical calculations, it has been demonstrated that PS-*b*-poly(2-vinylpyridine *N*-oxide) (PS-*b*-P_2VPyNO) and PS-*b*-poly(4-vinylpyridine *N*-oxide) (PS-*b*-P_4VPyNO) exhibited high-χ characteristics [29]. Very recently, the χ parameter for poly(4-VPy-*b*-2,2,2-trifluoroethyl methacrylate) was measured to be 0.12 + 5.77 which was higher than that of poly(4-VPy-*b*-methyl methacrylate) ($\chi = 0.08$), suggesting that the presence of the fluorine atoms in the structure of poly(4-VPy-*b*-2,2,2-trifluoroethyl methacrylate) enhances the repulsive interactions [271]. It has been reported that the counteranion species can affect the χ

parameter and phase behavior [262]. From a theoretical study on the phase separation of linear PS-*b*-P_2VPy (A-*b*-B), it has been reported that with increasing f_A at a fixed $\chi_{AB}N$, four equilibrium morphologies, gyroid (G), cylindrical (C), spherical (S), and lamellar (L), were appeared, respectively [234]. The Flory's interaction parameter for a variety of PVPy-containing BCP systems are given in literature [38,264].

Various approaches have been reported for the self-assembly of BCPs containing PVPy segments and very well documented [38,264,272–276]. In the following discussion, I present some of these methods.

Solvent vapor annealing (SVA) [273] is an approach leading to microphase separation of BCPs and formation of ordered self-assembled thin films, which are broadly applicable in different fields [273,277–282]. Due to the high T_g values for PVPy polymers and hence less polymer melt diffusivity below the thermal limit of decomposition, SVA is a useful approach for their fast self-assembly in their corresponding BCPs [272]. In SVA, well-organized BCP thin films are formed when the disorganized structure of the preformed BCP films is exposed to vapors of one or more solvents, at temperatures below the bulk T_g of both blocks. Solvent evaporation rate has a critical role in the level of the self-organization of BCP. At more rapid solvent evaporation rates, a disordered arrangement is produced, because the solvent evaporation rate is faster compared to the macromolecule diffusion. However, at slow evaporation rates, well-ordered organizations are achieved in the polymer structure (e.g., vertical, cylinders, in-plane cylinders, etc.) [283]. The efficiency of different solvents in SVA of PVPy BCPs has been investigated, and for example, it has been shown that acetone and THF for PS-*b*-P_2VPy [282,284] and PS-*b*-P_4VPy [272,285], respectively, are superior solvents.

Multicompartment particles produced by the self-assembly of PVPy-containing BCPs with different morphologies have gained extensive consideration [228,286–293] and have been recently reviewed [228,294,295]. Two common approaches for preparing these particles have been employed: (1) stepwise self-assembly [228,296,297] and (2) phase separation in emulsions upon solvent evaporation [228,286,291,294,295,298,299]. Preparation of ternary BCP systems by the stepwise self-assembly approach is particularly attractive. First, a precursor micelle is formed by assembling three constituting blocks in an aqueous solution, wherein the two relatively hydrophilic blocks are as the corona and shell and the most hydrophobic segment is located at the core. Afterward, the collapses of the corona and the shell blocks (e.g., by varying the pH of medium) drive the assembly of the pristine precursor micelle into multicompartment particles with a variety of different morphologies. The cosolvent method has extensively been employed for the generation of the pristine precursor micelles. In this method, the slow addition of a bad solvent to the dissolved BCP in a good solvent leads to aggregation of BCPs into precursor micelles; the order of BCPs from core to corona is determined based on the type of interaction that occurs between the blocks and solvent. An example for stepwise self-assembly approach is poly(oligo(ethylene glycol) methacrylate)-*b*-poly(benzyl acrylate)-*b*-P_4VPy (POEGMA-*b*-PBzA-*b*-P_4VPy) BCPs, prepared by reversible addition-fragmentation chain transfer (RAFT) method (will be discussed later). BCPs were first assembled into precursor micelles with PBzA cores, and then the collapse of P_4VPy segments at high pH values generated patchy micelles consisting of PBzA patches, P_4VPy core, and a POEGMA corona [296] (Scheme 1.21).

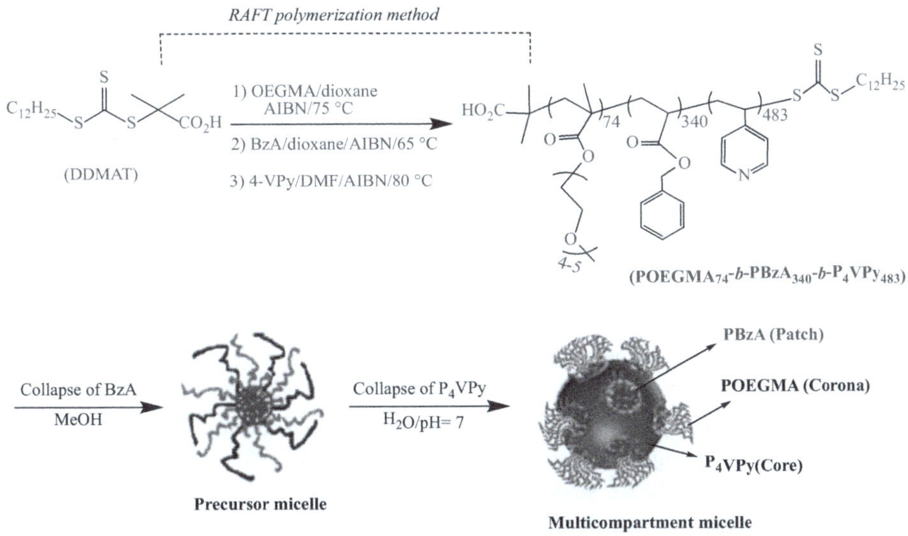

DDMAT: 2-(Dodecylthiocarbonothioylthio)-2-methylpropionic acid; OEGMA: Oligo(ethylene glycol) methacrylate monomethyl ether) ; BzA: Benzyl acrylate

SCHEME 1.21 Synthetic pathways for POEGMA$_{74}$-*b*-PBzA$_{340}$-*b*- P$_4$VPy$_{483}$ and its two-step self-assembly to multicompartment micelles via precursor micelles [296].

In the second approach of generation of multicompartment particles by self-assembly, BCPs are first dispersed in an organic solvent (Scheme 1.22). The solvent should dissolve BCP components and becomes volatile and immiscible with the continuous phase to produce a stable emulsion. Afterward, emulsification of BCPs/organic solvent phase into an aqueous surfactant solution leads to the formation of the emulsion droplets. Then, evaporation of the organic solvent continues until ordered BCP domains are formed near the interface between the surrounding aqueous solution and the droplet [294,297,299–301]. In fact, the assembly of BCPs was oriented by the emulsion droplets upon evaporation of the solvent [299]. For example, a shape- and color-switchable BCP system containing PS-*b*-P$_2$VPy segments [298] and also PS-*b*-P$_4$VPy with various morphologies (e.g., cylinder, dumbbells, triangles, snowmen, lamellae, and raspberry) assembled in an oil-in-water emulsion [289,290] has been prepared by this method.

The layer-by-layer (LbL) self-assembly is regarded as a crucial bottom-up approach for the design of multilayer films that have a range of applications [302]. In this regard, LbL assembly of P$_4$VP and a lignin derivative bearing carboxylic acid groups [303,304], carboxylated poly-(2,3,4,5,6-pentafluorostyrene) [305], and poly(acrylic acid) (PAA) [306] has been investigated for fabrication of LbL assembled films.

Self-assembly of an amphiphilic BCP, which bears solvophilic and solvophobic blocks, in block selective solvents is another method used for fabrication of various nano-assemblies. It can be done by fine-tuning the DP of the blocks and their chemical composition and the solvent character. Micellization of PVPy BCPs prepared by LbL method in block selective solvents is well documented [307] and will

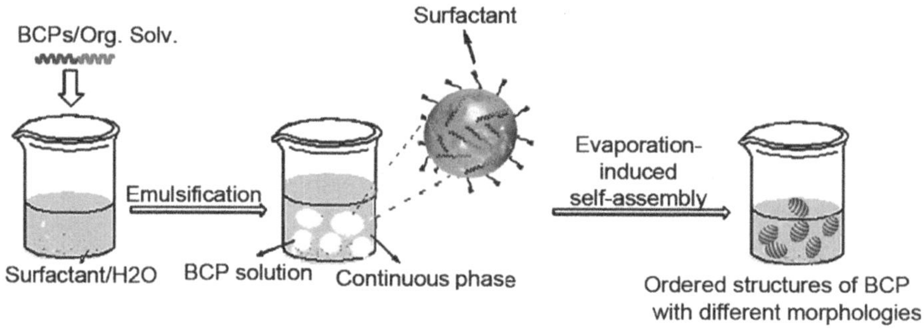

SCHEME 1.22 General concept of phase separation in emulsions upon solvent evaporation.

be discussed in the following section. For example, the star BCP of P_4VPy and PS, s-P_4VPy-*b*-PS, with multiple arms has been introduced [235]. The self-assembly of s-P_4VPy-*b*-PS in two block selective solvents, methanol and toluene, was surveyed. In methanol (a good solvent for P_4VP), s-P_4VP-*b*-PS produces core-corona nanospheres, while in toluene (a good solvent for PS), it forms s-P_4VPy-*b*-PS NP clusters in which P_4VPy creates dispersed nanodomains.

As mentioned in previous sections, P_2VPy and P_4VPy demonstrate relatively significant variations in solubility, T_g, acidity, hydrophilicity, due to the difference in the positioning of the pyridinyl nitrogen. Also, P_4VPy forms a stronger hydrogen bonding with compounds containing hydroxyl groups (e.g., phenolic resins) than P_2VPy [308,309]. Thus, when preparing BCPs of PVPy polymers bearing host sterically guests, further attention has to be paid to the use of P_2VPy or P_4VPy, depending on the employing of such BCPs [310–312]. In fact, the choice of isomers of PVPy polymers in BCPs has the strong effect on apparent solubility parameters and association constants; the first one is exemplified by the χ parameter between these segments. For example, it has been reported that microphase-separated domains of P_2VPy-*b*-P_4VPy preferentially absorb Au NPs into the P_4VPy segments, due to better availability of the pyridinyl nitrogen in P_4VPy (Scheme 1.23) [41]. In limited amounts of Au NPs, they are preferably bonded to the P_4VPy microdomains. On the other hand, with increasing amount of Au NPs, both P_4VPy and P_2VPy segments react with NPs. In that study, in order to check the performance of P_4VPy segments in absorbing more Au NPs compared to P_2VPy segments in P_2VPy-*b*-P_4VPy, two copolymers, PS-*b*-P_4VPy and PS-*b*-P_2VPy, were also tested in absorbing Au NPs. The results showed that P_4VPy segments easily bond to the

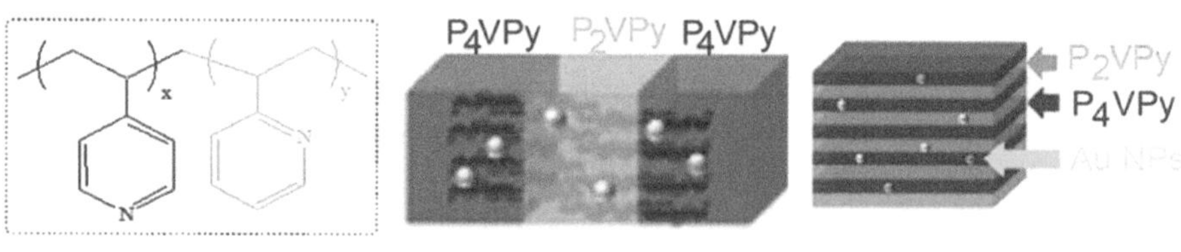

SCHEME 1.23 Preferential absorption of gold NPs onto P_4VPy blocks of P_4VPy-*b*-P_2VPy with a lamellar morphology [41]. (Reproduced with permission from Ref. [41]; Copyright 2017 American Chemical Society).

gold surface, induce density fluctuation near the surface, and hence reduce χ by limiting the interactions between PS and P_4VPy blocks. Interaction of P_2VPy with the gold surface is lower than that of P_4VPy blocks because of steric hindrance of the nitrogen in P_2VPy. The stretched P_2VPy chains cause χ to increase. Similar behavior was observed for P_2VPy-*b*-P_4VPy in absorbing Au NPs (in terms of changes in parameter χ). It is noteworthy that asymmetric sequestering of metal NPs into two microdomains in BCP systems containing PVPy segments is important for the fabrication of multicomponent nanopatterns by directed BCP self-assembly [313–315].

As mentioned above, self-assembled BCPs have many unique properties and therefore have potential applications in optical, electronic, catalytic, conductive, and biomedical nanomaterials. I refer to applications of some of self-assembled BCPs, bearing PVPy blocks, in the following discussion.

Ionic and protonic conducting materials, which are based on well-defined BCPs, have garnered considerable interest owing to their self-assembly characteristics [316,317]. These materials exhibit a variety of morphologies that demonstrate significant differences in conductivity values [316,318–325]. In this regard, BCPs containing PVPy segments have gained increasing attention [318,326,327]. To obtain high mechanical and thermal performance of conductive materials based on BCPs, one strategy is designing of AB diblock copolymers with a low glass transition ion conductor (block B) and a glassy and rigid insulator (block A) [328]. For example, the binding of PS-*b*-P_4VPy to oligomer of EO bearing SA groups doped with lithium perchlorate ($LiClO_4$) forms a self-organized polymer electrolyte which is denoted as PS-*b*-P_4VPy-($(EO)_n$-SA)($LiClO_4$) [326]. The obtained lamellar structure contains glassy reinforcing domains (PS) and ionically conducting nanochannels (P_4VPy, sulfonated EO, and $LiClO_4$) (Scheme 1.24). The conductivities of PS-*b*-P_4VPy-($(EO)_8$-SA)($LiClO_4$) and PS-*b*-P_4VPy-($(EO)_{13}$-SA)($LiClO_4$) were found to be 10^{-7} and 10^{-6} S/cm, respectively, at room temperature and rose slightly at 80°C (10^{-5} to 10^{-4} S/cm).

PS-*b*-P_4VPy containing alkyl-sulfonic acid groups and sulfate counteranions with cylindrical ionic nanochannel has been synthesized, and its proton conductivity was investigated [327]. Scheme 1.25 shows the synthetic pathways for this polyelectrolyte in which PS:P_4VPy ratio is 4. Films of this copolymer were prepared, and their ionic conductivity was measured at 25°C and 45% relative humidity. The vertical and lateral conductivities were found to be 0.015 and less than 10^{-15} S/cm, respectively. The preparation of protonically conducting self-assembled hierarchical nanostructure based on PS-*b*-P_4VPy has also been reported [318]. For the preparation of these nanomaterials, the P_4VPy block in PS-*b*-P_4VPy was first protonated with methane sulfonic acid (MSA) and then the resultant polyvinylpyridinium mesylate segments were H-bonded with pentadecylphenol (PDP) as surfactant (Scheme 1.26). Based on small-angle X-ray scattering measurements (SAXS), the self-organized structures of PS-*b*-P_4VPy(MSA)(PDP) as a function of temperature are as follows: (1) below 100°C a lamellar morphology; (2) between 100°C and 150°C, an order–disorder transition into the polymer layers; and (3) above 150°C, a hexagonal cylindrical structure formed from the disordered P_4VPy(MSA)(PDP) segments. Based on electrical conductivity measurements, at <100°C, 100°C–150°C, and >150°C, a small macroscopic conductivity in one direction, an increase of conducting lamellae in two dimensions, and reduction in the conductivity in one direction were observed,

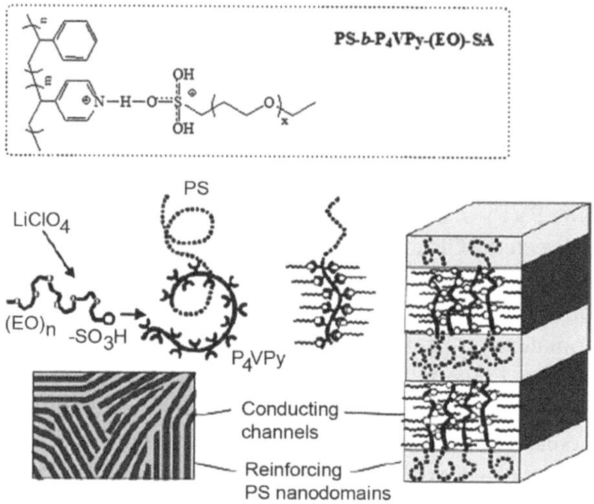

SCHEME 1.24 Immobilization of $(EO)_n$-SA oligomers to P_4VPy nitrogens of PS-*b*-P_4VPy and formation of a lamellar structure [326]. (Reproduced with permission from Ref. [326]; Copyright 2002 American Chemical Society).

respectively. At temperatures above ~195°C, the conductivity starts to increase again, where the order–order and the order–disorder transitions within the BCP length scale and the P_4VPy(MSA)(PDP) domain, respectively, both play a distinct role in conductivity. Such order–disorder transitions in P_4VPy(PDP) structures due to dynamic coupling between polymer and surfactant have been well studied [329–331]. Likewise, a lamellar-within-lamellar structure based on PS-*b*-P_4VPy in which P_4VPy blocks were reacted with PTSA, which are H-bonded with 3-*n*-pentadecylphenol (PDP), has been introduced [325,332]. The obtained copolymer is denoted as PS-*b*-P_4VPy(TSA)(PDP) (Scheme 1.27). A shear-flow field is utilized to minimize the number of domain boundaries, resulting in the successful formation of conducting nanowires. The conductivity exhibited anisotropic characteristics, being greatest along the aligned nanostructures [325].

The assembly of BCPs can be useful protocol in fabricating a range of highly ordered structures which may be directly employed as photonic devices or are utilized in nanostructures as templates. In the following discussion, I want to introduce several types of nanostructured assemblies of BCPs bearing PVPy segments for optical applications. For example, it has been demonstrated that in PS-*b*-PVPy systems, the hydrophobicity nature and $T_g \approx 100°C$ of PS leads to the selective swelling of PVPy domains with polar solvents at lower temperatures (while the morphology of PS is maintained). Therefore, this provides the use of such BCPs in systems that act as a photonic crystal with the ability to reflect different colors of light within the visible spectrum. Upon selective adsorption of ethanol by the P_2VPy phase of gyroid-forming PS-*b*-P_2VPy thin films, Bragg reflection of visible light would occur. As illustrated in Scheme 1.28, trapping and subsequent swelling of the domain at different stages cause the photonic films to undergo a reversible color change from colorless to red and then to green. [333]. They can be utilized as displays and optical sensors.

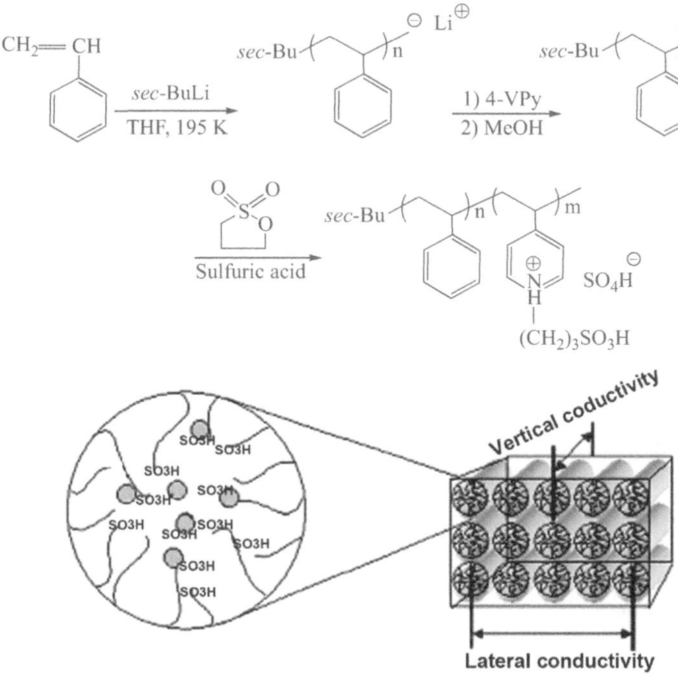

SCHEME 1.25 Attaching sulfonate groups to the pyridine units in PS-*b*-P_4VPy and cylindrically arranged ionic channels as nano highway to proton transfer [327]. (Reproduced with permission from Ref. [327] Copyright 2002 Elsevier).

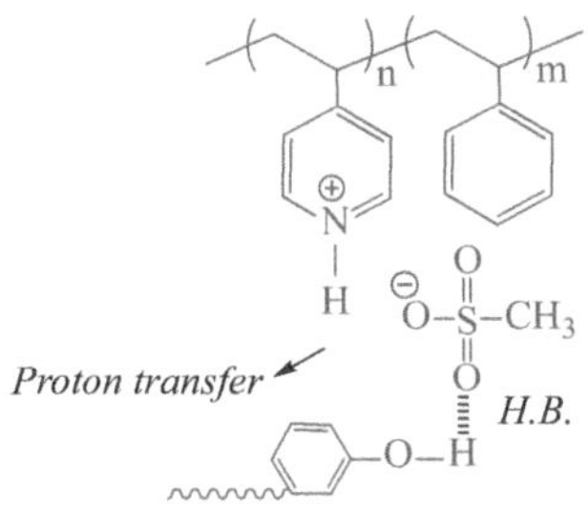

PS-*b*-P_4VPy/MSA/PDP

SCHEME 1.26 Schematic illustration of PS-*b*-P_4VPy/MSA/PDP [318].

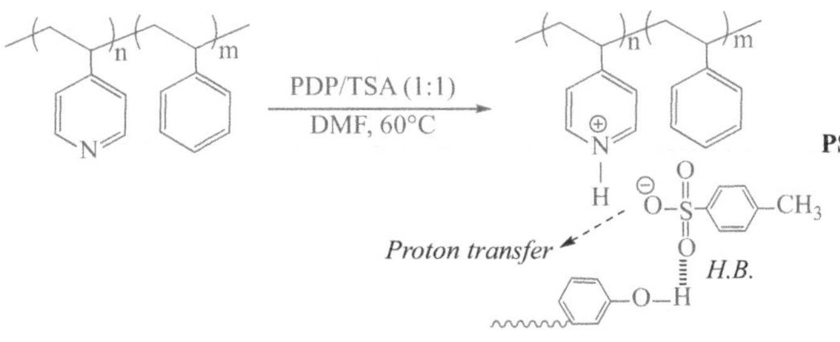

PS-*b*-P_4VPy/TSA/PDP

SCHEME 1.27 Preparation of PS-*b*-P_4VPy/MSA/PDP [325].

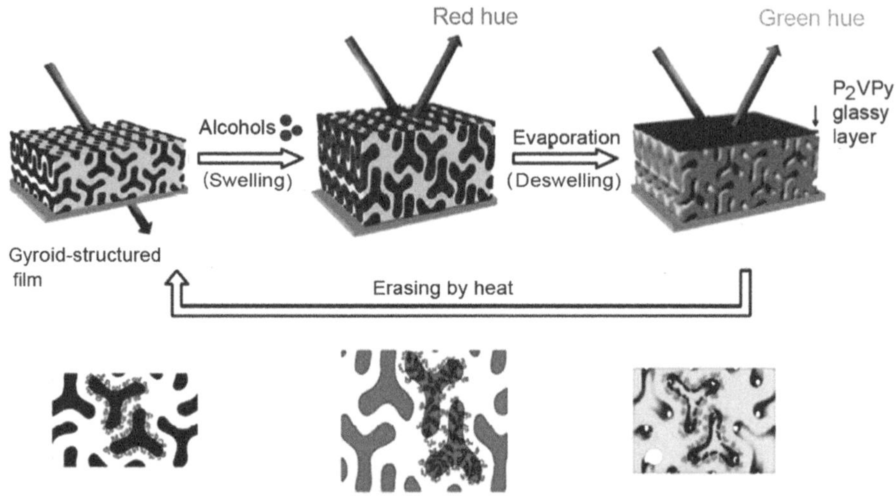

SCHEME 1.28 Selective adsorption of ethanol into the P_2VPy phase of gyroid-forming PS-*b*-P_2VPy thin films [333]. (Reproduced with permission from Ref. [333]; Copyright 2017 American Chemical Society).

Another example of the multicolor-emission phenomenon, like the one discussed above, is self-assembly of BCP of 2-VPy and ST monomers, in which 2-VPy blocks are cross-linked and quaternized with DBB and 1-bromoethane, yielding tunable periodic lamellar structure (as a photonic crystal) (Scheme 1.29) [334]. Swelling/deswelling of gel layers (by decreasing/adding salt (NH_4Cl)) causes increasing the refractive index contrast (n) as well as the domain spacing (Scheme 1.29a). The PS layers restrict expansion of QP_2VPy (the quaternized P_2VPy) gel layers to the direction normal to the layers. Pleasingly, through this chemical event, large and reversible real-time stop-band tenability (with huge shifts of stop-band, >1,250 nm, ca. 575%) was observed. In dry films, n_{QP2VPy} and n_{PS} are almost equal to 1.6, not sufficiently large to exhibit a stop band in visible region. Thus, they are transparent to the eye. When films are immersed in an aqueous solution, they exhibit a color that varies according to the degrees of cross-linking of DBB. For instance, at f_{DBB} (the DBB mole fraction) of 0.85, the films are blue ($\lambda_{peak}=421$ nm) and at $f_{DBB}=0.5$ they are red ($\lambda_{peak}=658$ nm). According to these findings, it is feasible to create multicolored patterns through the sequential application of coatings and the cross-linking of films. For example, first, a blue film of PS-*co*-QP_2VPy was constructed from PS-*co*-QP_2VPy ($f^1_{DBB}=0.85$). Afterward, a solution of the PS-*co*-P_2VPy was sprayed through a mask to make a second film, forming the letters "GEL" (Scheme 1.29b). The obtained layer was then cross-linked with DBB ($f^2_{DBB}=0.5$). Because $f^2_{DBB}<f^1_{DBB}$, the pattern displays the red-colored "GEL" on a blue background. This research demonstrates that these photonic gels may serve as an outstanding platform for the advancement of electronic display devices, colorimetric sensors, and more. The effect of salts on the tunability of photonic gels of PS-*co*-QP_2VPy lamellar films has also been investigated [335]. Self-assembly of diblock copolymer PS-*b*-P_2VPy fixed with SiO_2 has been introduced as photonic gel films with potential applications for fabricating hybrid organic/inorganic photonic crystals [336].

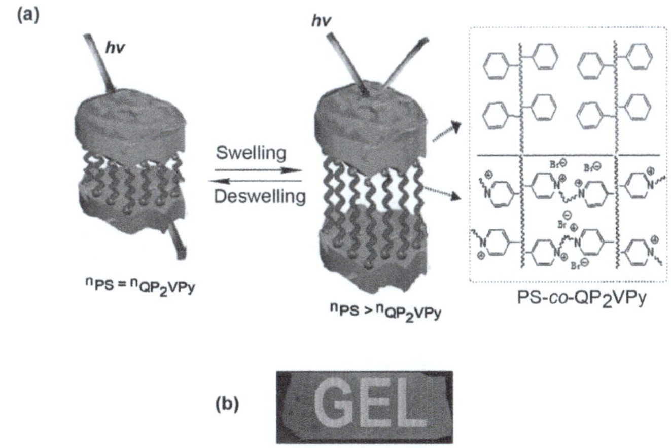

SCHEME 1.29 "(a) Swelling/deswelling of the QP_2VPy gel layers in aqueous medium and (b) photograph of the photonic gel film in water on a black background under fluorescent light [334].

Amine-containing BCPs are of interest for antimicrobial surfaces, biosensors, and cell scaffolds due to their tendency to functionalization with different chemical and biological moieties. However, the synthesis of microphase-separated morphologies of them seems still a challenge due to the aggregation of their charged sites or the destruction of their morphology during functionalization processes. Recently, PS-*b*-P_2VPy films were synthesized with a lamellar morphology and then functionalized by 3-bromo-*N*,*N*-bis(trimethylsilyl)propan-1-amine (BTMSPA) (Scheme 1.30) [337]. Deprotection of amine groups was done by HCl(aq). Afterward, the films were immersed in an aqueous solution of acetic acid (pH ~5), whereby protonated P_2VPy segments are formed, resulting in the swelling of films with the appearance of an orange color because of the lamellar structure acting as Bragg stack [337]. Subsequently, the swelled films were exposed to glutaraldehyde, leading to the formation of the cross-linked films (green in color). The findings indicated that the lamellar structure of the produced films remains unchanged following the processes of functionalization, deprotection, and cross-linking. Contrary to prevalent amine-protecting groups, e.g., di-*tert*-butyl dicarbonate (Boc) which can solubilize PS-*b*-P_2VPy or alter the original morphology, trimethylsilyl (TMS)-protecting groups do not result in the solubilization of copolymer or delamination of films. The incorporation of primary amine functionality into lamellar PS-*b*-P_2VPy films offers a sophisticated approach to integrate additional moieties such as oligonucleotides, enzymes, and others.

Ferroelectric polymers constructed from the self-assembly of a triblock copolymer consisting of P_2VPy and poly(vinylidene fluoride-*co*-trifluoroethylene) blocks, titled P_2VPy-*b*-P(VDF-TrFE)-*b*-P_2VPy, have been reported [338,339]. P_2VPy-*b*-P(VDF-TrFE)-*b*-P_2VPy were prepared by free radical copolymerization of TrFE and VDF (Scheme 1.31a) and via RAFT polymerization of 2-VPy (Scheme 1.31b), followed by click coupling of these polymers with each other (Scheme 1.31c). Various morphologies are obtained by modifying the block ratios as detailed below: (1) at high P(VDF-TrFE) content, a disordered melt structure and upon crystallization,

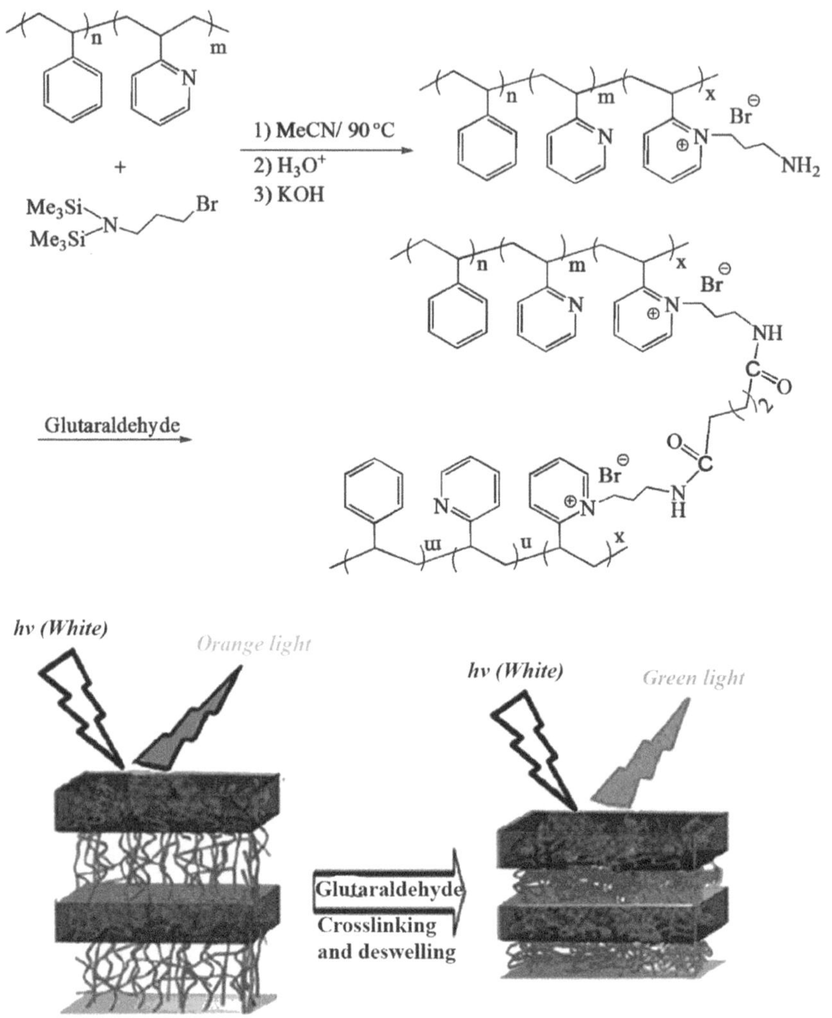

SCHEME 1.30 (a) Schematic illustration of the synthesis of microphase-separated PS-*b*-P_2VPy films containing primary amine functionalities, cross-linked by glutaraldehyde, and (b) the swollen lamellar PS-*b*-P_2VPy forms a photonic crystal due to its periodicity; its reflected color is a function of the degree of swelling [337].

a lamellar structure appear; (2) at higher content of P_2VPy the phase separation results a lamellar morphology; (3) with increasing the volume fraction of P_2VPy to 0.58, the cylindrical structure was observed; and (4) at 27 vol.% P(VDF-TrFE), disordered spherical structure was obtained (Scheme 1.32). Samples manifested as lamellar structures exhibit ferroelectric switching, whereas samples with cylindrical and spherical morphologies show no evidence of dipole switching.

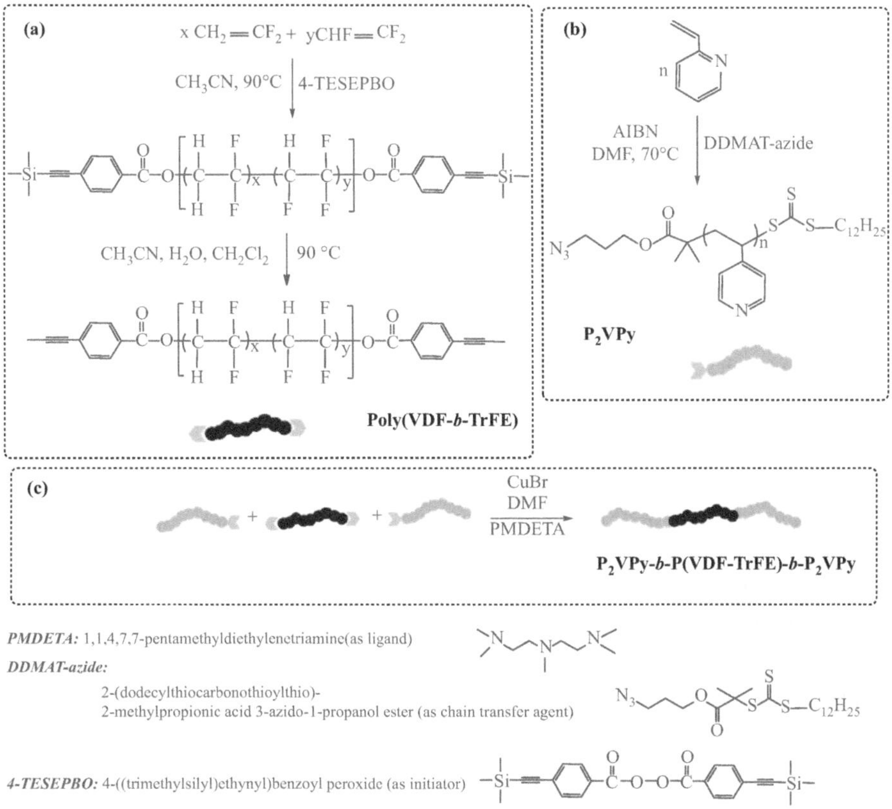

SCHEME 1.31 The manufacturing of P_2VPy-*b*-P(VDF-TrFE)-*b*-P_2VPy via CuAAc "(copper(I)-catalyzed azide-alkyne cycloaddition) click coupling of azide-terminated P_2VPy and alkyne-terminated P(VDF-TrFE) [338]. (Reproduced with permission from Ref. [338]; Copyright 2018 American Chemical Society).

1.2.5.1 Micelle Formation

As mentioned above, the self-assembly of BCPs leads to a variety of morphologies. Among them, micelle-like structures have gained a progressively increasing interest due to the facile and accurate control over the structure, shape, size, and functional tenability [340,341]. They are spherical but can appear in other morphologies such as vesicles, complex micellar aggregates, cylindrical, and platelike micelles [228].

In surfactant molecules, the morphology type can be anticipated from the packing parameter (p) according to the following equation: $p = \mathrm{v}/a_0 l_c$, where v, a_0, and l_c refer to volume of the hydrophobic segment, head group area, and hydrophobic tail length, respectively [342,276,343]. When $p \leq 1/3$, $1/3 < p \leq 1/2$, or $1/2 < p \leq 1$, spherical micelles, cylinders (worm-like or rod-like micelles), or vesicles (polymersomes) are formed, respectively. Although this equation is related to the low-molecular weight surfactants, it can also be applied for BCPs where the hydrophobic and hydrophilic blocks play the role of the surfactant tail and the polar head group, respectively.

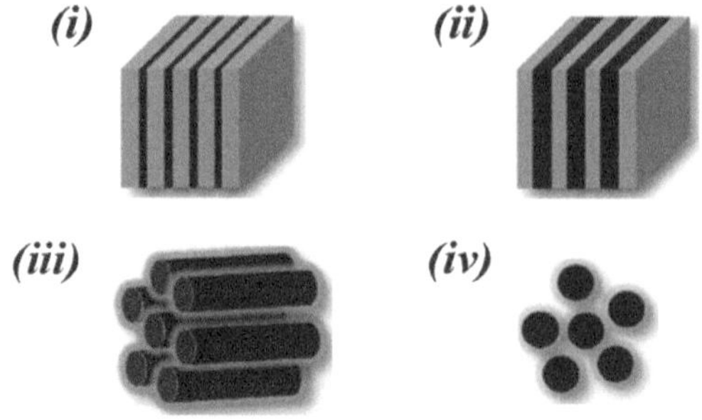

SCHEME 1.32 Different morphologies appeared in P_2VPy-*b*-P(VDF-TrFE)-*b*-P_2VPy [338]. (Reproduced with permission from Ref. [338]; Copyright 2018 American Chemical Society).

Spherical micelles were the first to be explored as nano-assemblies with various applications [276]. When dissolving AB-diblock copolymers, a well-defined micelle is formed if selective diluent is a good solvent for one block but precipitant for the other block. In this case, micelles consist of a glassy core (the insoluble blocks) and a polyelectrolyte shell or corona (the soluble blocks) [340,344]. Micelles with a small core and a large corona are known as "star-like" or "hairy" micelles, and conversely, those with a large core and a small corona are called "crew-cut" micelles [276]. The area of micellization of BCPs in aqueous media [345] or in nonaqueous media [346] strongly depends on their chemical structure. It has been demonstrated that the size of micelles is affected or changed by the solvent and the length of the blocks [347,348]. They respond to external stimuli such as the ionic strength, the pH value, the concentration, or temperature [349]. PVPy and its copolymers have been used widely and efficiently for the preparation of micelle structures [340,341,350]. They have been used as "nanoreactors" for metal NP formation [266,351,352] and as potential carriers for drug delivery [353–357].

A method to prepare polymeric micelles in water is micellization stimulated by controlling the solution pH. Mostly, these systems, which are recognized as double-hydrophilic BCPs, include a pH-sensitive acidic or basic block in the micelle core or corona. Two blocks are water-soluble, at least in some pH range, but upon an appropriate change of pH, aggregation of one block starts and then being insoluble in water. The dissociation constant pK_a is a commonly used parameter to characterize the pH-sensitive polymers, which can be relevant to the pH at which a conformational transition of polymer chains occurs. Both P_4VPy and P_2VPy, as pH-sensitive polymers, have been incorporated into amphiphilic BCPs [358–360], which are known to form various micelle structures. In this regard, PVPy-PEO-containing BCPs have attracted considerable attention [128,351,361–373]. These BCPs have the ability to create micellar structures in aqueous solution by adjusting the solution pH. PVPy is largely insoluble in water at neutral to high pH, while PEO is soluble in water at all pH levels. Although $\chi_{P2VPy\text{-}PEO}$ is small [353,363,374–376], microphase separation between these segments can be also promoted via complexation of PVPy with amphiphiles [377]. As a typical example, the micelle formation mechanism of P_2VPy-*b*-PEO is outlined in (Figure 1.9). As it is seen from the figure, the micellization of P_2VPy-*b*-PEO was started in the pH range of 4–5, correlated to the pK_a value

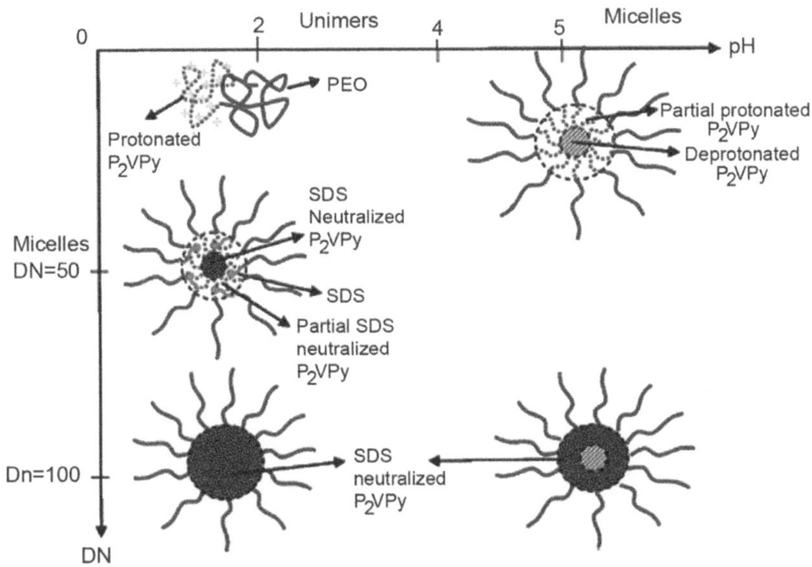

FIGURE 1.9 The micelle formation mechanism of P_2VPy-*b*-PEO [378]. [Reproduced with permission from Ref. [378]; Copyright 2013 Elsevier.]

of P_2VPy [378]. In this region of pH, both protonated and deprotonated P_2VPy blocks exist in the core and the PEO block makes the corona. At pH < 4–5, which leads to the protonation of the P_2VPy and its water solubility, double hydrophilic block-copolymer unimers were found. Experiments showed that sodium dodecylsulfate (SDS) as an anionic surfactant tends to induce the micellization of P_2VPy-*b*-PEO copolymers through the neutralization of protonated unimers at around pH 2–3 and at a given degree of neutralization (DN) of 40%–50%. It was observed that by adding more amount of SDS (about DN = 100), the particle size is partially increased. By increasing pH, the neutralization of the remaining protonated P_2VPy core of the already present micelles is continued along with a decrease in the particle size. The pH-induced release of drugs [379,380] and hydrophilic dyes from PEO-*b*-P_2VPy vesicles has been surveyed [366]. pH-sensitive poly(glycidol)-*b*-P_4VPy (PGl-*b*-P_4VPy), which is a functional analogue of P_2VPy-*b*-PEO (Scheme 1.33), has also been reported [196]. Like P_2VPy-*b*-PEO systems, the results revealed that the linkage of PGl block to P_4VPy influenced the pK_a value of P_4VPy, which can be regulated to facilitate the creation of polymeric micelle particles. Surfactants have also been used to induce micelle formation in other PVPy BCPs. For example, surfactant-mediated microphase separation in poly(4-VPy-*b*-hydroxybutylacrylate), P_4VPy-*b*-PHBA, as a model of miscible disordered BCPs, has been investigated [381]. The microphase separation including spherical and cylindrical morphologies was observed and the results showed that the ordered microstructures are affected by the miscible BCPs–surfactant interactions.

The potential of triblock copolymers composed of PVPy and PEO segments to form micellar structures has also been examined. For example, self-assembled monolayer of triblock copolymers of PS-*b*-P_2VPy-*b*-PEO (Scheme 1.34) has been

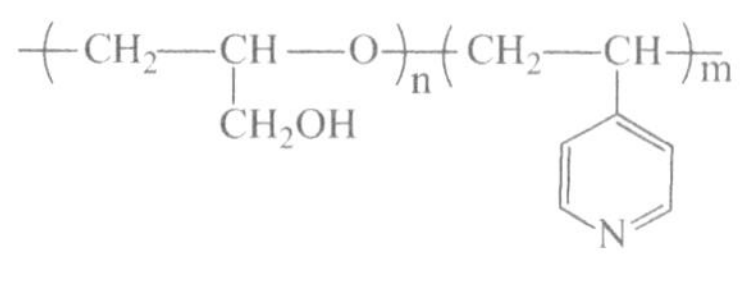

SCHEME 1.33 Structure of poly(glycidol)-*b*-P_4VPy [196].

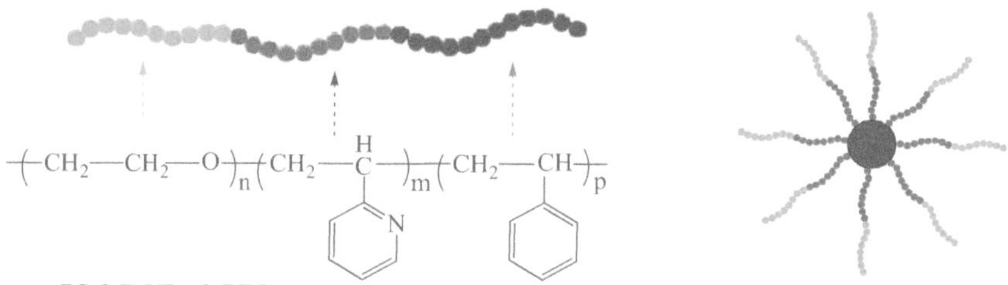

PS-*b*-P_2VPy-*b*-PEO

SCHEME 1.34 Micelle structure of PS-*b*-P_2VPy-*b*-PEO [310].

employed for the selective binding of guest metals; $AgNO_3$ and $HAuCl_4$ diffused inside the PEO and P_2VPy phases, respectively. Then, the resulting micelle structure was utilized for nanoscale patterning of metals on silicon by an aqueous galvanic displacement reaction [310]. Also, the pH-dependent micellization behavior of triblock copolymer PCL-*b*-PEO-*b*-P_2VPy (PCL: poly(ε-caprolactone), as a drug delivery system, has been investigated [382]. These flower-type micelles were prepared through sequential anionic and ring-opening polymerizations in the presence of stannous 2-ethylhexanoate (Sn(Oct)$_2$as catalyst (Scheme 1.35) [382]. They include a PCL core surrounded by a mixed corona of positively charged P_2VPy and neutral PEO. The copolymer is able to bind with biotin guests (as a nonspecific ligand) via the P_2VPy phase. PCL acts as a biocompatible core block. When P_2VPy blocks are unprotonated (neutral pH), PCL-*b*-PEO-*b*-P_2VPy collapses into flower-like micelles, and through protonation of P_2VPy, the micelle structure expands. The biotin ligand can be easily bonded to P_2VPy segments, and thus shielded in the micelles at neutral pH and exposed at the surface at low pH (down to pH 5). PS-*b*-P_2VPy-*b*-PEO triblock terpolymers having micellar structure have been synthesized [383]. All micelles possess a bright core (PS) and a dark corona, richer in PEO and P_2VPy. They exhibited various morphologies including spherical and hexagonal cylinders and lamellae, and the results showed that the composition of $PS_{80.5k}$-*b*-$P_2VP_{64.4k}$-*b*-$PEO_{16.1k}$ with a regular, isoporous morphology, is suitable for membrane application.

Three-layered micelles, namely "onion-skin" or "onion-type" micelles, can be considered as a one of the best examples to understand the formation process of micelles. They are formed by a two-step process using two diblock copolymers XY and YZ consisting of a common block Y. The first stage of synthesis includes the micellization of the copolymer that will make the micellar core (XY). The resulting micelles are immersed in a diluent which is as a solvent for the copolymer YZ. The YZ copolymer is poured into the above mixture. Then, by changing of the solvent the insolubilization of the block Y (common block) occurs. Probably,

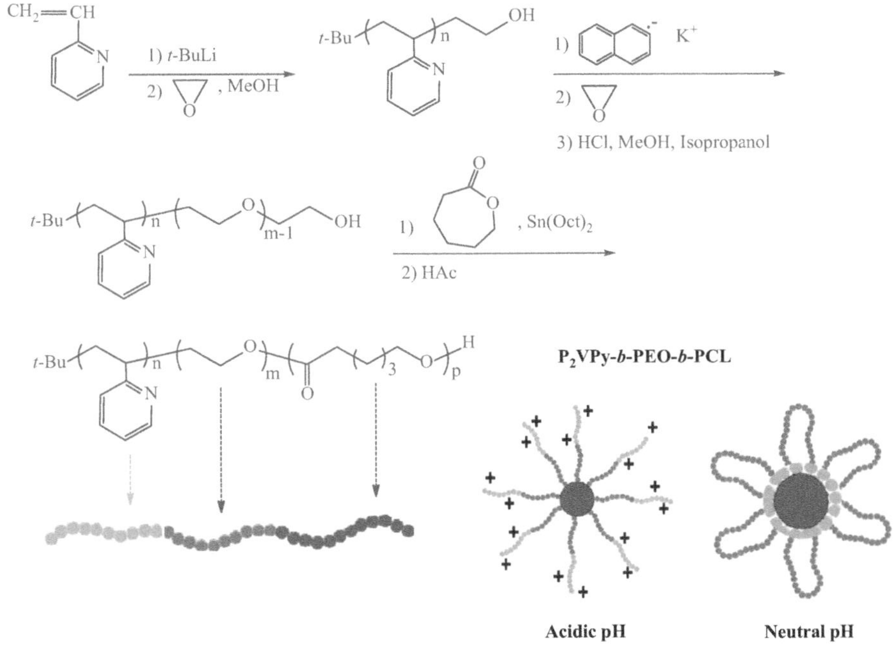

SCHEME 1.35 Synthetic routes of PCL-*b*-PEO-*b*-P_2VPy [382].

interpenetration of the Y blocks in each other leads to the formation a shell around the core X. Finally, the block Z forms the outer corona of the final micelle. For example, the synthesis of onion-type micelles from PS-*b*-P_2VPy and P_2VPy-*b*-PEO has been reported [384]. First, micellization of PS-*b*-P_2VPy was done after immersing in methanol:dioxane:water mixture (45:50:5 in volume), which was subjected to dialysis into an aqueous solution of HCl (0.1 M). The resulting micelles are comprised of the protonated P_2VPH^+ micellar corona and the PS core (Scheme 1.36). Afterward, P_2VPy-*b*-PEO was added to the mixture and the solution was adjusted to pH value of 10 using 1 M sodium hydroxide, where a coprecipitation of the common blocks P_2VPy is occurred and "onion micelles" are formed. Likely, onion-type micelles

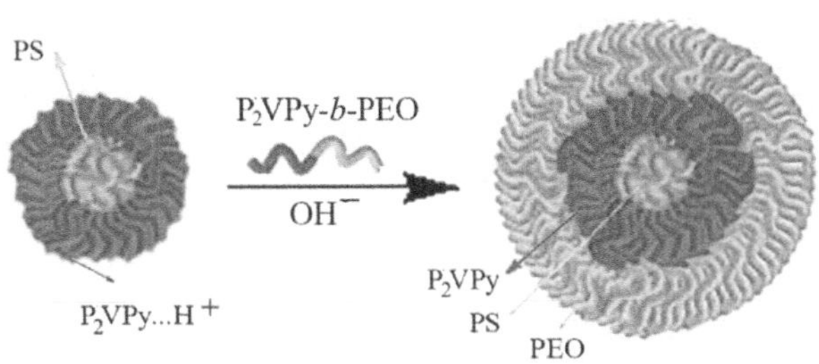

SCHEME 1.36 Schematic representations of the formation of onion-type micelles from P_2VPy-*b*-PEO and PS-*b*-P_2VPy [384]. (Reproduced with permission from Ref. [384]; Copyright 1999 American Chemical Society).

of poly(*tert*-butyl acrylate)-*b*-P_2VPy (PtBA-*b*-P_2VPy) and P_2VPy-*b*-PEO has been introduced [385].

As mentioned in previous sections, among many known amphiphilic BCPs, PS-*b*-P_2VPy and PS-*b*-P_4VPy have received considerable attention [386,387]. They have been used for micelle formations, due to their high quality, easy availability, and very low critical micelle concentration (cmc), so that the dispersion efficiency is kept, even at high dilution. In addition, the adequate difference between solution properties of P_4VPy and P_2VPy with PS in BCP favors micelle formation in various solvents [262,388–395]. The very different basicities of PVPy and PS allows acidic species to be selectively assembled on PVPy domains. For example, it has been shown that carboxylic acid-functionalized C_{60} fullerene molecules can be attached to the PS-*b*-P_4VPy micelles (PS corona and P_4VPy cores) by hydrogen bonding between the pyridyl group in the BCP micelles and carboxylic acid groups in C_{60} molecules [396]. Following this work, complexation between spherical micelles of PS-*b*-P_4VPy [397] with C_{60} molecules in xylene solutions was also investigated, and it was demonstrated that in fact charge-transfer complexation is mainly responsible in these interactions. Due to the good solubility of C_{60} in xylene molecules, it is expected that they are found in the PS micellar corona and not in the P_4VPy cores. However, the spectral data exhibited that the C_{60} molecules are able to penetrate into the P_4VPy cores to make charge-transfer interactions with them (Scheme 1.37a). Such interactions have been observed in the preparation of CdS NPs/C_{60} amphiphilic diblock copolymer PS-*b*-P_2VPy [398]. In this investigation, it was observed that C_{60} molecules

(a)

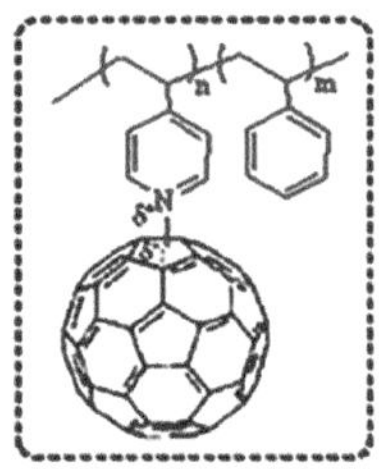

(b)

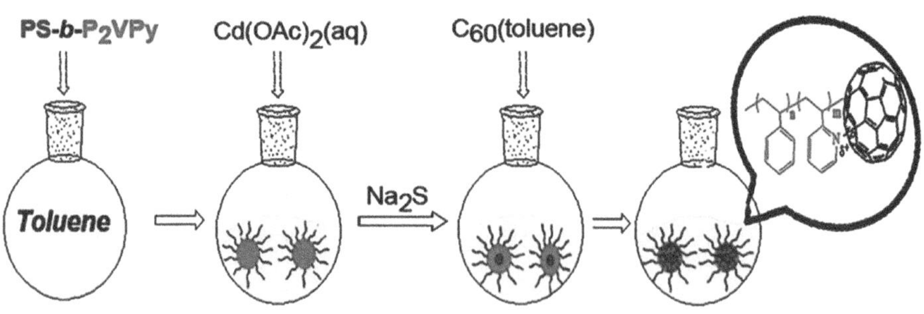

SCHEME 1.37 (a) Charge-transfer complexation between PS-*b*-P_4VPy and C_{60} [397]. (Reproduced with permission from Ref. [397]; Copyright 2006 American Chemical Society). (b) Preparation pathway of CdS/C_{60} NPs using PS-*b*-P_2VPy micellar template [398].

tend to coordinate to the pyridine rings of the P_2VPy cores ($C_{60}^{\delta-}$-$P_2VPy^{\delta+}$) without aggregations (Scheme 1.37b). Also, fluorescent quenching study revealed that there is an energy transfer between CdS (as electron donor) and C_{60} molecules (as electron-acceptor), useful for sensing or photoelectrochemical applications. Micelles loaded with different NPs as catalyst precursor for the synthesis of single-walled carbon nanotubes (SWCNTs) [399–403], functional organic/inorganic composite particles with different emission wavelengths [404], and nanocomposite with the spin crossover (SCO) properties [405–408] are other reported examples of PS-*b*-PVPy-based micelles. SCO materials show stimuli responses to light, pressure, or temperature between a low spin and a high spin state. Switching between these two states changes physical properties such as color, structure, or magnetism, thus rendering possibilities for sensor applications, display devices, or as functional contrast agents. Scheme 1.38 shows the synthetic pathways for an example of these systems [405]. Subsequent addition of the bridging ligand 4,4′-bipyridine (bipy) and an iron complex (Fe(L)$(MeOH)_2$) to PS-*b*-P_4VPy micelles leads to the crystallization of the SCO coordination polymer $[FeL(bipy)]_n$ inside the micellar core (P_4VPy blocks). Very recently, the inverse micelles of PS-*b*-P_2VPy through addition of D-tartaric acid (D-TA) to BCP structure in the non-solvent were prepared [409]. The results showed that various morphologies can be obtained with the change of D-TA concentration. Inverse micellar nanostructures, particularly those with open channels, have gained considerable attention in various fields like drug delivery and catalyst templating.

Another BCP of PVPy used in the construction of micelles and responsive hydrogels is the one having poly(*tert*-butylmethacrylate) (PtBMA) and poly(*tert*-butyl acrylate) (PtBA) in their structures [410–419]. PtBMA and PtBA are readily converted into PMAA and PAA, respectively, via ester hydrolysis. An interesting

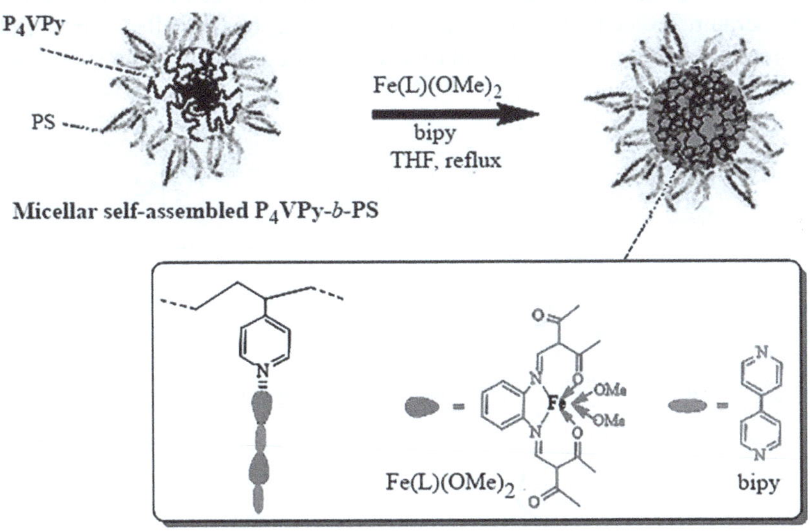

SCHEME 1.38 Synthesis of $[FeL(bipy)]_n$@P_4VPy-*b*-PS [405].

sample of these systems is A_n(B-*b*-C)$_n$ heteroarm star BCP containing cross-linked poly(divinylbenzene) (PDVB) core and PS and PAA-*b*-P_2VPy arms (Scheme 1.39) [38,420,421]. The multistep synthetic procedure involves the following steps: (1) synthesis of "living" linear PS chains by reacting *s*-BuLi with styrene, following by reaction with DVB monomer (cross-linking agent), resulting in a "living" PS star-shaped polymer with cross-linked core; (2) living P_2VPy arms were grown on the core upon the addition of 2-VPy monomers; (3) PtBA blocks were grown at the terminus of each living P_2VPy arm following the incorporation of t-BA monomers; and (4) acid hydrolysis of the ester groups resulting in the PS_n-PDVB-[P_2VPy-*b*-PAA]$_n$ star terpolymer [421]. PS_n-PDVB-[P_2VPy-*b*-PAA]$_n$ exhibits a rich phase behavior as a function of pH. At pH < 4, P_2VPy became protonated and the star terpolymers act as a reversible hydrogel. PS arms collapsed in the core, PAA forming the external corona, and P_2VPy extended blocks forming the intermediate shell. At pH > 7, PAA deprotonated. At pH ~5.5, a PAA/P_2VPy complex exists, and an isoelectric point (IEP) can be identified, where the negative and positive charges on the block polymer are equal (P_2VPy^+-H = PAA^-). Also, very likely, it has been reported that the PS_n-PDVB-[P_2VPy-*b*-PAA]$_n$ star terpolymer shows three main conformations at different pH regimes [420].

Among the stimulus-responsive copolymers [422], BCPs consisting of thermo-responsive block such as poly(*N*-isopropylacrylamide) (PNIPAM) and pH-responsive block such as PVPy have received remarkable attention because of their smart response to stimulus [423–428]. BCPs containing both thermo- and pH-responsive blocks are expected to respond to combined stimulus. In these systems, PVPy at pH < 5 is soluble and exists as unimers in water, while at pH > 5, it precipitates [127]. PNIPAM is recognized for its lower critical solution temperature (LCST, between 30°C and 35°C). Below LCST, the polymer chains stay adequately hydrated as a result of hydrogen bonding with water molecules. Conversely, when the temperature exceeds the LCST, the PNIPAM segments collapse into globular structures, resulting in a loss of solubility, which is attributed to the suppression of

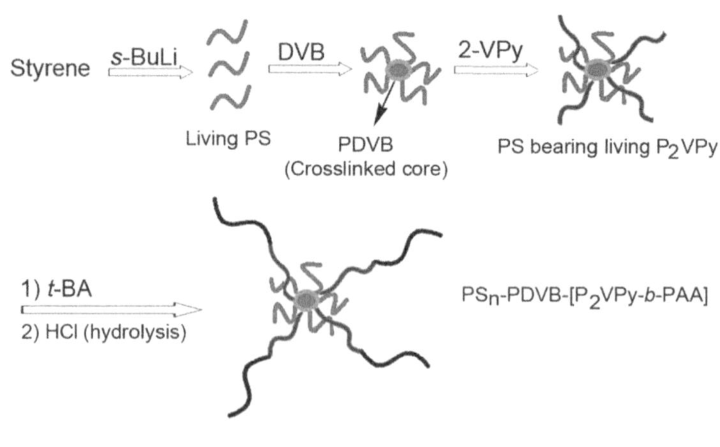

SCHEME 1.39 Synthetic pathways for PS_n-PDVB-[P_2VPy-*b*-PAA]$_n$ [421].

hydrogen bonding [429]. For example, nanogels with temperature-responsive core (photo-cross-linked of PNIPAM) and pH-responsive arms (P_2VPy) have been synthesized [430]. PEG-*b*-P_4VPy-*b*-PNIPAM (PEG: poly(ethylene glycol)) can form micelles with various morphologies during the combined stimulus of temperature and pH changes (Scheme 1.40) [129]. This triblock copolymer associates to form microphase-separated spherical core-corona micelles at pH 2–6.5 at 25°C, while exists as unimers at pH 2. The LCST of the PEG-*b*-P_4VPy-*b*-PNIPAM was found to be 34.6°C, which is higher than that of PNIPAM. When the copolymer solution was heated above its LCST, it first self-assembled into core–corona micelle structure or micellar cluster and then converted into core–shell–corona micelles with pH increasing from 2 to 6.5. The morphology of the copolymer transitioned from the core–shell–corona micelles into core–corona micelles with decreasing temperature from 50°C to 25°C. Also, very recently, the micellar characteristics of P_2VPy-*b*-poly(cyclohexyl methacrylate) in nonaqueous media with their potential applications as colloidal stabilizers as well as gasoline additives have been reported [431].

In contrast to linear PNIPAM, which is a water-soluble polymer, colloidal PNIPAM microgels are intramolecularly cross-linked polymeric particles which can be dispersed in an aqueous medium, exhibiting hydrodynamic diameters ranging from approximately 100 nm to 1 μm [432]. The incorporation of ionizable groups into the polymer network greatly improves both the swelling ratio and the temperature at which the phase transition of PNIPAM gels occurs. For example, the synthesis and thermo-responsive swelling/deswelling characteristics of a PNIPAM-*co*-P_4VPy gel cross-linked with MBAAM were detailed, alongside an examination of the electrochemical properties of this hydrogel when coordinated with ruthenium [130]. It was observed that the gel collapses at temperatures >40°C. Also, cross-linked PNIPAM-*co*-P_4VPy microgel particles with varying proportions of 4-VPy were produced, and their swelling/deswelling behaviors in relation to temperature, pH, and ionic strength were analyzed [433,434].

Photoinduced formation of micelles has attracted increasing attention, in which photo-stimulatory response polymers, containing photosensitive groups, are used [435,436]. Photocontrollable self-assembly of a polymer pair containing

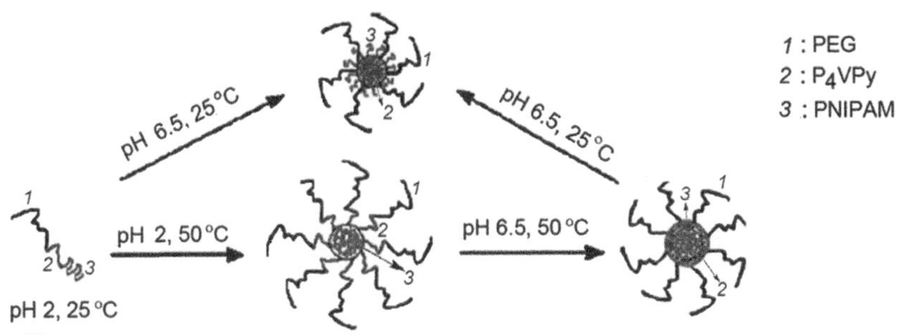

SCHEME 1.40 The thermo- and pH-responsive illustration of PEG-*b*-P_4VPy-*b*-PNIPAM [129]. (Reproduced with permission from Ref. [129]; Copyright 2005 American Chemical Society).

polybutadiene with a terminal carboxy group (CPB) and poly(4-phenylazomaleinanil-*co*-4-VPy) (AzoMI-VPy) has been investigated (Scheme 1.41a) [437]. Upon exposure to photo-irradiation via UV and VS light, azobenzene as a photosensitive group can undergoes reversible isomerization. The carboxyl group and pyridine unit act as a proton donor and a proton acceptor, respectively, resulting in the formation of a photosensitive micelle through hydrogen bonding. As it can be seen in Scheme 1.41b, through the conversion between VS and UV light, reversible isomerization of the *trans*-azobenzene group into *cis*-azobenzene occurs. AzoMI-VPy/CPB can act as a carrier in delivery systems.

1.2.6 Polyzwitterions based on PVPy Polymers

PVPy derivatives have been used in the preparation of a type of polyampholytes, namely polyzwitterions (PZs), also referred to "zwitterionic polyampholytes" or "zwitterionic polyelectrolytes", in which both cationic and anionic moieties exist within the same monomer unit, and their solution properties depends on the nature of both ionic groups, the ionic strength, and the distance between charges [105,438–450]. They have attracted particular attention due to their unique structures and excellent properties and can be used as components of thermally reversible gels, and some of them exhibited anti-polyelectrolyte behavior [451,452].

Typical studies on zwitterionic polyampholytes based on PVPy polymers can be found in an early report, where polysulfobetaines were prepared from the reaction of 2-VPy or 4-VPy with butanesultone, i.e., internal esters of hydroxysulfonic acids, and subsequent polymerization of the monomeric sulfobetaines [453] (Scheme 1.42a). Polysulfobetaines can also be prepared by direct reaction of the corresponding PVPy polymers with sultones (Scheme 1.42b). Noting that the preparation of totally sulfoalkylated polymers is very difficult, due to the precipitation of the polymers during the reaction. The $P_2VPy\text{-}BuSO_3$ and $P_4VPy\text{-}BuSO_3$ are soluble in water over the whole pH range. $P_4VPy\text{-}BuSO_3$, has been used as a zwitterionic polyelectrolyte for cathode modification of polymer solar cells [446]. It has been noted that 4-VPy carboxybetaine demonstrates effective blood-inert function in adhesion tests with different human blood cells [454].

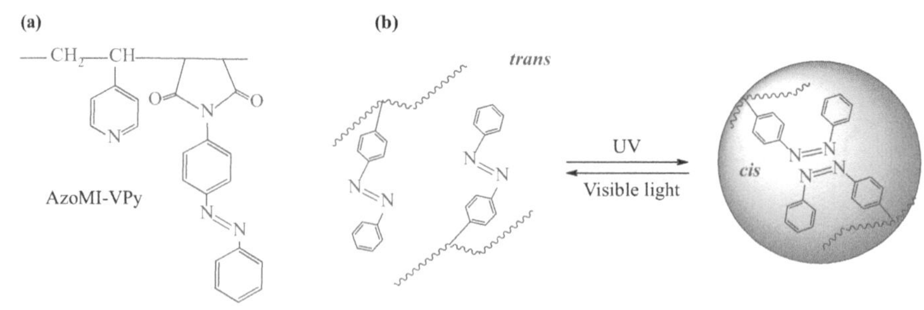

SCHEME 1.41 Illustration of (a) AzoMI-VPy and (b) the reversible photoinduced micellization [437].

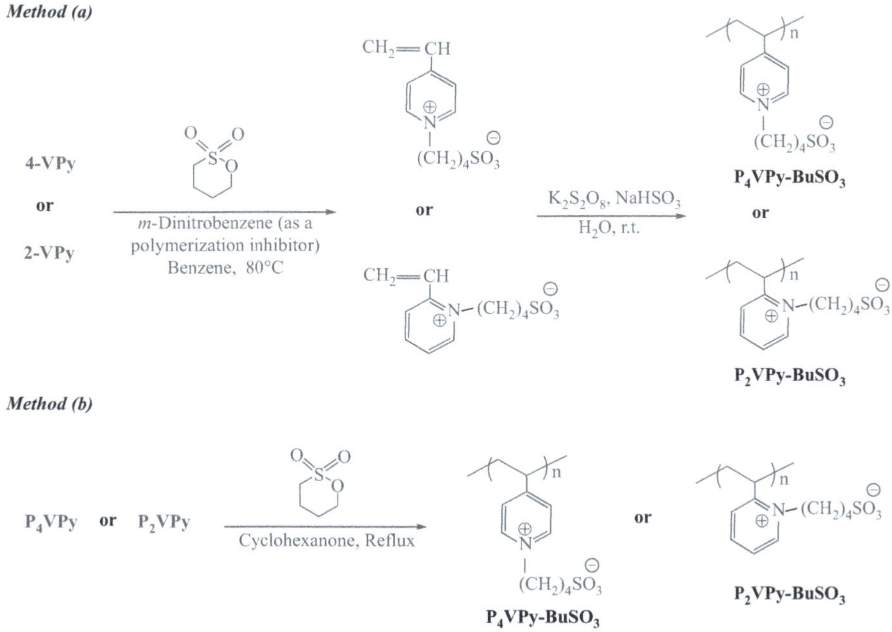

SCHEME 1.42 Two synthetic routes for polysulfobetaines.

Recently, zwitterionic copolymers based on P_2VPy-*co*-PS have been synthesized, and their solubility depending on pH was surveyed [445]. All polymers were synthesized via RAFT polymerization method (discussed later) using 4-cyano-4-(phenylcarbonothioylthio)pentanoic acid (CPeDB, as the chain transfer agent) and AIBN (as initiator). Scheme 1.43a shows the synthetic pathways for P_2VPy-*co*-PS. As shown in Scheme 1.43b, in order to synthesize the corresponding zwitterionic copolymers, the P_2VPy-*co*-PS was quaternized by 1,3-propanesultone (denoted by P_2VPys-*co*-PS) or ethyl-2-bromoacetate, followed by ester hydrolysis (denoted by P_2VPyc-*co*-PS). Likewise, zwitterionic homopolymers based on P_2VPy were prepared from the reaction of P_2VPy with 1,3-propanesultone (denoted by P_2VPys) or ethyl-2-bromoacetate, followed by ester hydrolysis (denoted by P_2VPyc). pH-dependent charge for P_2VPys and P_2VPyc are presented in Scheme 1.44. In the case of P_2VPys at pH < 4, non-quaternized P_2VPy segments are protonated. Raising the pH toward the pK_a of P_2VPy (4.7–5.1), i.e., between 4 and 8, the P_2VPyH^+ units are deprotonated gradually, and at pH > 8 the P_2VPyH^+ units are completely converted to uncharged P_2VPy, resulting in (overall) charge neutral polymer. Unlike P_2VPys-*co*-PS, P_2VPys was soluble over the whole range of pH. The solubility of P_2VPys-*co*-PS in water decreases with increasing content of the hydrophobic part in the P_2VPys-*co*-PS, i.e., PS blocks. In the case of P_2VPyc, at pH < 4, both non-quaternized P_2VPy units as well as the carboxylic acid moieties are protonated. Raising the pH from 4 to 7 will cause deprotonation of P_2VPyH^+ units and carboxylic acid functionalities. At pH > 7 the P_2VPyH^+ units and carboxylic acid groups were completely deprotonated. Interestingly, P_2VPyc

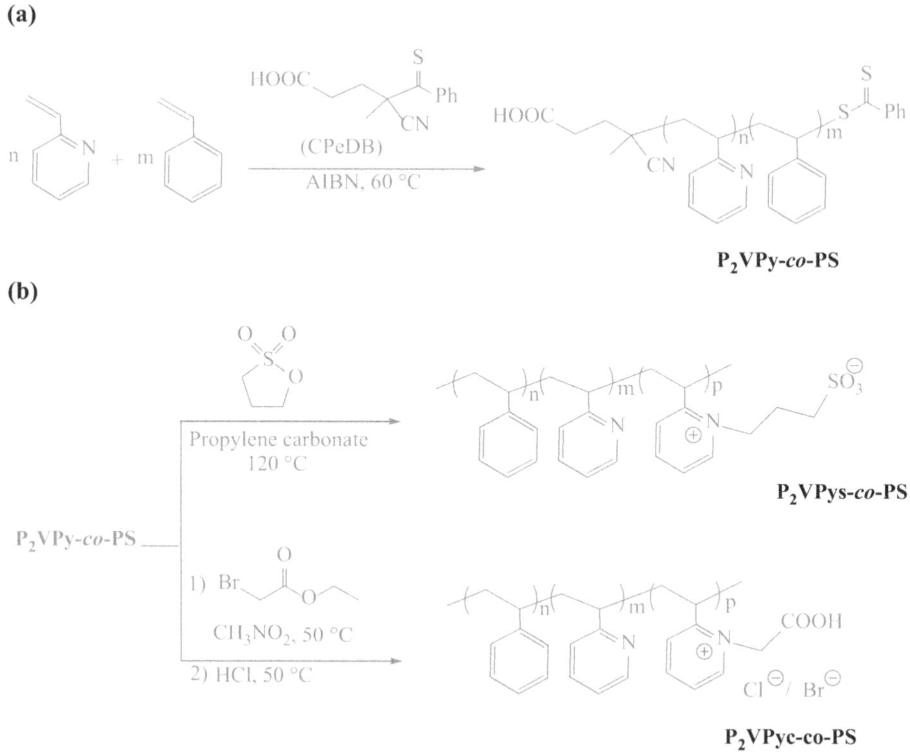

SCHEME 1.43 Synthetic pathways for (a) P_2VPy-*co*-PS and (b) its corresponding polyzwitterionic polymers [445].

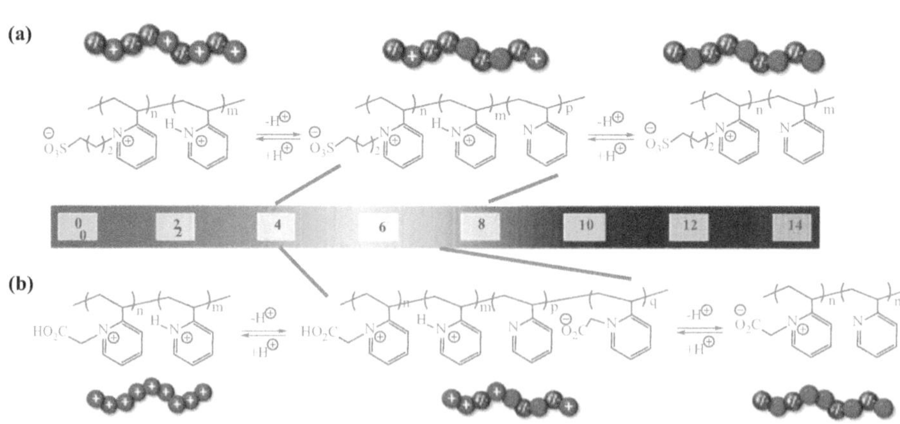

SCHEME 1.44 pH-dependent charge for (a) P_2VPys and (b) P_2VPyc [445]. (Reproduced with permission from Ref. [445]; Copyright 2016 Elsevier).

and P_2VPyc-*co*-PS were soluble over the whole pH range. A neutral zeta potential was observed in the pH range of ca. 6–9 and 5–7 for P_2VPys and P_2VPyc, respectively. Around pH 6 an IEP can be determined. The tunable charge of these homopolymers and copolymers make them potential as selective layers in polymer membranes.

Very recently, a zwitterionic polymer, namely P_4VPyPC-*co*-PDMAPC, was prepared from the reaction of poly(4-VPy-*co*-2-(*N,N′*-dimethylamino)ethyl methacrylate) (P_4VPy-*co*-PDMAEMA) with *β*-propiolactone via RAFT polymerization process (Scheme 1.45) [455]. Then, it was affixed to an ITO glass substrate through spin-casting, and its antifouling characteristics of the resulting polymer film were examined. The obtained polymer film was demonstrated to display high nonfouling activity against bovine serum albumin (BSA), lysozyme (Lys) proteins and human kidney cells. Zwitterionic copolymers of acrylamide (AAM) and 4-VPy propylsulfobetaine [456], (2-hydroxyethyl methacrylate) and 2-vinylpyridine propylsulfobetaine [457], and tertiary amino-methacrylates) and 4-vinylpyridine propylsulfobetaine [107], and also zwitterionic hydrogel of 2-vinylpyridine propylsulfobetaine and *N,N*'-methylene-bis-acrylamide as cross-linker [458] have also been reported. A zwitterionic polymer made from gamma radiation of poly(vinyl chloride) (PVC) catheters with 4-VPy and subsequent functionalization with 1,3-propane sultone was introduced and its biomedical applications, such as drug delivery and inhibit bacterial growth, were investigated [459].

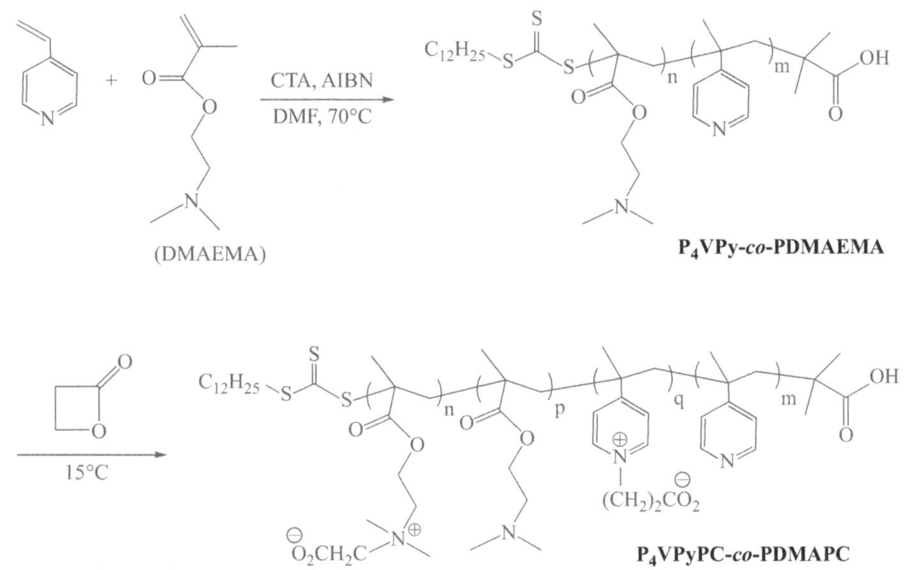

DMAEMA: (*N,N′*-Dimethylamino)ethyl methacrylate
CTA: 2-(Dodecylthiocarbonothioylthio)-2-methylpropionic acid as a chain transfer agent

SCHEME 1.45 Synthetic pathways for P_4VPyPC-*co*-PDMAPC as a zwitterionic polymer [455].

1.2.7 Polymer Blends based on PVPy Polymers

Although, as mentioned before the self-assembly of diblock copolymers (DBCPs) is an excellent approach for the preparation of nanostructures with well-defined morphologies and properties, sometimes this method requires complicated procedures and is time-consuming or expensive. To overcome such problems, blending a homopolymer or a second copolymer into the pristine copolymer, contributing to a stabilization of unusual self-assembled nanostructures through interpolymer specific interactions (e.g., hydrogen bonding), is thought to be an effective way [115,460]. The solution properties of blends of PVPy polymers with some polymers and copolymers have been investigated [171,461–471].

Typically, Figure 1.10 shows three self-assembled systems which can be produced from the blending of an immiscible A-*b*-B copolymer with a homopolymer C or C-*b*-D copolymer [472–474]. The first one is A-*b*-B/C blends, formed from blending immiscible A-*b*-B copolymer with a homopolymer C, where the B and C blocks are miscible via intermolecular hydrogen bonding (Figure 1.10a), e.g., PS-*b*-PVPh blended with PVPy [115]; PS-*b*-P_2VPy blended with PVPh [475,476]; poly(isoprene)-*b*-P_2VPy (PI-*b*-P_2VPy) blended with phenolic thermosets [477]; poly(caprolactone-*b*-2-VPy) (PCL-*b*-P_2VPy) blended with PAA [478], P_4VPy [479], and with poly(hydroxyether of bisphenol A) [480]; blends of P_2VPy-*b*-PMMA and

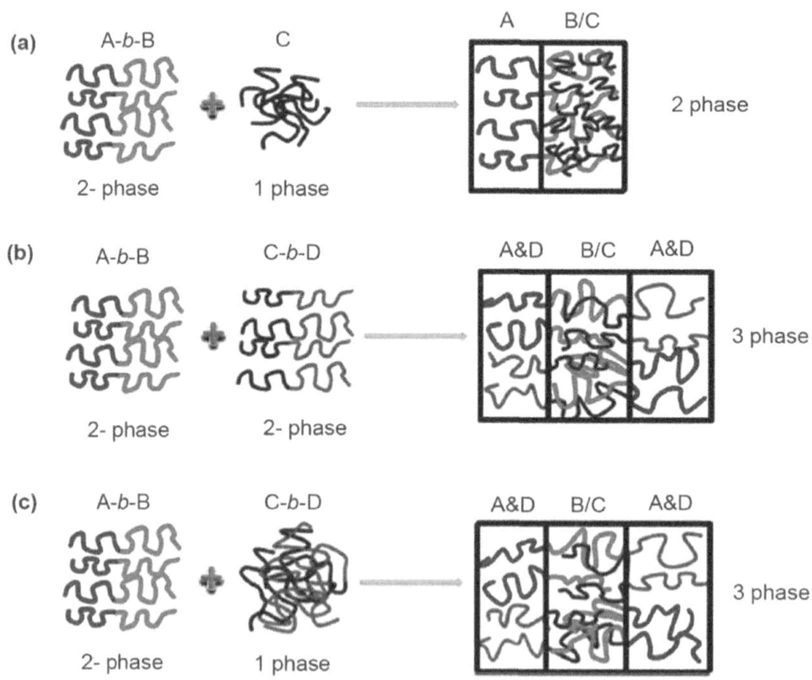

FIGURE 1.10 Schematic representation of three different systems for diblock copolymer blends [486]. (Reproduced with permission from Ref. [486]; Copyright 2018 American Chemical Society).

poly(hydroxyether of bisphenol A) [481]; and P_2VPy-*b*-PMMA blended with PVPh [482]. The second one is A-*b*-B/C-*b*-D blends, formed from blending immiscible A-*b*-B copolymers with immiscible C-*b*-D DBCPs, where the A and D blocks are immiscible and the B and C blocks are miscible by intermolecular hydrogen bonding (Figure 1.10b), e.g., blends of PI-*b*-P_2VPy and PS-*b*-PVPh [483], PCL-*b*-P_4VPy and PVPh-*b*-PS [484], and PS-*co*-PVPh an PMMA-*co*-P_4VPy [485]. The third case of the above-mentioned systems are A-*b*-B/C-*b*-D blends, formed from blending immiscible A-*b*-B copolymers with miscible C-*b*-D copolymers (Figure 1.10c), e.g., the immiscible PS-*b*-P_4VPy with miscible PVPh-*b*-PMMA providing hierarchical self-assembled nanostructures, including cylinder and three lamellar morphologies [486].

Noticeably, the blending of A-*b*-B copolymer with A or B homopolymers can also cause order–order structures or macrophase separation [487,488]. Three distinct possibilities arise based on the molecular weight ratios of the A homopolymer and the A block segment ($a = M_{h-A}/M_{b-A}$): (1) a "wet-brush" system ($a < 1$), (2) a "dry brush" system ($a \sim 1$), and (3) complete macrophase separation ($a >> 1$).

T_g of the blends is commonly employed for the determination of polymer–polymer miscibility. For example, in binary blends, if a composition-dependent shift in T_g of the blend is observed, along with a singular glass transition, the blend may be regarded as highly miscible. Kwei proposed an empirical Equation 1.1 (where w_1 and w_2 ($=1 - w_1$) are the weight fractions of the component polymers; k and q are the interaction parameters to exhibit the composition dependence of T_g for a binary blend of polymers 1 and 2, with specific interaction parameters [489]. Very recently, glass transition temperatures as functions of the molecular weight and the blend composition in P_2VPy/PVPh and P_4VPy/PVPh blends were introduced [468]. A weak deviation from the Kwei equation for P_2VPy/PVPh blends but not for P_4VPy/PVPh blends was observed, which can be due to difference in their hydrogen bonding capabilities.

$$T_g = \frac{w_1 T_{g1} + k w_2 T_{g2}}{w_1 + k w_2} + q w_1 w_2 \tag{1.1}$$

Miscibility in blends of poly(benzimidazole) (PBI) and P_4VPy has been studied and the results showed that they are miscible over entire composition range, due to hydrogen bonding [490,491]. The T_g values of the samples were measured and compared with those obtained from another equation used to describe the T_g of blend materials, namely the Fox equation (Equation 1.2, where *w is* the weight fraction) [492]. The experimental and calculated values were in good agreement with each other [491]. The T_g value of PBI/P_4VPy blend (50:50, w/w) was 538°C, which was between the T_g values of PBI (683°C) and P_4VPy (423°C) alone.

$$\frac{1}{T_g} = \frac{w_{\mathrm{PBI}}}{w_{\mathrm{g,PBI}}} + \frac{w_{\mathrm{P_4VPy}}}{w_{\mathrm{gP_4VPy}}} \tag{1.2}$$

Interpolymer complexes (polymer blends) made by hydrogen bonding interactions have attracted particular attention due to their modified physicochemical properties with respect to the individual constituent polymers [462,493]. The interactions that occur between PVPy polymers and PVPh through hydrogen bonding serve as an excellent illustration for comprehending the fundamental behavior of BCP/blend [476,494]. Usually, the miscibility is raised by the intermolecular interaction through hydrogen bonding between the pyridine moieties of the PVPy and other existing polar groups such as amide and hydroxyl groups in the polymer blend [166,476,495]. Such inter-associated hydrogen bonds in the blends eventually lead to the formation of miscible materials with highly unique performance. The addition of PVPh to PS-*b*-PVPy systems (or the addition of PVPy to PS-*b*-PVPh systems) is very similar to the addition of molecule phenols to this BCP. In fact, the association of PVPh and PVPy improves their miscibility with increasing volume fraction of the PVPh/PVPy phase [38]. For such blend systems, the choice of P_4VPy or P_2VPy is important. For example, PS-*b*-PVPh/P_2VPy and PS-*b*-PVPh/P_4VPy blends displayed different morphologies and solution properties, due to the higher steric demand for PVPh to hydrogen bond with P_2VPy over P_4VPy [115].

The PS-*co*-P_2VPy and specially PS-*co*-P_4VPy are most frequently present in various blends with good physicochemical properties [472,486,496–501]. For instance, PS-*co*-P_4VPy has been used to improve miscibility of blends of syndiotactic polystyrene (sPS)/thermoplastic polyurethane (TPU), one of the famous chemicals with low-temperature elasticity and wear and tear resistance [502]. The compatibilization mechanism of the diblock copolymer in sPS/TPU blends is due to miscibility of the noncrystalline regions of the sPS with its PS blocks, and interaction of its P_4VPy blocks with TPU via intermolecular hydrogen bonding. A series of polymer blends were constructed from sulfonated poly(etheretherketone) or poly(ethersulfone) with P_4VPy and used in membrane fuel cells [503]. A binary blend of PI-*b*-PS-*b*-P_2VPy has been used for the preparation of thin films with exciting potential for nanopatterning [504–507]. Blends of the PS-*b*-P_2VPy and the homopolymer PS have been proved to be successful in dispersion of multiwalled carbon nanotubes (MWCNTs) by up to 263% [488].

1.2.8 Magnetic Behavior of PVPy Derivatives

Magnetic behavior of complexes of both P_4VPy and P_2VPy with some metal chlorides ($M(II)Cl_2$, M = Zn, Co, Ni, Cu) has been investigated [508]. The results showed that P_4VPy prefers intermolecular complexation of the metal cation with the nitrogen atoms of pyridine rings, resulting in cross-linked polymer (Scheme 1.46). In contrast, P_2VPy undergoes complexation with metal ions through intramolecular coordination. All PVPy-$ZnCl_2$ complexes were diamagnetic. The magnetic moments for PVPy-$CuCl_2$ and PVPy-$NiCl_2$ complexes ranged from 1.59 to 1.97 and 3.14 to 3.63 BM, respectively. PVPy-$CoCl_2$ complexes exhibited magnetic moments in the range of 4.43–4.60 BM. In a similar study, it was demonstrated that the magnetic moments for P_2VPy-$CuCl_2$ complexes were decreased from 1.64 to 1.66 to 0.1 μB upon heating the complexes to 200°C, indicating reduction of paramagnetic Cu(II) to diamagnetic forms (Cu(I) or Cu[0]) [509]. Based on another investigations, the magnetic moment values of P_4VPy-$CuCl_2$, P_4VPy-$CoCl_2$, and P_4VPy-$NiCl_2$ were determined to be 1.90, 5.20, and 1.87 BM, respectively [142]. The results revealed the square planar and octahedral structures for P_4VPy-$CuCl_2$ and P_4VPy-$CoCl_2$ complexes, respectively, and P_4VPy-$NiCl_2$ forms a mixture of both configurations. Superparamagnetic behavior of a series of PVPy-Co(II) and

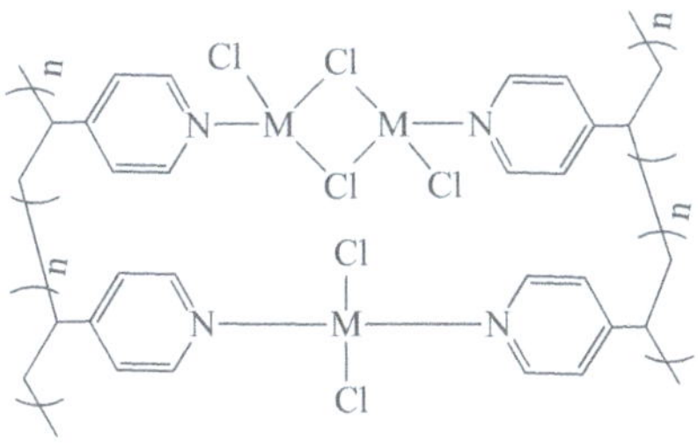

P_4VPy-M(II)Cl_2 M= Zn, Co, Ni, Cu

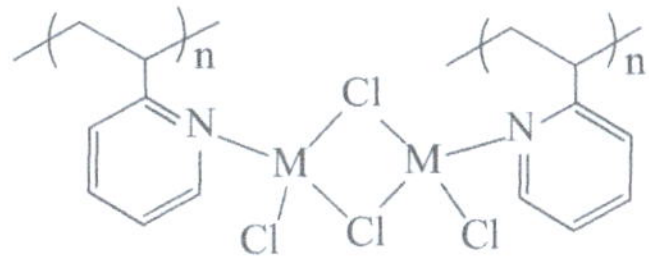

P_2VPy-M(II)Cl_2

SCHEME 1.46 Representation of the coordination of metal chlorides to PVPy. (a) P_4VPy cross-linked by metal chlorides, (b) P_2VPy intramolecularly coordinated by metal chlorides [508].

PVPy-Fe(II) complexes has also been emphasized by magnetic susceptibility measurements [510]. In another study, it was revealed that PVPy-$FeCl_2$, PVPy-$FeCl_3$, and PVPy-$CoCl_2$ exhibited superparamagnetic behavior [511], while PVPy-$FeSO_4$, PVPy-$CoSO_4$, and PVPy-$NiCl_2$ had non-superparamagnetic behavior.

Based on ^{57}Fe Mössbauer spectra [512], P_4VPy-$FeCl_3$ and P_4VPy-$Fe(NO_3)_3$ complexes, prepared from the reaction of PVPy with $FeCl_3$ and $Fe(NO_3)_3{\cdot}9H_2O$, respectively, are consisted of amorphous ferric oxide and hydroxide clusters and showed superparamagnetic behavior at low temperature and undergo collective magnetic excitations [513]. From the measured Mössbauer parameters and IR data, structures such as (I) and (II) were proposed for P_4VPy-$FeCl_3$ complexes but not for P_4VPy-$Fe(NO_3)_3$ (Scheme 1.47). The size of the clusters for the PVPy-$Fe(NO_3)_3$ complexes is smaller than that for the PVPy-$FeCl_3$ complexes and increases with increasing iron content of samples.

Fe_3O_4 NPs-loaded poly(ethylene glycol-*b*-4-VPy) copolymer (PEG-*b*-P_4VPy) [362], the Fe_3O_4@P_4VPy composites [514], Fe_2O_3 NPs-loaded polybutadiene-*b*-P_2VPy, Fe_3O_4@P_4VPy-*b*-PS composites [515], and core–shell Fe_3O_4@polypyrrole@4-VPy composites [516] with superparamagnetic behavior have been reported. Polybutadiene-*b*-P_2VPy was applied as template (structure-directing agent) for superparamagnetic Fe_2O_3 NPs into nanotubes [517].

(I) **(II)**

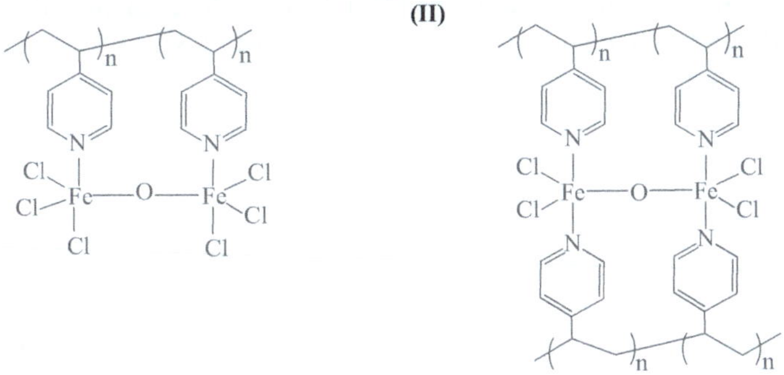

SCHEME 1.47 Oxygen-bridged ferric dimers in P_4VPy-$FeCl_3$ [513].

1.2.9 Oxidized Forms of PVPy Polymers

Poly(4-vinylpyridine-*N*-oxide) (P_4VPyNO) and Poly(2-vinylpyridine-*N*-oxide) (P_2VPyNO) are two oxidized forms of PVPy polymers which can easily be created by oxidation of P_4VPy and P_2VPy using hydrogen peroxide (The Ochiai method) [518–520] (Scheme 1.48). They are soluble in water, trifluoroethanol, and mixed solvents of acetone:water and p-dioxane:water [29,521]. P_2VPyNO is also soluble in methanol and isopropanol. No T_g was observed for the P_2VPyNO and P_4VPyNO at temperatures below their decomposition temperature. Because such behavior was also observed for polyzwitterions, it can be imagined that the zwitterionic pendent groups of (PVPyNO)s impart polar interactions, resulting in an enhancement of T_g which is undetectable for the standard DSC measurement [29]. Both P_2VPyNO and P_4VPyNO exhibited good thermal stability up to 220°C [29]. Self-assembly of BCPs of P_2VPyNO and P_4VPyNO with PS has been investigated [29]. They created well-organized nanostructures with a high χ value, which may have significant applications in nanolithography and microelectronics. In contrast to P_2VPyNO and P_4VPyNO, P_2VPyNO-*b*-PS, P_4VPyNO-*b*-PS, P_2VPy-*b*-PS, and P_4VPy-*b*-PS showed a T_g around 90°C–100°C.

The positioning of the pyridinyl nitrogen of P_4VPyNO and P_2VPyNO causes these polymers to exhibit different chemical properties. For instance, although the reduced viscosity dependence on concentration for P_2VPyNO is similar to that for nonionic polymers (Figure 1.11a), P_4VPyNO behaves like polyelectrolytes, e.g., its reduced viscosity increased at polymer concentrations smaller than 1.0 g/dL (Figure 1.11b) [521,522], due to chain expansion. These observations were postulated based on the difference in the position of the N–O group in these polymers. Complexation of P_4VPyNO with water converts the N–O group to a week acidic species; hence, the resulting complex behaves like a polyelectrolyte. It is important to note that, despite P_4VPyNO exhibiting a charge separation across the $N^+–O^-$ bond, it cannot be classified as a polyelectrolyte since it lacks ionizable groups [521]. On the other hand, upon the addition of terbium chloride hexahydrate ($TbCl_3{\cdot}6H_2O$) to the aqueous solutions of P_4VPyNO and P_2VPyNO, the reduced viscosity decreased in both isomeric polymers (Figure 1.11a and b). In this regard, P_4VPyNO behaves like polyelectrolyte solutions, but the P_2VPyNO does not exhibit such behavior and its viscosity decreases at concentrations < 1 g·dL^{-1}, which is attributed to a complexation of Tb^{3+} with the ligands resulting in chain contraction [521]. It is worth mentioning that aqueous solution of P_4VPyNO in deionized water has acidic pH of 3.4–4.5, due to the formation of polymer-water complex [28,521]. In the case of P_2VPyNO due to the proximity of

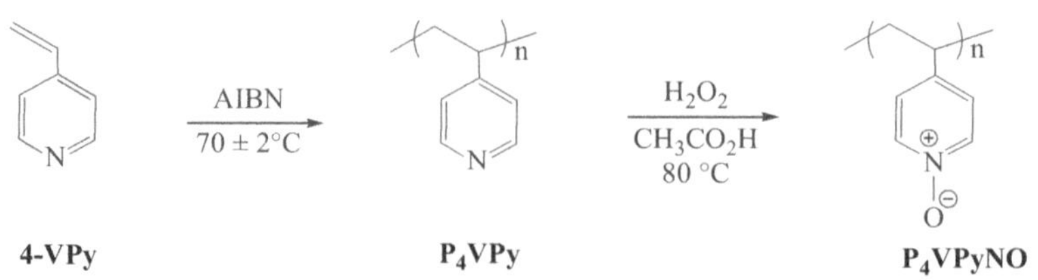

SCHEME 1.48 Preparation of P_4VPyNO [520].

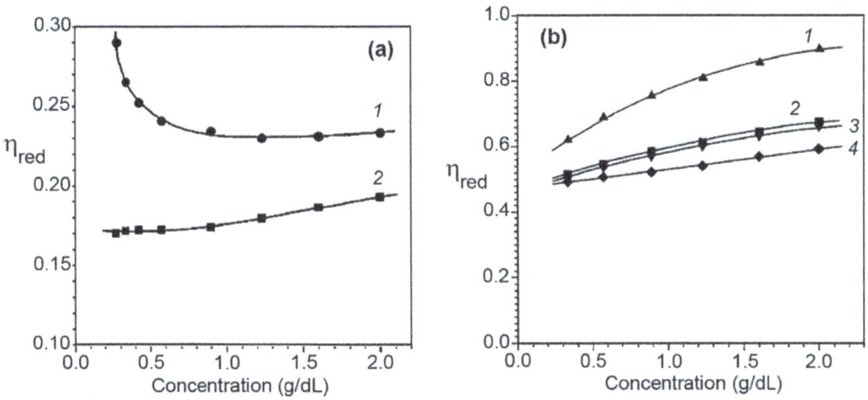

FIGURE 1.11 Reduced viscosity of (a) P_4VPyNO [1] and $P_4VPyNO\text{-}Tb^{3+}$ [2, molar ratio 5:1) and (b) P_2VPyNO [1] and $P_2VPyNO\text{-}Tb^{3+}$ (2, molar ratio 10:1), (3, molar ratio 5:1), and (4, molar ratio 3:1) as a function of concentration [521]. (Reproduced with permission from Ref. [521]; Copyright 1998 American Chemical Society).

the N–O group to the polymer backbone, the possibility of forming complex with water molecules is very low, due to steric repulsion. The remarkable charge separation across the N–O group also affects electrical conductivity of (PVPyNO)s so that while P_4VPyNO solutions in water exhibit a nonzero electrical conductivity (2.86×10^{-7}–3.46×10^{-6} S/cm), the corresponding non-oxidated P_4VPy solutions have an electrical conductivity of $<10^{-14}$ S/cm [28].

Titration of atactic PVPyNO aqueous solutions with HCl revealed that P_2VPyNO and P_4VPyNO behave as base at pH <5 and <3, respectively, and as acids at pH >10 [31]. It is proposed that the difference in behavior of these polymers is due to the interaction between the *N*-oxide group and CH of the polyvinyl chain in P_2VPyNO (Scheme 1.49), while such CH–ON bonding is absent in the P_4VPyNO. Influence of pH on the viscosity of P_2VPyNO and P_4VPyNO has also been investigated [523–525]. The intrinsic viscosity of P_2VPyNO between pH 4 and 8.5 is approximately constant, but below pH 4 and above pH 8 it increases and decreases, respectively. In the case of P_4VPyNO, the viscosity remains constant until a pH of 3 and then it increases between pH 3 and 9 and after that decreases. Again, it is suggested that an interaction between oxygen of the NO group and CH of the polymer chain in this range of pH (ca. 4–8.5) causes a tightly packed structure in P_2VPyNO (but not in P_4VPyNO structure) [524,526]

(I)

SCHEME 1.49 Proposed α-H···O interaction in P_2VPyNO [524].

There are three different types of tacticity in the P_2VPyNO, atactic, isotactic, and syndiotactic. Scheme 1.50 shows the synthesis of isotactic and syndiotactic P_2VPyNO [527,528]. Studies indicate that the tacticity of P_2VPyNO significantly influences the viscosity number of the resulting polymers [31]. Relation between viscosity of the stereospecific forms of P_2VPyNO in aqueous solution has been surveyed and the results do not resemble that for P_4VPyNO [31]. The lowest viscosity was noted within the pH range of 4–8, while a rapid increase in viscosity was observed beyond this range. In this respect, viscosity behavior of the atactic polymer is a combination of the behavior of isotactic and syndiotactic polymers.

The *N*-oxide group of PVPyNO can act as an electron donor, which is coordinated with various metals. For example, complexation of both P_2VPyNO and P_4VPyNO with $CuCl_2$, $HgCl_2$, $NiCl_2$, $CoCl_2$, $FeCl_3$, and $ZnCl_2$ has been investigated [529,530]. The results revealed that P_4VPyNO is coordinated with these metal chlorides to produce the corresponding metal complexes mostly by intermolecular cross-linking reactions in which the ratio of *N*-oxide group to metal is 2:1 (Scheme 1.51, I). When the feed molar ratio of metal chloride to P_4VPyNO was 2:1, another structure of metal complexes could be produced (Scheme 1.51, structure II). On the other hand, in P_2VPyNO, *N*-oxide groups on adjacent rings are closely associated with each other and halide chloride is shared between them to produce metal complex through

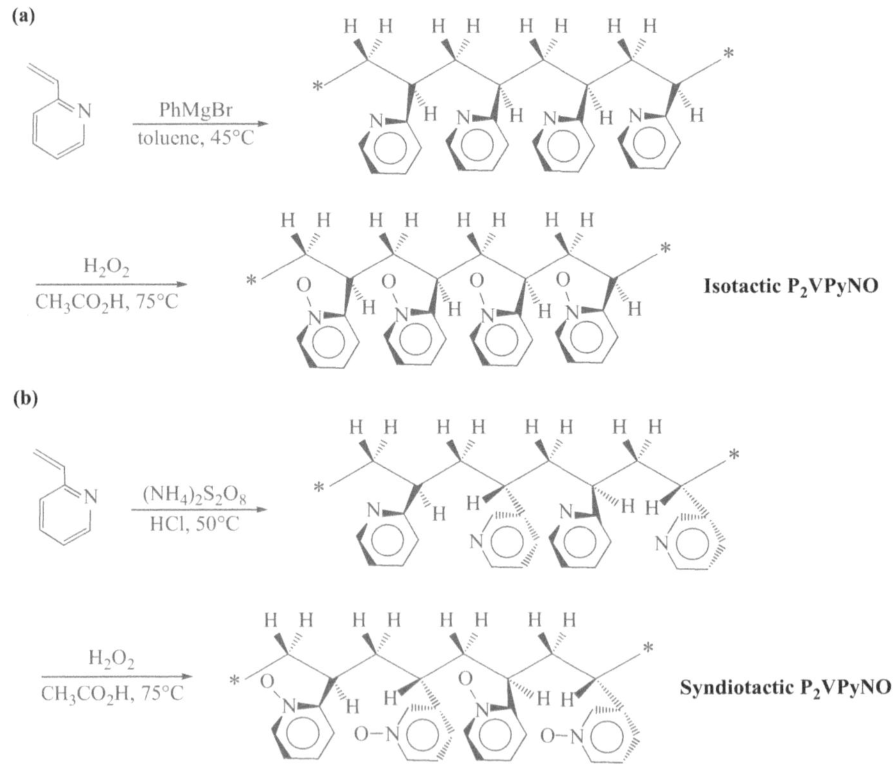

SCHEME 1.50 Synthetic pathways for (a) isotactic and (b) syndiotactic P_2VPyNO.

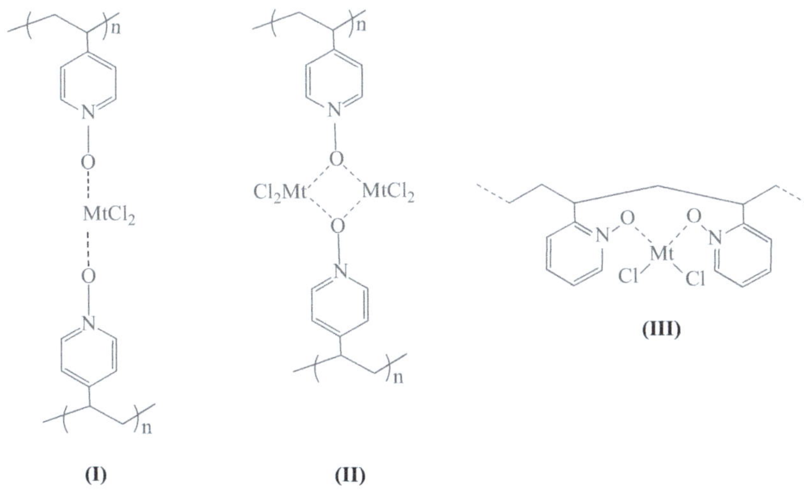

SCHEME 1.51 Complexing metal chlorides with P_4VPyNO (structures I and II) and P_2VPyNO (structure III) [529].

intramolecular interactions (Scheme 1.51, structure III). In contrast to structure (III), which decomposes in water because the *N*-oxide oxygen is hydrated, structures (I) and (II) dissolve unchanged in water. Following the above results, a series of alkyl substituted P_2VPyNO were subjected to complexation with $CuCl_2$, $HgCl_2$, $NiCl_2$, $CoCl_2$, $FeCl_3$, and $ZnCl_2$ to determine the influence of alkyl substituents on complexation of the pyridine rings (Scheme 1.52a) [531,532]. The synthetic pathway for one of them is given in Scheme 1.52b as a typical. Similar results were found for the complexation of metal halide on all alkyl substituted P_2VPyNO as described in the above studies, meaning that polymers (I) and (II) form their corresponding metal complexes in which each metal halide is shared between two *N*-oxide units. Regarding polymer (III), although the *N*-oxide group is situated far from the polymer chain and is expected to function similarly to P_4VPyNO (which undergoes cross-linking reactions), this polymer, like polymers (I) and (II), formed complex metals with a ratio of *N*-oxide group to metal of 2:1. It is likely that the steric hindrance caused by the methyl group, along with the hydrogen bonding interaction between the negatively charged oxygen of the *N*-oxide group and the hydrogen of the methyl group, inhibits the approach of a second metal ion to the active sites of polymer (III). Relations between pH and viscosity of the polymers (I–III) have also been investigated [533].

1.3 SYNTHESIS OF PVPY POLYMERS AND THEIR COPOLYMERS

PVPy and its derivatives are attractive polymeric materials due to exhibiting excellent performance in fundamental science and technological applications. Thus, the precise synthesis of this useful polymer and its copolymers with predictable molecular weights and tailormade architectures is attracting considerable attention from the view point of laboratory synthesis and industrial manufacturing. To my knowledge, polymerization of vinylpyridine was first reported in the 1950s, and the

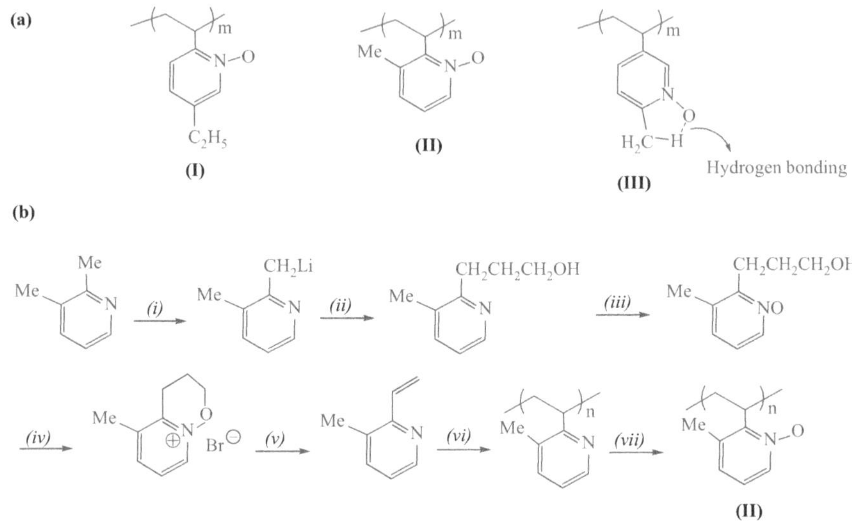

(i): PhLi
(ii): a: $(CH_2)_2O$, b: hydrolysis
(iii): H_2O_2-HOAc, 80°C
(iv): 48 % HBr
(v): NaOH (aq)
(vi): AIBN;
(vii): H_2O_2, CH_3CO_2H, 80°C

SCHEME 1.52 (a) Structure of poly(3-ethyl-6-vinylpyridine l-oxide) (I), poly(3-methyl-2-vinylpyridine l-oxide) (II), and poly(4-methyl-2-vinylpyridine l-oxide) (III), and (b) the synthetic routes for polymer (III) [532].

obtained polymer was used for coating of photographic material [3]. It was reported many years ago that both 2-VPy and 4-VPy were converted to their corresponding polymers upon reaction with methyl iodide [9]. In the case of 4-VPy, an anionic mechanism was proposed in which a zwitterion (I) initiates chain propagation (Scheme 1.53) [534]. On the other hand, for the synthesis of P_2VPy a charge-transfer polymerization of radical nature between 2-VPy (as electron donor) and its quaternized derivative (as electron acceptor) was suggested [535].

Another early report on the polymerization of vinylpyridine monomers is electrochemical polymerization of 4-VPy [536,537]. Rapid polymerization occurs upon addition of monomers to a solution of Na metal in liquid NH_3 at −50°C and during the rapid disappearance of the dark blue color of the solution Na-NH_3, a red-orange solid, i.e., P_4VPy with high molecular weight was produced. The proposed mechanism for

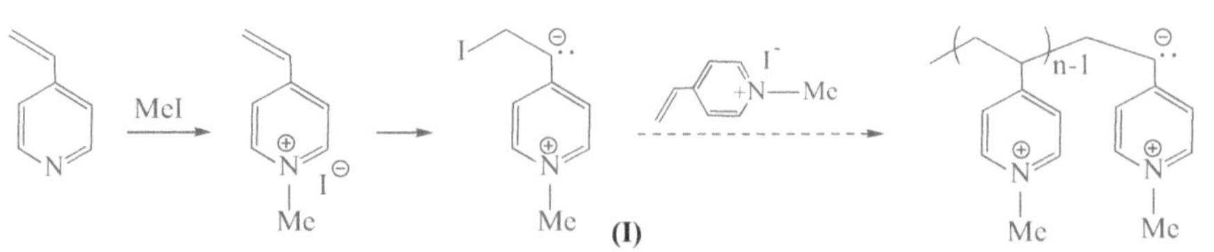

SCHEME 1.53 An anionic mechanism for the polymerization of 4-VPy in the presence of MeI [534].

polymerization involves the formation of an anion radical during the electron transfer process from Na to 4-VPy monomer, followed by anionic propagation (Scheme 1.54a). Also, the results showed that a small amount of termination occurs, in which the created amide ion initiates homogeneous polymerization [537] (Scheme 1.54b).

After the aforementioned reports, a variety of synthetic methods have been investigated for the preparation of PVPy family and its corresponding copolymers. The most frequently employed method for the preparation of PVPy and its derivatives is free radical polymerization [6,9] using an initiator such as AIBN [538,539], potassium persulfate ($K_2S_2O_8$) [540,541], ammonium persulfate (APS, $(NH_4)_2S_2O_8$) [542], and benzoylperoxide (BPO) [543]. The mechanism of chain initiation and chain transfer [544] and the kinetic studies [545–548] in the polymerization of vinylpyridine monomers have been reported. Photo-initiated polymerization of vinylpyridine monomers has also been investigated [549–552]. Industrial production of PVPy was mostly restricted to radical copolymerizations (typically in suspension or emulsion) for the manufacture of rubbers [553] or cross-linked ion-exchange resins [554]. Fortunately, with the coming of CRP or reversible deactivation radical polymerization (RDRP) methods, the synthesis of well-defined vinylpyridine-based polymers (especially BCPs) with different architectures and functions is possible [555]. Usually, unlike anionic polymerization, these techniques do not require glassware methods, and extensive purifications are not needed. In the following subsections, several methods which have been used for the synthesis of PVPy and its copolymers are discussed.

1.3.1 Suspension Polymerization

It is a kind of radical polymerization, divided into bead and powder polymerization. In the former system, the polymer is soluble in its corresponding monomer and in the later, the polymer is not soluble in its monomer. Mostly, the initiators such as

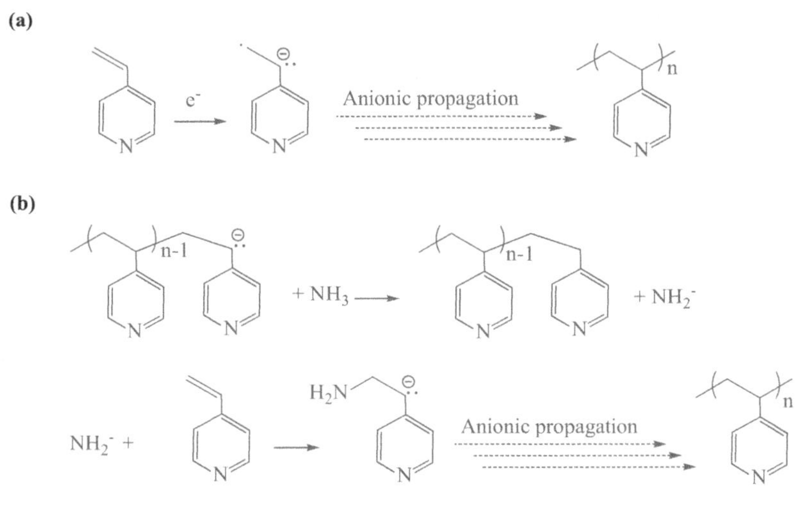

SCHEME 1.54 The electrochemical polymerization of 4-VPy in a solution of Na metal in liquid NH_3 [536,537].

tert-butyl peroxypivalate (Lupersol® 11), 2,5-dimethyl-2,5-di-(*tert*-butylperoxy) hexane (Lupersol® 101), and AIBN are located in the monomer phase. Usually, suspension agents such as PVA, SDS, and PVP are used. Suspension polymerization has frequently been employed for production of commercial resin beads from the corresponding vinylic monomers in the presence of a cross-linker such as DVB.

Suspension polymerization of vinylpyridine monomers is one of the most important methods for the synthesis of PVPy polymers [556–560]. Typical recipe for suspension polymerization of 4-VPy is as follows: 4-VPy (247 ppm)/BPO (4 ppm)/dioctyl phthalates (50 ppm)/styrene (62 ppm)/hydroxyethyl cellulose (4.5 ppm)/NaCl (234 ppm)/$NaNO_2$ (2.9 ppm)/H_2O (800 ppm)/9 h/80°C [6]. In the case of suspension polymerization of the quaternized vinylpyridine monomers, due to their water solubility, no suspension agent is needed [561]. Poly(4-VPy-*St*-DVB) copolymers of different compositions have been prepared via suspension polymerization in the presence of PVA and AIBN [562]. The synthesis of cross-linked P_4VPy beads (4, 8, and 12% DVB) via suspension polymerization method in the presence of hydroxyethyl cellulose and gelatin as suspension agents and BPO as initiator has been reported [563]. Preparation of copolymer beads of 4-VPy and *N*-vinyl pyrrolidone (NVP) via suspension polymerization has also been investigated [558]. Copolymerization of the monomers is accomplished through the dropwise addition of a solution of monomers and AIBN in cyclohexane to an aqueous solution of PVP (Scheme 1.55). A few drops of NH_3 along with NaCl were used in order to help to maintain NVP in organic phase.

The influence of the diluent systems (porogens) on the porous properties and the swollen state morphology of macroporous resins based on P_2VPy and P_4VPy during the suspension polymerization has been surveyed [49,564–570]. In fact, the porous property of these resins can be adjusted by phase separation, achieved using porogens. Among the porogens tested, *n*-heptane/methyl ethyl ketone, *n*-heptane/diethylphthalate, toluene/cyclohexanone, and *n*-heptane/ethyl acetate, are most important. For example, the effect of *n*-heptane/toluene (HEP/TOL) on the porous property of P_2VPy cross-linked with DVB was investigated [48,566]. The resin beads were synthesized via suspension polymerization. Ingredients distributed between the two phases as follows: aqueous phase including 2-hydroxyethyl cellulose and gelatin (as suspension agents) and NaCl, and organic phase consisting of AIBN initiator, 2-VPy and DVB monomers, and diluent mixture (HEP/TOL). The results showed that with decreasing the proportion of HEP in the diluent system (e.g., HEP:TOL, 25/75) or in TOL alone, the obtained resin was gel-type like material and transparent. On the contrary, at high amounts of HEP (e.g., HEP:TOL, 75/25), the synthesized resin was opaque. Furthermore, it was observed that as the proportion of HEP in the diluent system increases, the pore volumes rise while the apparent density decreases across all percentages of DVB. In many cases, an increase in pore volumes correlates with a reduction in the swelling of

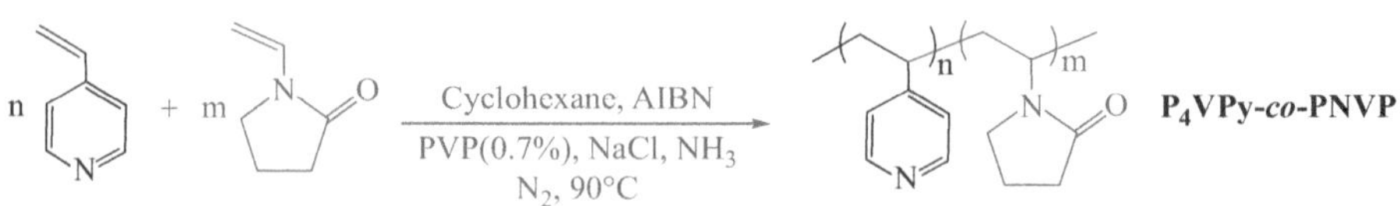

SCHEME 1.55 Synthesis of P_4VPy-*co*-PVP [558].

samples in methanol. This phenomenon can be attributed to the occurrence of phase separation at elevated levels of HEP (a non-solvent diluent), which leads to the development of more rigid structures and ultimately a decrease in swelling. This behavior was also observed in the synthesis of the resins constructed from the suspension polymerization of 2-VPy and DVB in the presence of *n*-heptane/diethylphthalate [566], heptane [567], and *n*-heptane/methyl ethyl ketone [564] as porogens.

1.3.2 Emulsion Polymerization

In this method, the polymer is produced through the dispersion of the corresponding monomers in an aqueous phase in the presence of a surfactant such as the soap micelle and an initiator. Monomer and generated initiator radical migrate into micelles and the polymerization starts. Then, with migration of more monomers into the micelle, the polymerization continues. Some water-soluble polymers such as PVA or hydroxyethyl celluloses can also be used to act as emulsifiers/stabilizers. PVPy and its copolymers have been synthesized by emulsion polymerization [70,379,571–577]. Typical recipe for emulsion polymerization of 4-VPy is as follows: 4-VPy (100 parts)/BPO (0.1–2.5 parts depending on molecular weight desired)/Nekal AEMA (general dyestuff corporation) as an emulsifier (0.25 part)/H_2O (200 parts)/4 h/50°C [575]. One-step synthesis of P_4VPy hydrogel NPs using an oil-in-water emulsion system including DVB and ethylene glycol dimethacrylate (EGDMA) as cross-linking agents in the presence of SDS has been studied [72].

Despite its good performance in the preparation of polymers, the emulsion polymerization still suffers from some shortcomings, such as high cost, the need for large amounts of surfactants, toxicity, foaming, and the separation of surfactants after polymerization [578]. One strategy to overcome some of these limitations is the use of solid nano- or microparticles as new types of very stable emulsions, namely Pickering or Mickering emulsion [579,580]. The adsorption of solid particles accumulated at the interface between two immiscible liquids leads to the formation of a stable emulsion. For example, very recently, fabrication of pH-sensitive P_4VPy nanospheres stabilized by Mickering microgel emulsions based on PAMPS has been reported (Scheme 1.56) [580]. The microgel was prepared via the cross-linking radical polymerization method in the presence of *N,N′*-methylenebisacrylamide (MBAM), APS, tetramethylethylenediamine (TEMED) as cross-linker, initiator, and activator, respectively, and the nonionic surfactant Span 85. Self-emulsion polymerization as a type of surfactant-free emulsion polymerization (SFEP) methods [581] have been reported for the preparation of P_4VPy [582]. As the polymerization begins and the polymer chains reach a certain length, they behave as a surfactant. Polymerization carried out in the presence of only monomer and a water-soluble cationic initiator, [(2-imidazolin-2-yl)propane] dihydrochloride.

1.3.3 Dispersion Polymerization

In this kind of polymerization, the corresponding monomer and the initiator are soluble in the reaction solvent, but the obtained polymer is insoluble in the dispersion medium. Generally, a stabilizer is used to prevent the aggregation of colloidal

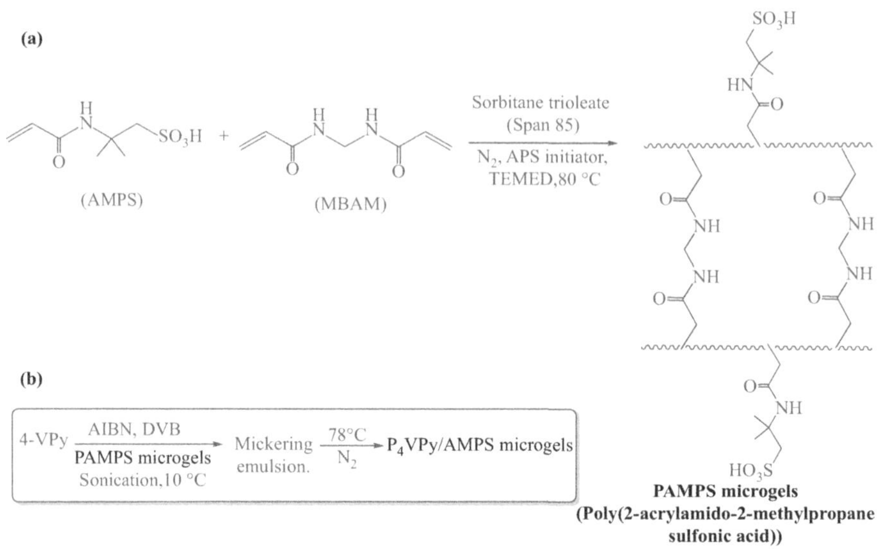

SCHEME 1.56 (a) Synthesis of PAMPS microgels and (b) preparation of P_4VPy nanospheres stabilized by Mickering microgel emulsions based on PAMPS [580].

particles. Dispersion polymerization is superior to emulsion or suspension polymerization with respect to its better control over the particle size of the synthesized beads and preparation of micrometer-scale monodisperse polymer particles. The dispersion polymerization of vinylpyridine monomers was first introduced in 1997 for the preparation of monodisperse PVPy polymers and P_4VPy-*co*-PMAA particles [583,584]. Polymerization was performed in a solvent pair of DMF and toluene (good and poor solvents for PVPy, respectively) using PS-*b*-polybutadiene as a stabilizer. Dispersion copolymerizations of 4-VPy with PS have been reported [388,585–587]. By employing small amounts of a lyophilic comonomer, 4-VPy and styrene were polymerized by dispersion polymerization [388]. The main advantage of this investigation is the possibility of the synthesis of polymers in the absence of stabilizers which usually cause hazardous pollution and high cost of the final polymer, as well as they can remain in the polymer particles.

1.3.4 Atom Transfer Radical Polymerization (ATRP)

Polymerization in which a reversible reaction between a low-oxidation state metal complex M^zL_m (catalyst–ligand) and an alkyl halide initiator produces a radical which causes subsequent propagation of the polymerization using the corresponding monomers. The use of copper-based catalysts in ATRP has attracted commercial interest due to easy availability and their ability to form complex with different kinds of ligands. The main disadvantage of this method is that the amount of the remained metal impurities is sometimes detrimental for optical and photovoltaic applications of the obtained polymers. This method has been used for the synthesis of PVPy family [588–598]. Typical recipe for the preparation of P_4VPy via ATRP is as follows:

4-VPy:PECl (95:1 molar ratio)/PECl:CuCl:Me_6TREN (1:1:1 molar ratio)/2-propanol/25 h/40°C, in which PECl (1-phenylethyl chloride) is as initiator and CuCl/Me_6TREN (tris[2-(dimethylamino)ethyl]amine) complex is as catalyst [593]. For example, a schematic representation of the synthetic approach for the P_4VPy and the P_4VPy-*b*-PS by ATRP method is depicted in Scheme 1.57 [594]. However, in contrast to many of vinylic monomers, the ATRP of vinylpyridine monomers is not a convenient route because of tendency of the pyridine units of both monomer and obtained polymer with the used copper catalyst [596]. A strategy to overcome this problem is the use of the multidentate ligands, which tend to form complexes with the catalyst in a competitive reaction with the pyridine rings [593]. In addition, to overcome the oxidation of the catalyst–ligand system in ATRP, a modified method of ATRP, denoted as reverse ATRP, has been reported in which the polymerization is accomplished in the presence of a radical initiator such as AIBN with transition metal complexes in their higher oxidation state [597,598]. In comparison to 4-VPy monomer, ATRP of 2-VPy is largely difficult, probably due to deactivation of 2-VPy monomer arising from the *ortho* positioning of the pyridinyl nitrogen. Recently, an activating agent known as 1,1,1,3,3,3-hexafluoro-2-propanol (HFIP), which can activate 2-VPy through complexation, has been effectively utilized in the synthesis of P_2VPy using the ATRP method; however, the dispersities remain insufficient yet [599]. The effect of alkyl halides (chlorides or bromides) as well as catalytic systems (the $Cu^{I}Cl$ complexes of tris(2-pyridylmethyl)amine (TPMA) or *N,N,N′,N″,N‴,N‴*-hexamethy ltriethylenetetramine (HMDETA)) on polydispersity of P_4VPy obtained by ATRP has been investigated [600]. "Grafting from" method for surface modification of Fe_3O_4 NPs with P_4VPy by ATRP has been reported [601]. As shown in Scheme 1.58, Fe_3O_4-BTPAM macroinitiator, which is prepared by thermal decomposition of $Fe(acac)_3$ following by reaction with BTPAm, in the presence of PMDETA/CuBr system initiates ATRP of 4-VPy monomers, resulting in P_4VPy-coated magnetite NPs.

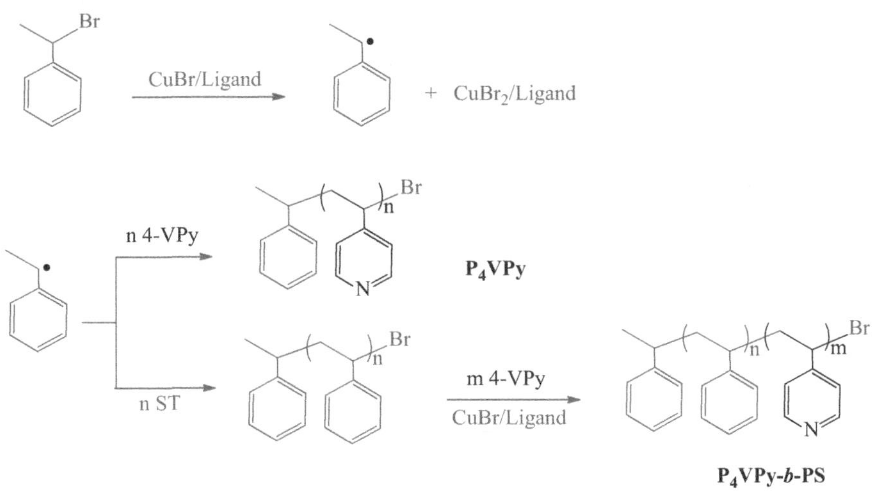

SCHEME 1.57 The synthetic approach for P_4VPy and P_4VPy-*b*-PS via ATRP method (ligand: *N,N,N′,N″,N″*-pentamethyldiethylenetriamine (PMDETA) [594].

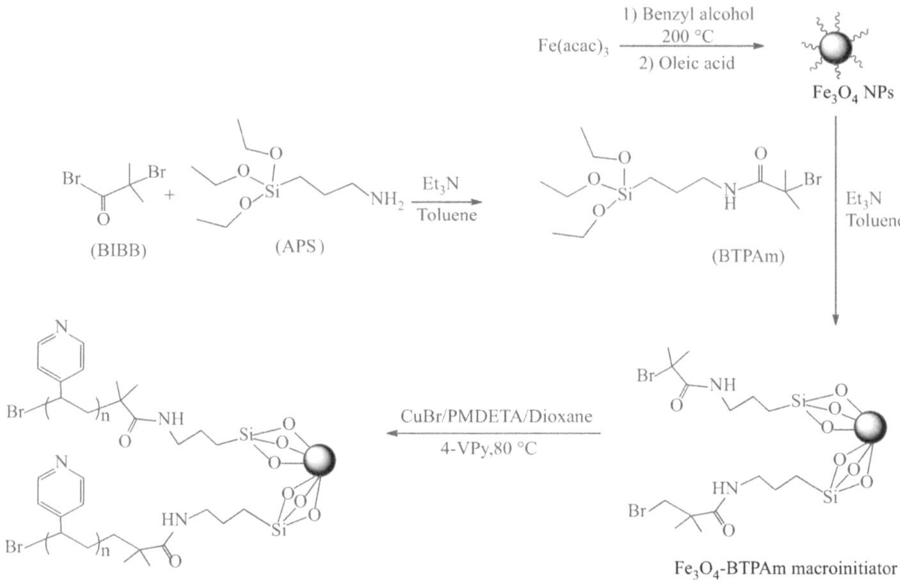

SCHEME 1.58 Synthetic pathways for P$_4$VPy-coated magnetite NPs. APS: 3-aminopropyl triethoxysilane; BIBB: 2-bromoisobutyryl bromide; BTPAm: 2-bromo-2-methyl-*N*-(3-(trieth oxysilyl) propanamide; PMDETA: *N,N,N',N",N"*-pentamethyldiethylenetriamine; Fe(acac)$_3$: iron(III) acetylacetonate [601].

1.3.5 Reversible Addition-Fragmentation Chain Transfer (RAFT) Polymerization

In this method, the control of polymerization can be smoothly achieved by employing an appropriate reversible addition-fragmentation chain transfer agent (CTA) often possessing a thiocarbonylthio moiety (e.g., dithiocarbamates, trithiocarbonates) or xanthates [602–604]. Of the CRP techniques, RAFT is superior from the view of the compatibility with different monomers and less sensitive to the choice of the reaction conditions (temperature and solvent). Although RAFT method has many advantages, one notable disadvantage is the remained CTAs on the polymer chain ends, resulting in unpleasant odor, toxicity, limited thermal stability, and coloration [605]. To overcome this issue, various strategies were employed, including replacing the CTAs with a strong and inert C–H bond by using trialkylstannane reagents as effective hydrogen sources [606,607], radical-induced reduction strategies by using reducing agents like tris(trimethylsilyl)silane (TTMSS) and *N*-ethylpiperidine hypophosphite (EPHP) [605], and light-mediated removal of CTAs [608,609].

Homopolymers of P$_2$VPy and P$_4$VPy and their copolymers have been synthesized by RAFT [445,610–618]. Typical recipe for the preparation of P$_2$VPy or P$_4$VPy via RAFT is as follows: CDB:2-VPy or 4-VPy (1.374 molar ratio)/AIBN:CDB (1:4.75 molar ratio)/24 h/60°C under a nitrogen atmosphere in septa-sealed vials, in which CDB is cumyl dithiobenzoate as RAFT CTA [615]. For example, schematic representations of the synthetic approach for the preparation of BCPs of 2-VPy and 4-VPy via RAFT method, are depicted in Schemes 1.59 and 1.60 [615,619]. The use of BPO:CTA (1:1) in RAFT polymerization of 2-VPy monomer led to the synthesis of P_2VPy at lower temperature (25°C) with narrow dispersity (1.11) [620]. RAFT polymerization of 2-VPy using blue light initiation of S-1-dodecyl-S′-(α,α′-dimethyl-α″-acetic acid) trithiocarbonate (DDMAT) has been surveyed [602]. The possible reaction mechanism for this polymerization is depicted in Scheme 1.61. Liquid-crystalline triblock copolymer P_4VPy_{60}-*b*-poly(*p*-(6-acryloyloxycaproyloxy phenyl)-*p*-methoxybenzoate)$_{40}$-*b*-P_4VPy_{60} has been synthesized via RAFT polymerization using *S*,*S*′-bis(methyl-2-isobutyrate)trithiocarbonate as RAFT agent [213]. Also, poly[(2-methyl-2-oxazoline)-*random*-4-VPy] brush copolymer was prepared via RAFT method [621]. The copolymer could be attached on the surfaces of gold, silicon, glass, and polydimethylsiloxane. The surface-initiated RAFT polymerization of 4-VPy on silicon substrate by RAFT strategy has been investigated [622].

The combination of RDRP methods with dispersion or emulsion polymerization has gained considerable attention in recent years for the preparation of BCP architectures with predetermined molecular weights and low dispersities, owing to its advantages such as low viscosity during the reaction period, producing well-defined polymers, and heat transfer efficiency [623,624]. For instance, in RAFT dispersion polymerization, a pre-prepared soluble homopolymer as polymeric chain transfer agent (namely macroRAFT agents) is employed. Then, polymerization of the second monomer begins in the homogeneous phase stabilized by the aforementioned chain transfer agent. As polymerization continues, the growing second block becomes insoluble and consequently BCPs self-assemble into NPs (so-called polymerization-induced self-assembly (PISA)) [625]. Different types of the used stabilizer segments in RAFT dispersion (or emulsion) polymerization are given in the recent reports [276,626]. P_4VPy-*b*-PS and P_2VPy-*b*-PS copolymers have been synthesized via RAFT dispersion polymerization which takes advantage of the PISA method [611,627]. For example, 4-cyano-4-[(dodecylsulfanylthiocarbonyl)-sulfanyl]pentanoic acid (CDTPA) and 2-cyano-2-propyl dodecyl trithiocarbonate (CPDTC) were used as trithiocarbonate RAFT agents in the preparation of P_4VPy-*b*-PS (Scheme 1.62) [611]. Very recently, Fe_3O_4@P_4VPy-*b*-PS magnetic core–shell–shell magnetic nanospheres were obtained via RAFT dispersion polymerization [515]. RAFT dispersion polymerization of ST monomers in methanol/water in the presence of macro-P_4VPy chain transfer agent produced a polymeric network template. With addition of $FeCl_3$, Fe_3O_4 grows on the

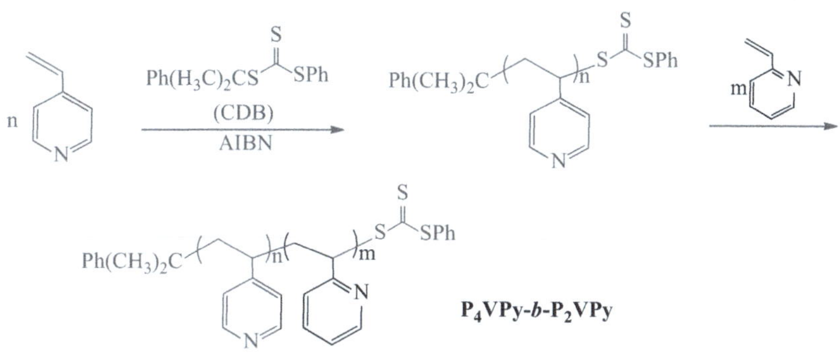

SCHEME 1.59 Synthetic pathway for P_4VPy-*b*-P_2VPy via RAFT method [615].

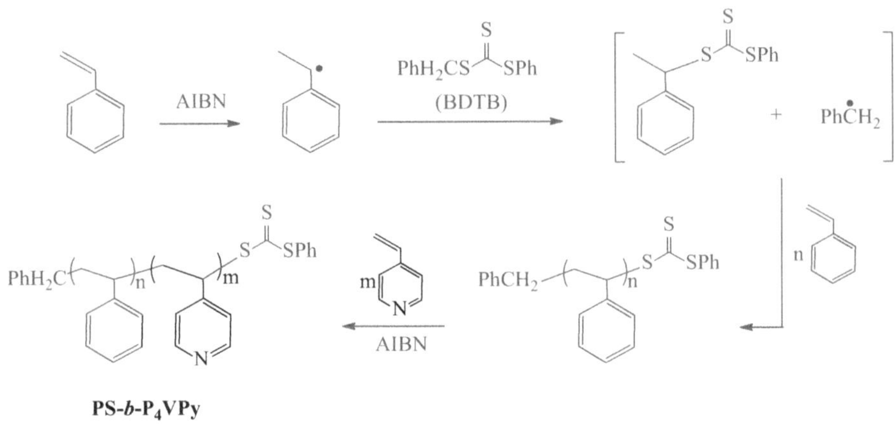

SCHEME 1.60 Synthetic pathway for PS-*b*-P$_4$VPy via RAFT method. BDTA: benzyl dithiobenzoate as CTA [619].

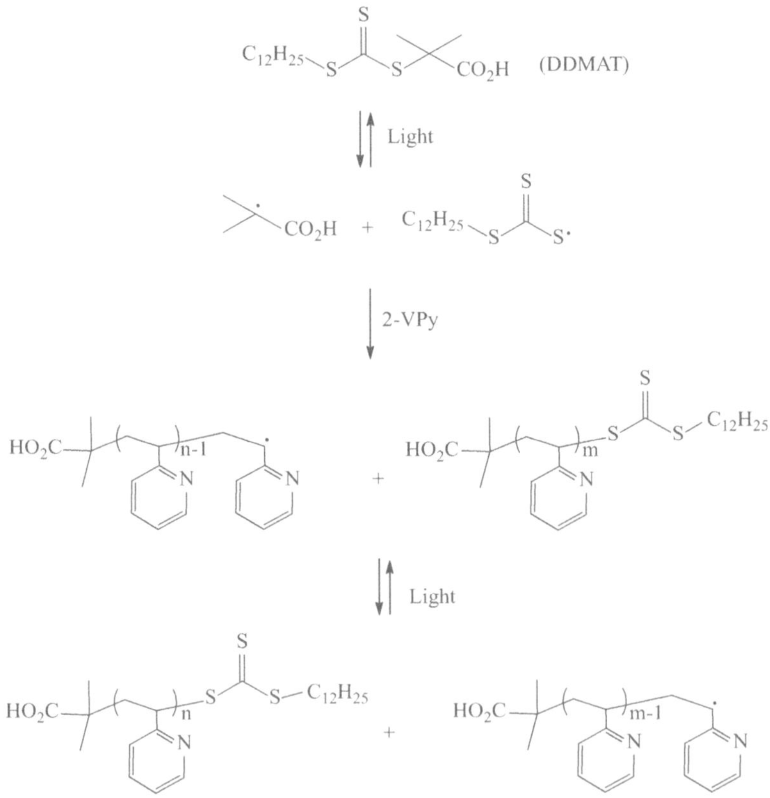

SCHEME 1.61 The proposed mechanism for the polymerization of 2-VPy using blue light initiation of (DDMAT) [602].

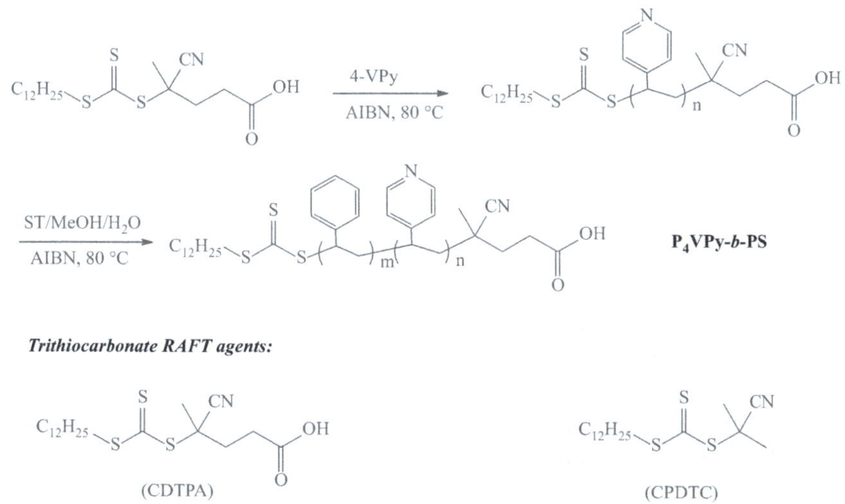

SCHEME 1.62 Synthesis of P_4VPy-*b*-PS via RAFT aqueous-alcoholic dispersion polymerization [611].

obtained templates and finally nanospheres appeared. The synthesis of cationic sterically stabilized P_4VPy-*b*-PS-*co*-poly(benzyl methacrylate) by RAFT-mediated PISA under aqueous emulsion polymerization conditions has also been reported [628].

The synthesis of P_3VPy-*b*-PS by surfactant-free RAFT emulsion polymerization has been reported [13]. The PISA method was proposed for this reaction which includes the in situ generation of micellar objects, followed by self-assembly of the growing diblock copolymers (Scheme 1.63). This method has the obvious advantages such as the use of water instead of organic solvent, and there is no need for the removal of surfactants from the polymer product which is problematic during common emulsion polymerization methods.

1.3.6 Nitroxide-Mediated Radical Polymerization (NMRP)

This type of polymerization includes the reversible terminations of the living chain end using the persistent nitroxide radicals. NMRP is a promising candidate for the

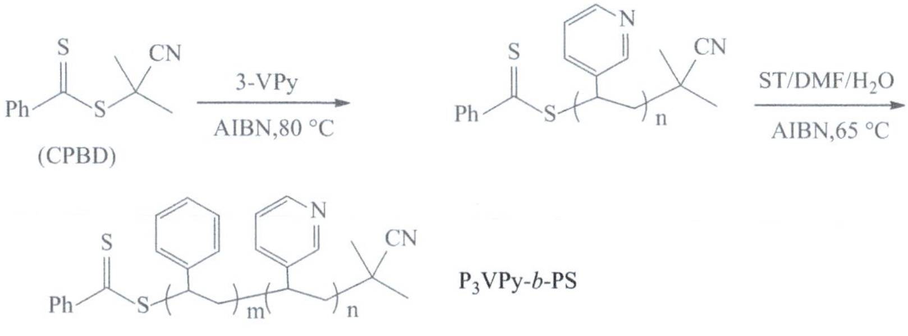

SCHEME 1.63 Synthesis of P_3VPy-*b*-PS via surfactant-free RAFT emulsion polymerization. CPBD: 2-cyano-2-propyl benzodithioate as RAFT agent [13].

synthesis of various polymers which exhibit much narrower molecular weight distributions as well as for better control over polymer chain characteristics. Of the CRP methods, NMRP is synthetically the simplest one; mostly only the addition of a nitroxide, an initiator, and heat is required. In addition, the obtained polymers are free of any contaminants such as sulfur or metal compounds, which are present when using ATRP or RAFT, leading to easy purification of products and tolerance to the presence of functional groups. However, the method suffers from the need for elevated reaction temperatures. The 2,2,6,6-tetramethylpiperidine-*N*-oxyl, known as TEMPO (Scheme 1.64), and other nitroxides have been used as initiators (capping agents) for polymerization of PVPy family via NMRP method [225,629–641]. Typical recipe for the preparation of P_2VPy, P_3VPy, and P_4VPy via NMRP, respectively, are as follows: 2-VPy/TEMPO:BPO (1:1 molar ratio)/$(CH_3CO)_2O$/22 h/95°C in a sealed glass tube degassed by three freeze–thaw cycle [642]; 3-VPy/TEMPO:BPO (1:3 molar ratio)/0.5 h/138°C in a reactor degassed by three freeze–thaw cycle [643]; 4-VPy/TEMPO:BPO (1:3 molar ratio)/dioxan/2 h/100°C in a reactor equipped with N_2 inlet [106]. In fact, the first report on the synthesis of PVPy by CRP methods is related to the work of Bohrisch et. al. [106]. They prepared P_4VPy and P_4VPy-*b*-PS by NMRP method using BPO initiator and TEMPO. Polymerization of 3-VPy monomer and the effect of solvent, the TEMPO concentration, and temperature on the reaction have also been studied [643,644]. Schemes 1.65 and 1.66 depict the synthesis of P_4VPy and poly(*N*,*N*-dimethylacrylamide-*b*-4-VPy), respectively, via NMRP method.

Although in the early reports of NMRP, polydispersity index (PDI) was often more than 1.2 and conversions were modest (~50%) [632,642], later, thanks to the use of other initiators such as BlocBuilder MAMA-SG1 2-({*tert*-butyl [1-(diethoxyphosphoryl)-2,2-dimetylpropyl]amino}oxy)-2-methylpropanoic acid, *N*-*tert*-butyl-*N*-(1-diethylphosphono-2,2-dimethylpropyl) nitroxide (DEPN), and 2,2,5-trimethyl-4-phenyl-3-azahexane-*N*-oxyl (TIPNO), conversions were significantly improved and PDI <1.2 were achieved [645].

Recently, different BCPs containing P_4VPy were synthesized by the desirable combination of CRP methods [646,647]. For example, BCPs of P_4VPy, PS, PEG, and PMMA have been prepared using successive NMP, ATRP, and SET-NRC (single-electron-transfer nitroxide-radical-coupling) [646]. Typically, Scheme 1.67 shows the synthesis of PMMA-*b*-P_4VPy via combination of NMP, ATRP, and SET-NRC methods.

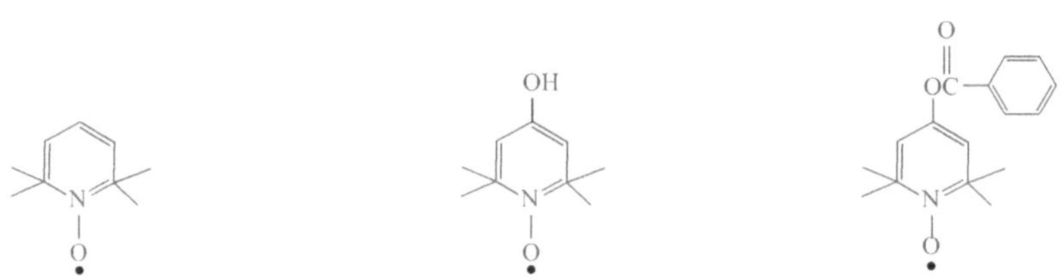

2,2,6,6-Tetramethylpiperidine-*N*-oxyl (TEMPO)

4-Hydroxy-2,2,6,6-tetramethylpiperidine-*N*-oxyl (HO-TEMPO)

2,2,6,6-Tetramethylpiperidine-*N*-oxyl-4-benzoate (G-TEMPO)

SCHEME 1.64 TEMPO derivatives used for NMRP.

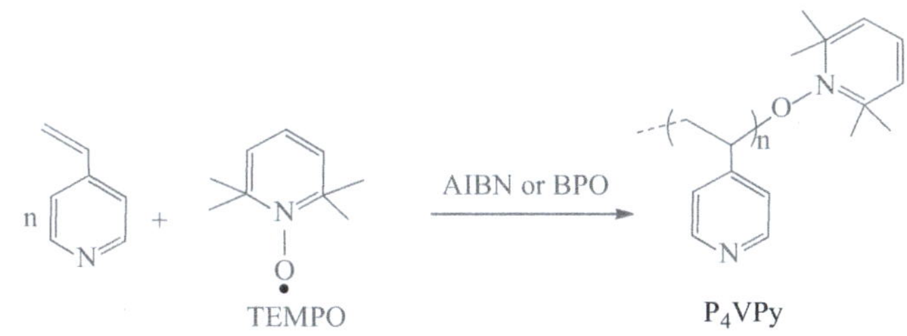

SCHEME 1.65 The polymerization of P_4VPy using BPO or AIBN as initiators and TEMPO as a capping agent [630].

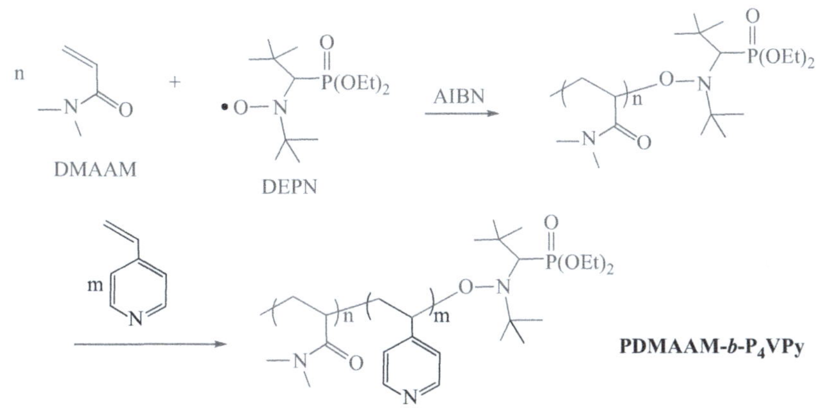

SCHEME 1.66 Synthesis of poly(*N,N*-dimethylacrylamide-*b*-4-vinylpyridine). DEPN: *N-tert*-butyl-*N*-(1-diethylphosphono-2,2-dimethylpropyl) nitroxide as capping agent, DMAA: *N,N*-dimethylacrylamide [632].

1.3.7 The Living Anionic Polymerization

It is a chain polymerization process which involves no formal termination step. It is another approach for the synthesis of PVPy polymers with narrow molecular weight distribution [114,243,360,383,648–669]. Typical recipe for the living anionic polymerization of 4-VPy is as follows: 4-VPy (0.041 mol)/ α-methylstyryllithium (0.9 mol)/ THF/15 min/–78°C in a flamed glass reactor under a nitrogen atmosphere [114]. P_2VPy homopolymer and poly(2-VPy-*b*-isoprene) (P_2VPy-*b*-PI) copolymer have been prepared by anionic polymerization method at low temperature (<–60°C) and in polar aprotic solvents (e.g., THF) [665]. Due to less solubility of P_4VPy, especially at cold temperatures, it is more liable to side reactions than P_2VPy and thus dilution of the monomer prior to addition using a cosolvent (e.g., DMF, pyridine) may be needed to retain solubility [114,649,670,671]. In contrast to 4-VPy, anionic polymerization of 2-VPy is more convenient [649,672].

Occasionally, due to nucleophilic addition of the anionic chain end of the living PVPy to the –N=CH– bond in the pyridine ring (especially in less polar solvents and at warm temperatures), star-branched or cross-linked polymers as side products

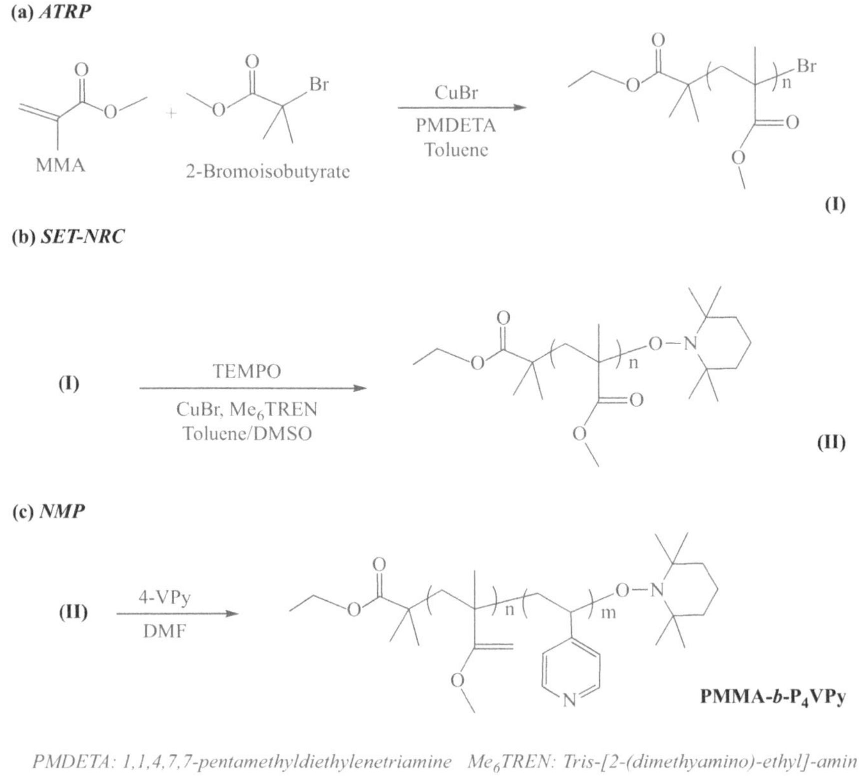

SCHEME 1.67 Synthesis route of PMMA-*b*-P_4VPy via successive (a) ATRP, (b) SET-NRC, and (c) NMP methods [646].

are formed [673,674]. This effect is more important for P_4VPy since the ring has two of these bonds [673]. Thus, there have been several attempts to overcome this problem. One strategy is the use of metallic salts, e.g., lithium chloride [189,675]. Probably, the formation of the mixed aggregates between the anionic chain end and LiCl decrease the chain end reactivity. Another strategy to suppress the side reactions is rapid deactivation of the living chain end using proper capping agents. For example, the synthesis of different types of PVPy-*b*-poly(dimethylsiloxane) copolymers (PVPy-*b*-PDMS) through end-capping with chlorosilanes and epoxides has been investigated [650]. As shown in Scheme 1.68, the method includes the polymerization of VPy monomer in the presence of a bromo-functionalized acetal (I), prepared from the reaction of the corresponding alkyl bromide with *n*-butyllithium. The hydrolysis of the resulting acetal and then the reaction of the formed hydroxyl groups with triphenyllithium produced a lithium alkoxide end group which was further employed (as a macroinitiator) in the polymerization of hexamethylcyclotrisiloxane. Finally, the resulting P_2VPy-*b*-PDMSLi blocks were coupled by dimethyldichlorosilane to produce P_2VPy-*b*-PDMS-SiM e_2-PDMS-*b*-P_2VPy. In this study, the ABA BCPs can be prepared as shown in Scheme 1.69. First, by the addition of 2-isopropenylpyridine (2-IPP, as capping agent) to living P_2VPyLi, followed by reaction with EO, a living P_2VPy

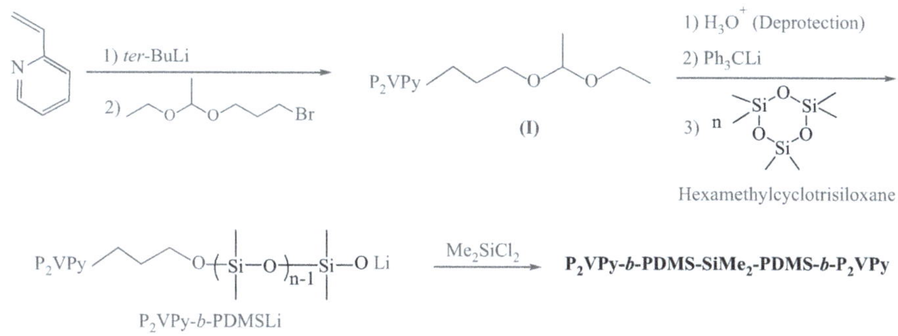

SCHEME 1.68 Synthesis of P_2VPy-*b*-PDMS-$SiMe_2$-PDMS-*b*-P_2VPy [650].

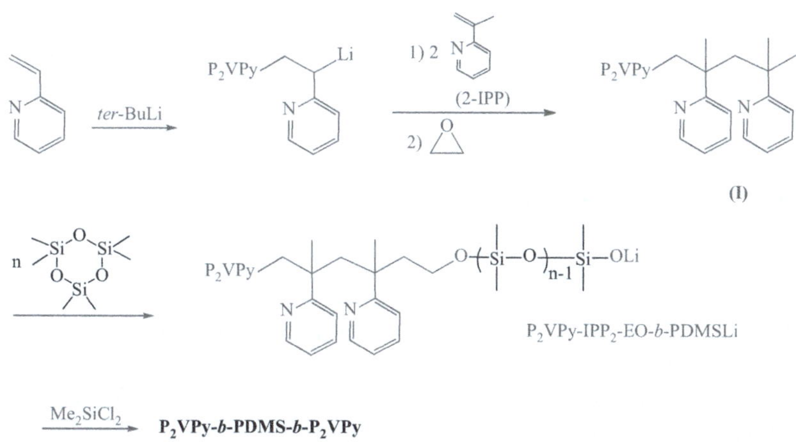

SCHEME 1.69 Synthesis of P_2VPy-*b*-PDMS-*b*-P_2VPy via living anionic polymerization [650].

with lithium alkoxide end group (I) was produced. Then, compound (I) reacts with hexamethylcyclotrisiloxane to afford the living P_2VPy-*b*-PDMSLi blocks. Afterward, the obtained BCP could be coupled with dimethyldichlorosilane to yield the final P_2VPy-*b*-PDMS-*b*-P_2VPy product. BCPs of 2-VPy and dimethylsiloxane have also been synthesized using *s*-BuLi [676].

Also, capping agents of diphenylethylene (DPE) derivatives were employed in anionic polymerization [412,677–682]. Recently, a combination of them with EO was used for the synthesis of amphiphilic copolymers of P_2VPy [360]. Scheme 1.70 shows the synthetic pathway for one of these polymers. First, living P_2VPy chain, prepared from the polymerization of 2-VPy monomers with *sec*-butyllithium, was end-capped with DPE to terminate the active ends of the anions. This approach effectively mitigated crossover steps and transfer reactions that could arise from excessive nucleophilicity during the addition of the third monomer. Then, the resulting polymer (I) reacted with ethoxy ethyl glycidyl ether (EEGE) to afford polymer II, which was subjected to hydrolysis to produce end-functional P_2VPy-$(OH)_2$. Finally, from the polymerization of EO using

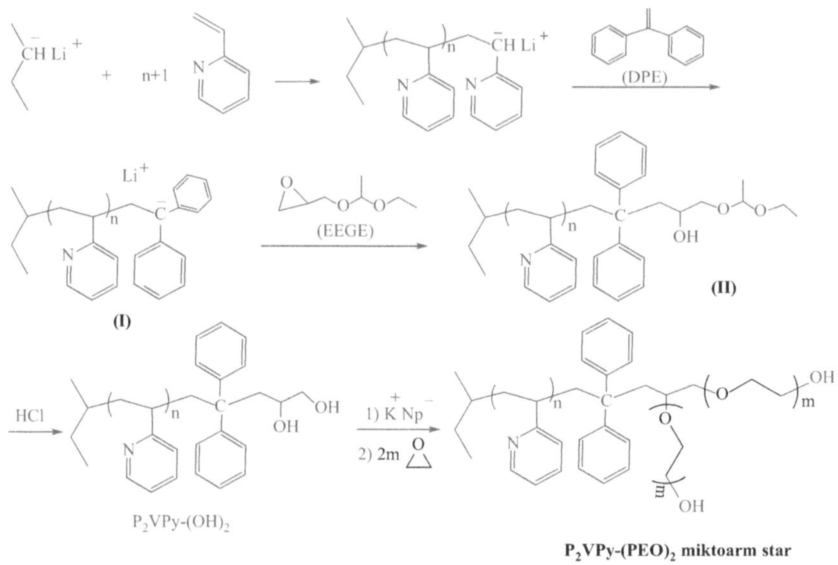

SCHEME 1.70 Synthesis of amphiphilic $P_2VPy\text{-}(PEO)_2$ miktoarm stars [360].

$P_2VPy\text{-}(OH)_2$ as a macroinitiator in the presence of potassium naphthalide (KNp), the desired amphiphilic AB_2-miktoarm stars were obtained.

In addition to DPE, other π-stabilized bulky initiators such as (1,1,4,4-tetraphenyl)butanedilyldilithium ($TPBLi_2$) and fluorenyllithium (FlLi) have been used in the anionic polymerization of vinylpyridine monomers (Scheme 1.71) [648,683,684]. $TPBLi_2$ is prepared from lithium naphthalenide and DPE.

Finally, the use of flow chemistry for the preparation of BCPs of PVPy using anionic initiators (e.g., *s*-BuLi) has been documented [660,685]. The results revealed that both the number-average molar mass and dispersity were affected by the flow rate.

1.3.8 Coordinative Anionic Polymerization and Rare Earth Metal-Mediated Group Transfer Polymerizations (REM-GTP)

Before CRP methods, coordinative anionic polymerization of 2-VPy and 4-VPy monomers using some organomagnesium initiators, e.g., Grignard reagents (such as phenyl magnesium bromide), halides of magnesium diethyl amide, and diethyl magnesium has been investigated [527,686–690]. Typical recipe for coordinative anionic polymerization of 2-VPy is as follows: 2-VPy (10 g)/$(C_2H_5)_2NMgBr$ (0.33 g)/toluene (40 mL)/5 h/45°C in a three-necked flask equipped with nitrogen inlet tube; the polymerization was stopped by adding 50 mL of 10% HCl [527]. Generally, for such polymerization, a complex formation is predicted between the magnesium and the pyridine rings (the last and penultimate rings) located in the living chain end (Scheme 1.72). Due to this complex formation, the anionic polymerization of ethylenic monomers initiated by organomagnesium compounds is restricted by the

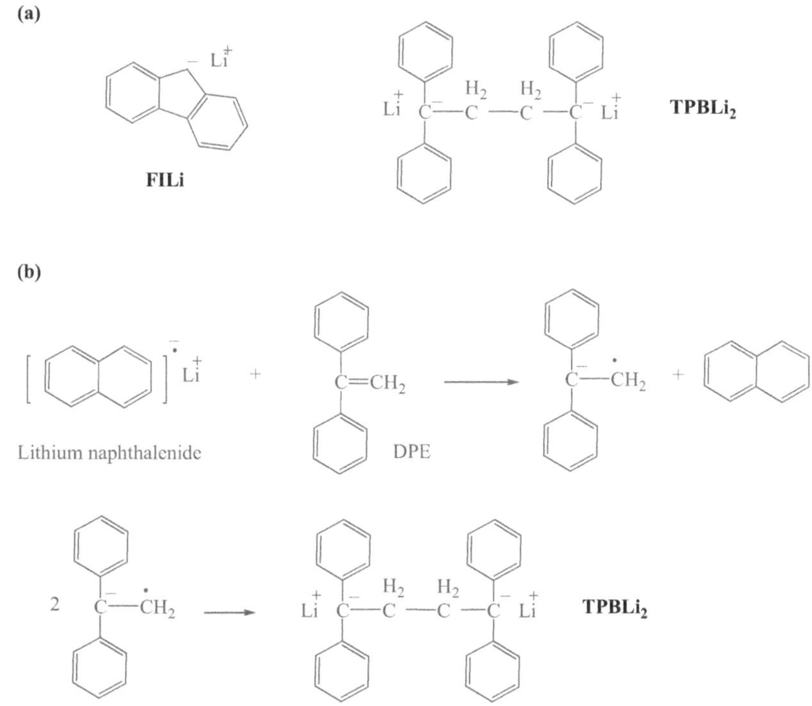

SCHEME 1.71 (a) Structures of $TPBLi_2$ and FlLi and (b) preparation of $TPBLi_2$ [683,684].

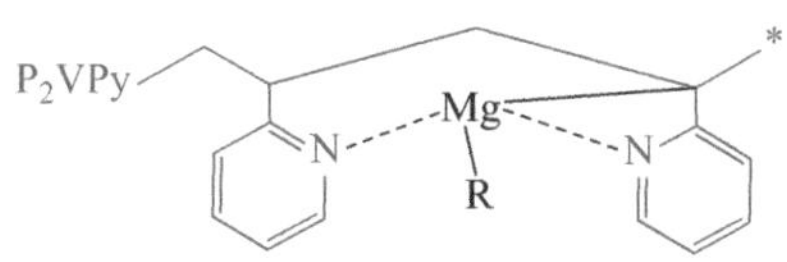

SCHEME 1.72 A possible intramolecular complexation during coordinative anionic polymerization of 2-VPy.

low reactivity of the living chain end toward double bonds, and it is very important in the polymerization of monomers containing strongly electroattractive substituents, such as acrylics or those monomers which are susceptible to the nucleophilic attack, such as vinylpyridine monomers. Organomagnesium compounds have been used for highly stereoregular polymerization of vinylpyridine monomers in nonpolar media or in noncoordinating solvents [527,528,687–689,691–694]. Investigations revealed that while P_2VPy obtained by radical polymerization is an atactic polymer, isotactic polymer is obtained with phenylmagnesium bromide (PhMgBr) in toluene, especially at higher temperatures [695]. At lower temperatures (−78°C), the fraction of isotactic polymer decreased. On the other hand, in polar solvents (e.g., THF), the fraction of isotactic polymer decreased and with decreasing temperature (−78°C), it increased. However, P_4VPy obtained by radical and anionic polymerizations using PhMgBr gave atactic polymers.

Along with the studies on coordinative polymerization methods, some researchers have been showing interest in the development and use of complexes of rare-earth metals (REM) such as yttrium as alternatives to conventional organomagnesium compounds for the iso-selective polymerization of vinylpyridine monomers [696–702]. Due to similarity of these types of polymerization reactions to silyl ketene acetal-initiated group transfer polymerization (GTP), therefore they are often called REM-GTP. The advantages of GTP over simple anionic polymerization methods include its ability to produce strictly linear homopolymers and BCPs with low PDI (<1.1) and increased operation temperature of the polymerization. REM-GTP of 2-VPy in the presence of yttrium complexes has been investigated [697]. The mechanistic studies revealed that REM-GTP has highly living character and follows a monometallic group transfer polymerization mechanism including an eight-membered ring chelate (I) which drives the chain growth by the consecutive conjugate addition of monomers (Scheme 1.73). Also, highly isotactic P_2VPy was prepared using the REM-GTP of 2-VPy with yttrium complex [703]. Isotacticities of up to $P_m = 0.92$ (the meso diads) were achieved by these catalytic systems (Scheme 1.74).

Syndio-specific polymerization of 2-VPy has very rarely been explored. *t*-BuLi/AlR_3 has been used for the syndio-specific polymerization of 2-VPy [704]. Recently, a scandium complex after activation with an organic borate, such as $[PhNMe_2H][B(C_6F_5)_4]$, has been reported for the synthesis of syndiotactic P_2VPy at room temperature (Scheme 1.75) [705]. However, in the absence of borate, isotactic P_2VPy was achieved.

Very recently, new pyrrolidinyl-ethylene fluorenyl REM (yttrium (Y), lutetium (Lu), or scandium (Sc)) complexes were introduced for the synthesis of P_2VPy [706]. The results showed that these complexes alone are able to catalyze the polymerization of 2-VPy, giving isotactic P_2VPy (Scheme 1.76). However, in the presence of

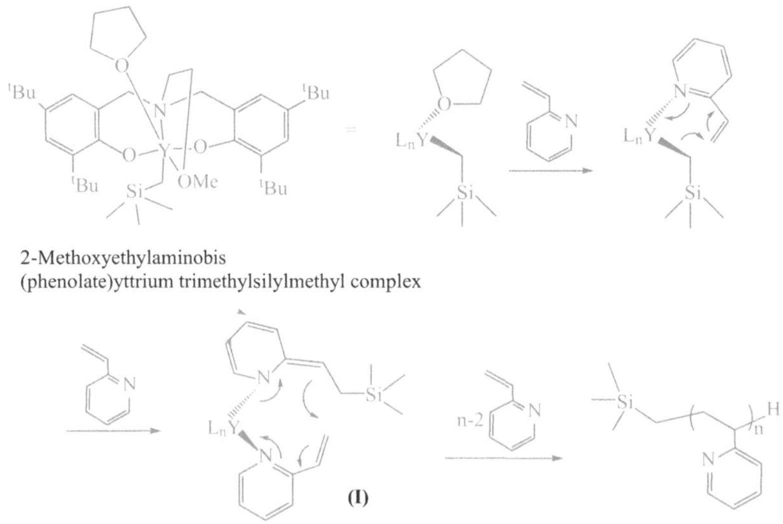

SCHEME 1.73 REM-GTP of 2-VPy in the presence of an yttrium complex [697].

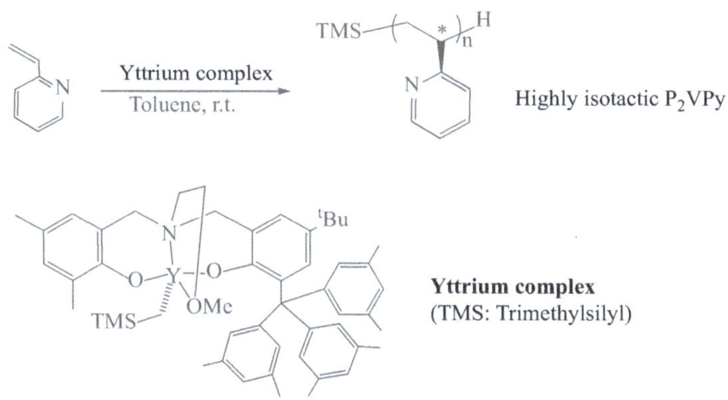

SCHEME 1.74 Synthesis of highly isotactic P_2VPy using REM-GTP [703].

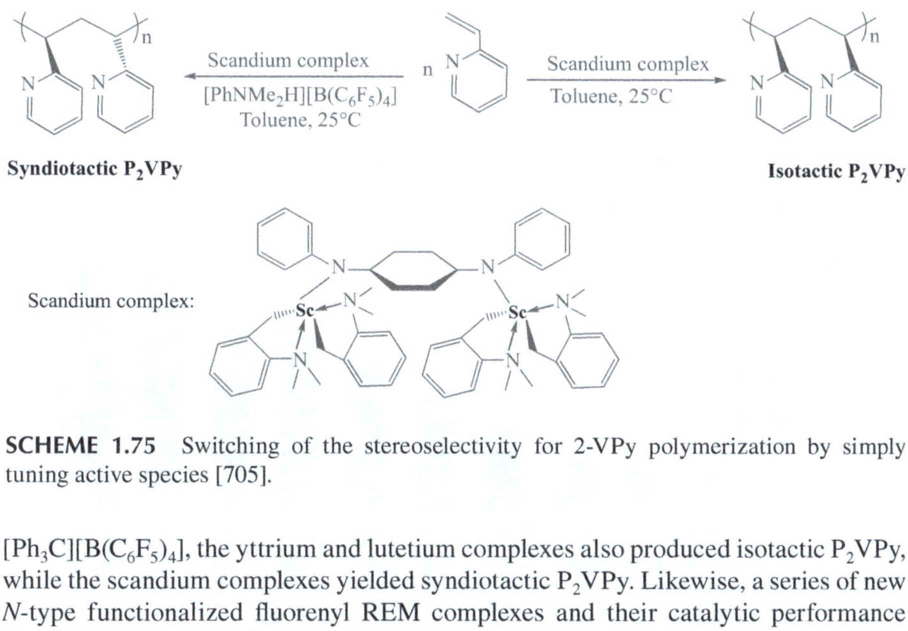

SCHEME 1.75 Switching of the stereoselectivity for 2-VPy polymerization by simply tuning active species [705].

$[Ph_3C][B(C_6F_5)_4]$, the yttrium and lutetium complexes also produced isotactic P_2VPy, while the scandium complexes yielded syndiotactic P_2VPy. Likewise, a series of new *N*-type functionalized fluorenyl REM complexes and their catalytic performance toward 2-VPy polymerization, giving isotactic P_2VPy, was studied [707].

1.4 APPLICATIONS OF PVPy POLYMERS AND THEIR DERIVATIVES

As mentioned previously, PVPy polymers with weakly basic pyridine rings have ideal properties such as good solubility in acidic media and some polar solvents; pH-responsive behavior; thermal resistance; chemical stability against alkalis, acids, oxidants, and reductors; and strong tendency of pyridine rings to metals (metal–ion affinity interactions). Accordingly, PVPy and its derivatives have gained interest in many different fields over the last decade. Figure 1.12 shows the increase in the number of publications on PVPy polymers since 1975 according to Google Scholar Citations based on the keyword "Poly(vinylpyridine)".

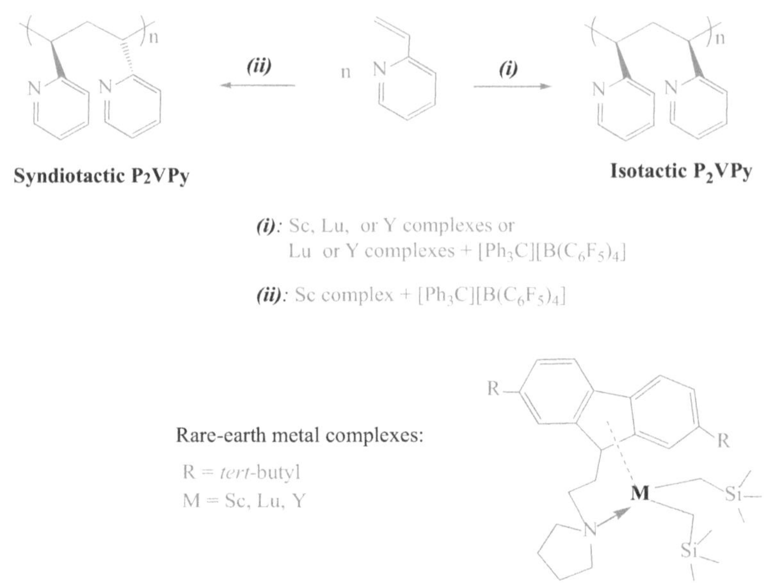

SCHEME 1.76 The stereospecific polymerization of 2-VPy [706].

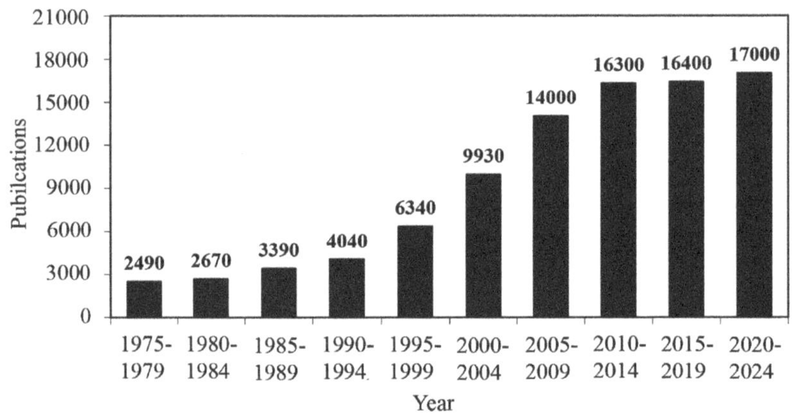

FIGURE 1.12 Growth in publications about PVPy polymers.

PVPy polymers have many applications in various fields of science such as electrochemistry, metal recovery, photonics, removal of acidic and neutral materials, medicine, electronics, catalysis, membrane technology, and so on. In the following subsections, I introduce various applications of PVPy polymers.

1.4.1 In Medicine

The use of PVPy and its derivatives in medicinal chemistry dates back to 1960s, when P_2VPyNO showed an ability to inhibit fibrogenesis (resulted from quartz powders) in the lungs of some animals [708–712]. In this regard, P_4VPyNO has little effect.

The antisilicostic action of P_2VPyNO was ascribed to its tendency of complex formation with monosilicic acid which is formed from the dissolution of the particles of quartz crystallites [708]. Therapy of silicosis in humans by P_2VPyNO has also been investigated [713–717]. The prophylactic effect of P_2VPyNO in pneumoconiosis has been suggested to be due to its ability to capture or prevent the formation of reactive oxygen species (ROS) [717]. Very recently, P_2VPyNO as an excipient affecting the dissolution behavior of amorphous dispersions carrying posaconazole has been introduced, which exhibited comparable or superior performance over commonly used excipients such as Eudragit EPO [718].

In addition to the aforementioned primary properties of P_2VPyNO in medicine, other derivatives of PVPy polymers have demonstrated promising biological characteristics in various medical fields, the representative examples are shown in Table 1.4.

TABLE 1.4
The Representative Biological Activity of PVPy Derivatives

Entry	Reagent	Activity
1	Electrospun PU nanofiber modified by quaternized P_4VPy	Antibacterial [719]
2	Cross-linked and quaternized P_4VPy-*co*-PS	Antibacterial [720]
3	P_4VPy-*co*-PAAM, grafted onto micron-sized silica particles	Antibacterial [721]
4	Quaternized P_4VPy/poly(vinylidene fluoride) (PVDF)	Antibacterial and antifungal [722]
5	Quaternized P_4VPy-*co*-PS	Antibacterial [723]
6	Poly(*N*-benzyl-4-vinylpyridinium chloride)	Removal of airborne bacteria [724]
7	P_4VPyNO/Cu and P_4VPyNO/Ag complexes	Antibacterial [725]
8	PVA-*g*-P_4VPy	Antibacterial [726]
9	P_4VPy-*g*-cellulose (after quaternization of P_4VPy units)	Antibacterial [727]
10	PU-*b*-P_4VPy copolymer	Antimicrobial (as binder in leather finishing) [728]
11	PS-*b*-P_4VPy-loaded fullerene and silver NPs	Antimicrobial [729]
12	Quaternized PVP-*co*-P_4VPy	Antibacterial/antiviral [730,731]
13	Cross-linked P_4VPy-*co*-PNIPAM	Antimicrobial [732]
14	Quaternized PVC-*g*-P_4VPy	Antibacterial [733]
15	Cross-linked poly(*N*-benzyl-4-vinylpyridinium chloride)	Removal of virus from air [734]
16	Cross-linked polv(*N*-benzyl-4-vinylpyridinium bromide)	Removal of pathogenic human viruses [735]
17	Films of quaternized P_4VPy on high-density polyethylene (HDPE)	Antimicrobial [736]
18	Ag NPs anchored on cotton fiber bearing PAA-*g*-P_4VPy (PAA: polyacrylic acid)	Antimicrobial [737]
19	Cationic P_4VPy nanospheres	Antibacterial coatings [738]
20	HDPE matrices grafted with P_4VPy loaded with silver	Antimicrobial [739]
21	Natural rubber latex films with surface-anchored quaternary P_4VPy polyelectrolyte	Antibacterial [740]
22	Lignocellulosic membrane grafted with P_4VPy	Antibacterial [741]
23	Films of spin-coated polyvinylpyrrolidone (PVP) and PVPy	Antibacterial [742]
24	Grafting zinc-coordinated fragments to P_4VPy matrix	Antibacterial [743]
25	Poly(cyclohexene carbonate)-*b*-P_2VPy/Ag	Antibacterial [744]

As mentioned previously, PVPy is a pH-sensitive polyelectrolyte [745,746] and therefore, its potential as a drug delivery carrier is widely investigated [379,747–750]. For example, P_4VPy hydrogel NPs have been synthesized by ATRP method, followed by nanoprecipitation in the presence of the surfactant PVA [746]. It was observed that with decreasing the solution pH from 7 to 5 (close to pK_a of the P_4VPy polymer), the hydrogel swells and the size of the particles increases, resulting in the drug release. It is due to the protonation of the vinylpyridine monomer in this pH range and the electrostatic repulsion inside the particles. Drug delivery capability of pH-sensitive biocompatible P_4VPy-*b*-PEO nanomicelles loaded with 5-fluorouracil and curcumin drugs has been investigated [751]. The preparation of acidic pH-responsive nanogels based on P_2VPy cross-linked with DVB, below 200 nm in size, has also been reported [747]. The nanogels were prepared via SFEP procedure [131] and employed as delivery systems for short oligonucleotides and iron oxide NPs (IONPs) [747]. By changing the pH from 7 to 3.5, the nanogel was swollen and both the magnetic NPs and oligonucleotides penetrate into the swollen nanogels (Scheme 1.77). On the other hand, by addition of NaOH and return of pH to 7, the nanogel collapses and closes down, resulting in the release of both payloads. The results suggest that if such nanogels accumulate into the extracellular tumor environment (pH ~6.5), they can be swelled and consequently the released IONPs produce for performing hyperthermia.

pH-sensitive H40-P_4VPy-PEG-FA has been used for transporting paclitaxel (PTX), an anticancer drug [379]. As shown in Scheme 1.78, it is a nanogel consisting of a core (H40-poly(ε-caprolactone), H40-PCL), a pH-sensitive cross-linker (P_4VPy), a corona (hydrophilic PEG), and a targeting agent (folic acid). The results showed that inside cytosol of cancer cells (pH 6.7), the drug release is highest (~60%). In cancer cells (pH 5) and physiological conditions (pH 7.4), the drug release was found to be approximately 30% and 20%, respectively. The biocompatibility and biodegradability are other unique characteristics of these nanogel particles. An amphiphilic pentablock terpolymer (PCL/PEO-P_2VPy-PEO/PCL) was synthesized, and its potential for delivery and release of calcein was investigated [353]. It possesses the central P_2VPy block (a pH-sensitive polymer), having at both ends PEO looping chains. PCL which is at the end of PEO segments can act as a stabilizer for the construction of layers. The release of the drug by this terpolymer

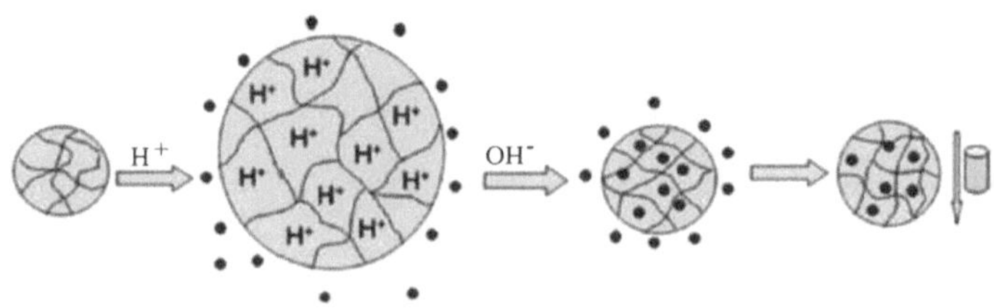

SCHEME 1.77 Swelling/deswelling behavior of nanogels based on cross-linked P_2VPy bearing short oligonucleotides and iron oxide NPs upon changes in pH [747]. (Reproduced with permission from Ref. [747]; Copyright 2010 American Chemical Society).

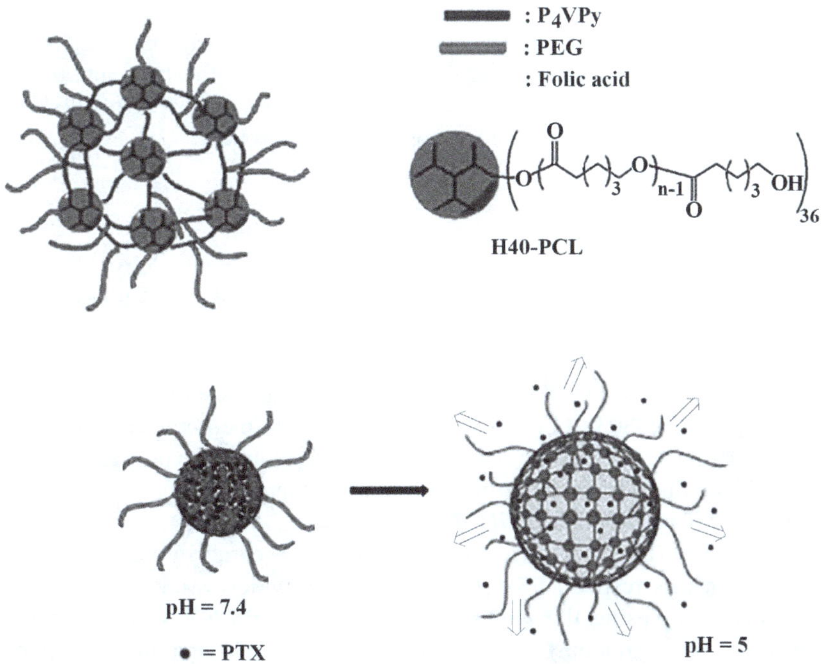

SCHEME 1.78 Structure of H40-P_4VPy-PEG-FA nanogels and mechanism of drug release [379]. (Reproduced with permission from Ref. [379]; Copyright 2014 Elsevier).

was examined at different pH levels, and the findings indicated that the highest release occurs in a medium with a pH value of approximately 5.5, which is similar to the lysosome environment.

Very recently, it has been reported that pH-responsive poly(2-VPy-*co*-butyl methacrylate) (P_2VPy-*co*-PBuMA) copolymers can be used as a targeting polymer coating [752]. pH-responsive coatings are used to prevent the release of pharmaceutical formulations (via the formation of protective coatings on drug) until the target environment with desirable pH is reached [753]. The biological properties of coatings composed of chitosan (CS)/Eudragit E 100 (EE100) or CS/P_4VPy on Ag NPs-decorated titanium substrate in contact with osteoblast cells has been investigated [754].

A semi-interpenetrating polymer network based on hydroxyethyl methacrylate and liner P_4VPy has been used as a hydrogel with the ability to release tetracycline in a controlled/sustained manner for use in dental applications [450]. Mesoporous silica NPs coated with P_2VPy for the adsorption and release of 5-fluorouracil as an anticancer drug [755] and temperature-responsive properties of P_4VPy spin-coated films, having temperature dependence of water contact angle with a well-defined transition at 13°C–14°C, for protein adsorption [756] have been reported.

BCPs composed of either P_2VPy or P_4VPy have been integrated into gold-labeled micelles [351,390,757,758], which serve as visualizing agents. For example, P_4VPy-*b*-PEO micelles consisting of Au NPs have been synthesized, and their visualization in human lung and kidney cells was investigated [759]. The results revealed that the micelles are mainly inside lysosomes and endosomes.

1.4.2 Membrane Materials

An important application of PVPy polymers is in membrane technology [188,255,760–762]. Representative examples of membranes based on PVPy derivatives are shown in Table 1.5.

PVPy polymers have garnered significant attention in the field of membrane fabrication due to their self-assembly behavior as well as their several stimuli-responsive characters [760,799–803]. For example, microfiltration membranes have been prepared by grafting of 4-vinylpyridine onto polyethylene and polypropylene [549] and poly(vinylidene fluoride) [765]. Stimuli-responsive membranes based on a thin film of P_4VPy have been introduced [760].

BCPs of PVPy have also attracted tremendous research interest in membrane technology. In this respect, PS-*b*-P_4VPy has been widely studied [242,803,804]. The possibility of using the membranes made from PS-*b*-P_4VPy in drug delivery, for filtration of aqueous solutions, and in biomolecule separations depends on careful choice of their preparation methods along with highly ordered morphologies. For this purpose, different solvents and mixed solvent systems and additives have been used for the preparation of PS-*b*-P_4VPy membranes with well-defined morphology [805–807]. Nonetheless, these methods are hindered by their inadequate biocompatibility. Carbohydrates have been identified as biocompatible substitutes for chemical solvents and additives in the fabrication of PS-*b*-P_4VPy membranes [808]. For example, highly porous PS-*b*-P_4VPy membranes were synthesized in the presence of some carbohydrates (e.g., α-(D)-glucose, saccharose, and α-cyclodextrin) as additives [808]. They can reversibly interact with the pyridine units located in BCP chains via hydrogen bonding [809]. The synthesized membranes exhibited increased porosity along with a well-structured porous layer when compared to the membranes produced without carbohydrates. In addition, they exhibited higher water flux value.

Double stimuli-responsive membranes composed of pH-sensitive PS-*b*-P_4VPy membranes and temperature-responsive PNIPAM have been studied [803]. PS-*b*-P_4VPy membranes are coated with polydopamine (PDA, as an interlayer), which subsequently reacts with amine-terminated PNIPAM. The pH-dependent water fluxes of the membranes were evaluated at temperatures ranging from 25°C to 45°C, including the LCST of PNIPAM. The water flux significantly diminishes below and rises above the LCST of PNIPAM (30°C–35°C). At all tested temperatures, a reduction in water flux is observed between pH 3.4 and 3.8. It is noted that at lower temperatures, the variation in water flux is less significant owing to the swelling of PNIPAM, resulting in decreasing of pore sizes.

TABLE 1.5
Representative Examples of Membranes Based on PVPy Derivatives

Entry	Components of Membrane	Application
1	P_2VPy and P_4VPy-*g*-PVDF copolymers	Microfiltration [763–765]
2	PU/PVPy composite	Desulfurizing gasoline [766]
3	Quaternized P_2VPy with 1,4-diiodobutane (DIB)	As polymer gel with pH-controllable permeability [767]
4	PI-*b*-PS-*b*-P_4VPy	Ultrafiltration [768]
5	P_4VPy- and poly(vinyl acetate-*co*-4-VPy)-grafted PVC filled with organo clay	Removal of THF from water [769]
6	PVC-grafted P_4VPy	Anion-exchange processes [770,771]
7	P_4VPy-*g*-polyethylene and P_4VPy-*g*-polypropylene	Ion exchange [549–551,761,772–776]
8	$LiClO_4$-doped PS-*b*-P_2VPy	Ionic conductivity and microfiltration [598,777]
9	P_4VPy/(hexafluoro propane dianhydride)-durene (6FDA)/PAN composite	Air separation [141]
10	Silicone rubber/P_4VPy/polysulfone	Gas separation [778,779]
11	P_4VPy/poly(4-methyl-petene-1)	O_2/N_2 separation [780]
12	P_4VPy-*co*-PAN/Cu(II) complex	Sorption and permeation of CO_2 [781]
13	P_4VPy or P_2VPy/natural rubber or poly(dimethylsiloxane)	Permeabilities for O_2 and N_2 [782]
14	P_2VPy-containing oxadiazole	CO_2 capture [783]
15	Nanocages/P_4VPy	Gas separation [784]
16	Silicone rubber/P_4VPy/polyetherimide	Gas separation [785]
17	Poly(*N*-ethyl-4-vinylpyridinium bromide) and sodium carboxymethyl cellulose	Pervaporation dehydration of ethanol [786,787]
18	Poly(ST-*b*-4-VPy-*b*-solketal methacrylate), PS-*b*-P_4VPy-*b*-PSMA	The protein adsorption [788]
19	Poly(ionic liquid)-based quaternized P_4VPy an P_2VPy supported on polyamide	Fractionation of dyes and sugar [789]
20	PS-*b*-P_4VPy/polyquaternium-10	Anion exchange [790]
21	PAN/P_4VPy-Au NPs	Separation of nitrophenols and methylene blue [791]
22	Polyethylene film containing P_4VPy and poly(imidazolium)	Anion exchanger in fuel cells [792]
23	P_4VPy as a capture layer on PVDF	Rapid purification of antibodies [793]
24	P_4VPy as a capture layer on PVC	Proton exchange membranes for fuel cells [794]
25	Cellulose acetate/polyethylene glycol/PS-*b*-P_4VPy/ TiO_2 NPs	Wastewater treatment [795]
26	Highly cross-linked butene-grafted PVA-*co*-P_4VPy	For acid recovery via diffusion dialysis and desalination via electrodialysis [796]
27	Composite P_4VPy and polytetrafluoroethylene (PTFE)	Protection against phosphonate-based cholinesterase inhibitors [797]
28	Cross-linked and quaternized P_4VPy	Electrodialysis desalination [798]

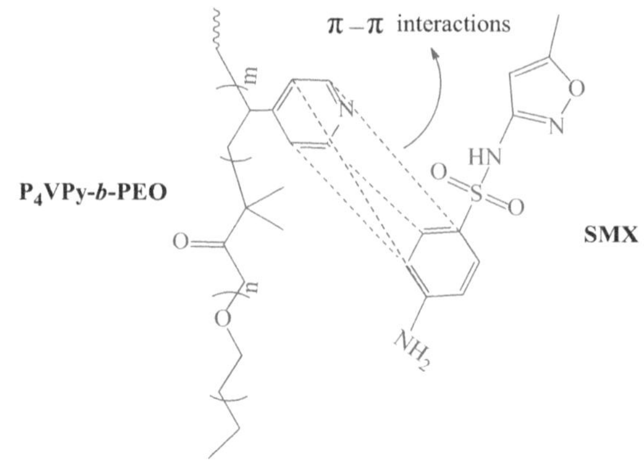

SCHEME 1.79 A schematic of π–π interactions that take place on the aromatic rings of the SMX and the pyridine ring of P_4VPy-*b*-PEO [810].

BCPs bearing PVPy and PEO blocks are other examples of PVPy-based membranes. Recently, the fabrication of membranes modified with the P_4VPy-*b*-PEO for the removal of sulfadiazine (SDZ), sulfamethoxazole (SMX), and omeprazole (OMZ), a class of electron-deficient pharmaceuticals, from aqueous media has been reported [810]. P_4VPy segments adsorb contaminants through electrostatic π–π interactions (Scheme 1.79) [810]. Such interactions were also observed in nanocellulose-P_4VPy-*b*-PEO films utilized for the removal of sulfamethoxazole [811]. The permeability of membranes of P_2VPy-*b*-PEO vesicles has been studied [812].

1.4.3 Chemosensors and Biosensors

Stimuli-responsive polymers have gained considerable attention in the field of artificial sensors owing to their physicochemical characteristics, high selectivity, and rapid measurements [813–817]. They have been used to fabricate various types of sensors (e.g., humidity sensors, pH sensors, biosensors) through mainly coating the electrode surface. PVPy with stimuli-responsive characters is one of the building blocks of copolymers used in sensor devices due to its modifiable basic and acidic properties, flexible polymeric chains, and own hydrophobic/hydrophilic indices. The applications of PVPy polymers in electrode materials have been recently reviewed [101]. The coating of electrodes by the electrochemical polymerization of vinylpyridine complexes has been reported [818–823].

Humidity sensors are one of the most important types of sensors. A practical humidity sensor should exhibit important properties such as quick response and high sensitivity, small hysteresis, water durability, and small temperature coefficient. Among humidity sensors, those containing PVPy polymers have attracted considerable interest due to the above-mentioned reasons [824]. For example, it has been shown that cross-linked form of quaternized P_4VPy as a weakly hydrophilic polymer is insoluble in water with electrical resistivity varied over a wide humidity range, which can be used as humidity-sensitive materials [101,825–831]. It exhibits a response linearity in semilogarithmic scale, a very small hysteresis (<1%

relative humidity), and a resistance to ethanol vapor. A composite of quaternized and cross-linked P_4VPy with processable polypyrrole has been used for the fabrication of thin-film humidity sensors [827]. The prepared sensor exhibited small hysteresis (~1% relative humidity), high sensitivity with impedance changes from 10^7 to 10^3 Ω over the range of 11–95% relative humidity.

PVPy-based BCPs have been used in the structure of humidity sensors [829,832]. For instance, silicon-based nanoporous membranes as the micellar films based on Ps-*b*-P_2VPy were found to be quite sensitive to the humidity of the environment [833]. Gold electrodes with deposited sulfonated PS-*b*-P_4VPy were employed as humidity sensors [834]. They exhibited relatively small hysteresis, good sensing repeatability, and quick response. Cross-linked and quaternized P_4VPy-*co*-PBuMA (Scheme 1.80) [94] has been used for fabrication of humidity sensors with very small hysteresis (<1% RH).

PVPy polymers have also been used in the structure of sensors for detecting various chemicals. Representative example of this type sensors is shown in Table 1.6. For example, a series of functional polymer microspheres are constructed from radical copolymerization of fluorescent *N*-vinylcarbazole (NVCz) monomers with different nitrogen-containing vinyl monomers including 4-VPy, 1-vinylimidazole (VIM), and AAM (Table 1.6, entry 1) [835]. Reaction was accomplished in the presence of DVB and AIBN as cross-linker and initiator, respectively. Scheme 1.81 shows the preparation of one of these microspheres, namely poly(DVB-*co*-NVCz-*co*-4-VPy). These microspheres were used as a sensor for detecting Cu^{2+} based on the quenching effect of Cu^{2+} on the fluorescence intensity of carbazole units. The lowest detectable concentrations of copper cation for poly(DVB-*co*-NVCz-*co*-AM), poly(DVB-*co*-NVCz-*co*-4-VPy), and poly(DVB-*co*-NVCz-*co*-VIM) were determined to be 0.028, 0.030, and 0.028 μM, respectively. Very recently, PS-*b*-P_2VPy micelle bearing Ag NPs has been investigated as a colorimetric sensor for the determination of cartap (Table 1.6, entry 2) [836]. As shown in Scheme 1.82, PS-*b*-P_2VPy/Ag NPs was prepared by micellization of BCPs in toluene (non-solvent for P_2VPy blocks) followed by addition of the formed micelles to $AgNO_3$ (aq) and subsequent reaction with $NaBH_4$. With the formation of Ag NPs, the reaction medium became yellow in color. Using UV–Vis spectroscopic studies, it was found that among the various pesticides (e.g., carbofuran, cartap, alpha-cypermethrin,

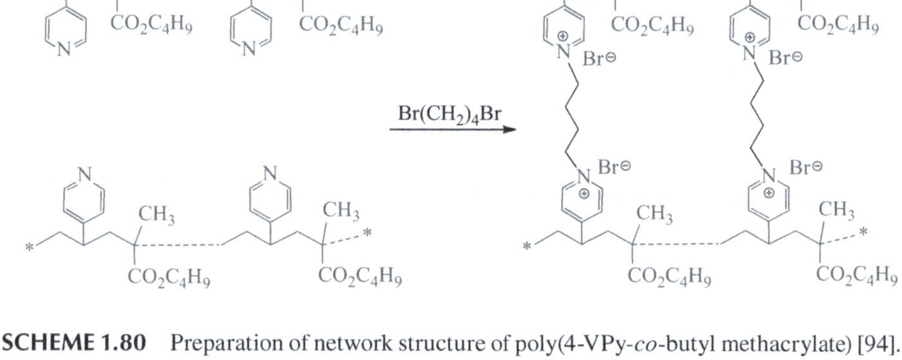

SCHEME 1.80 Preparation of network structure of poly(4-VPy-*co*-butyl methacrylate) [94].

TABLE 1.6
Representative Examples of Sensors Based on PVPy for Detecting Various Chemicals

Entry	Components of Sensor	Detected Chemicals
1	Poly(DVB-*co*-NVCz-*co*-4-VPy)	Cu^{2+} [835]
2	PS-*b*-P_2VPy micelle bearing Ag NPs	Cartap in aqueous media and blood plasma [836]
3	Cross-linked poly(FcMMA[a]-*co*-4-VPy) incorporated in SPCE[b]	Bisphenol A [837]
4	GCE[c] modified by quaternized P_4VPy/ $[Ag_4(CN)_6]$	Silver ions [838]
5	GCE modified by P_4VPy/MWCNT	Aspirin and caffeine [839] and Paracetamol [840]
6	GCE modified with composite film of P_4VPy and graphene sheet	Acetaminophen [841]
7	GCE loaded with MWCNT/P_4VPy/complex copper ions	L-cysteine [842]
8	GCE modified with $Mo(CN)_8^{4-}$-doped-P_4VPy	L-ascorbic acid (vitamin C) [843,844]
9	[Osmium(2,2′-bipyridyl)$_2$-(P_4VPy)$_{10}$Cl] Cl-modified electrodes	Ascorbic acid [845,846]
10	MnO_2 nanorods/carbon NPs/P_4VPy	2-Propanol, acetone, ethanol, methanol, and mesitylene vapors [847]
11	Core–shell hybrids of P_2VPy@barium titanate oxide ($BaTiO_3$)	CO_2 [848]
12	Titanium carbide/carboxylated GO-P_4VPy	Aflatoxin B1 [849]
13	Graphite electrode modified with P_4VPy-osmium complex	CdS NPs [850]
14	P_4VPy-*g*-PVDF nanoporous electrodes	Monitoring mercury in water [817]
15	GCE[c] modified by Au NPs stabilized by P_2VPy-*b*-PMMA	Nicotine [851]

[a] Ferrocenylmethyl methacrylate; [b] screen-printed carbon electrodes; [c] glassy carbon electrode.

deltamethrin, etc.), only cartap has an interaction with PS-*b*-P_2VPy/Ag NPs. The nanosensor identified cartap within analytical concentrations ranging from 0.036 to 0.36 μg/L, with a limit of detection (LOD) of 0.06 μg/L.

Another application of PVPy polymers in the field of artificial sensors is their use in designing biosensors [846,852–855]. Among these, the production of glucose-sensitive biosensors using amperometric enzyme electrodes in the treatment of diabetes mellitus, one of the leading causes of death and disability in the world, is of great importance [856–861]. The electrical communication between metal electrodes and the active site of the redox enzyme is crucial in enzyme electrodes. However, transfer of electrons was disturbed by protein layers that function as a thick insulator material. To address this limitation, some mediators such as ferrocenes, components of organic metals, and ruthenium complexes capable of permeating the

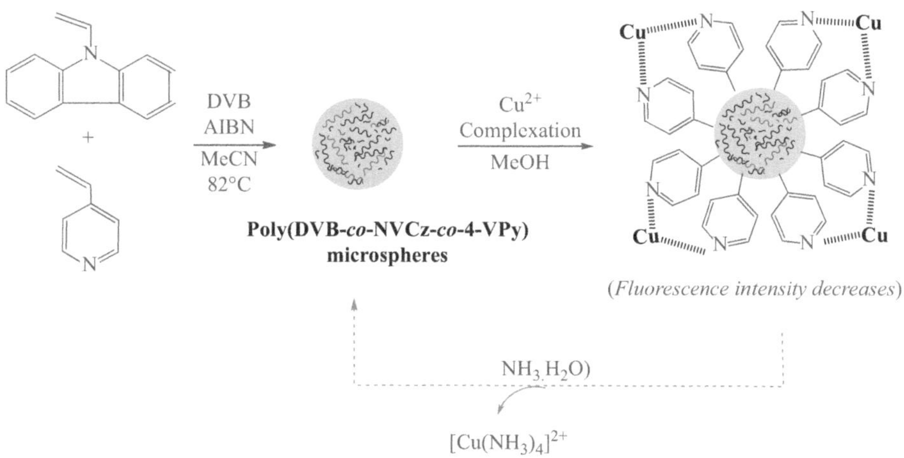

SCHEME 1.81 Illustration of the synthetic pathways for poly(DVB-*co*-NVCz-*co*-4-VPy) microsphere doped with Cu^{2+} and its regeneration using ammonium hydroxide [835].

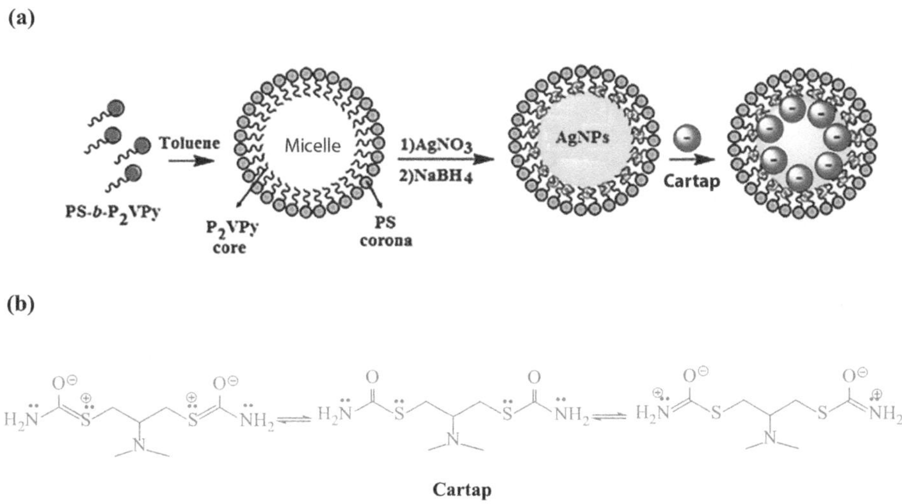

SCHEME 1.82 Illustration of (a) synthesis of P_2VPy-*b*-PS/Ag NPs micelles and entrapment of cartap into micelles and (b) the resonance structures of cartap [836]. (Reproduced with permission from Ref. [836]; Copyright 2018 Elsevier).

enzyme have been used [862]. Another strategy is the use of a polycationic redox polymer as a membrane which causes a reduction in the electron-transfer distance via its complexation of the polyanionic enzyme. For example, P_4VPy complex of $Os(bipyridyl)_2Cl$ ($Os(bpy)_2Cl$) has been used as a polymeric redox for manufacturing of glucose biosensors [863–865]. As shown in Scheme 1.83, P_4VPy blocks are quaternized with bromoethanol or 3-bromopropionic acid (for promoting hydrophilicity) as well as an ester of *N*-hydroxysuccinimide. Electron transfer occurs in a complex made from P_4VPy-Os(bpy)$_2$Cl and glucose oxidase (GOX). Also, one type of glucose biosensors from gold electrodes containing P_4VPy units made via self-assembly layer by layer process has been introduced (Scheme 1.84) [866]. The first step involves the immobilization of 11-mercaptoundecanoic acid (MUA) on gold electrode. Next, this surface undergoes further reactions with P_4VPy-Os(bpy)$_2$Cl and then with GOX-SWNTs, alternatively. The obtained sensors exhibited a sensitivity of 56 μA/mM·cm^2 as well as a current density of 440 μA/cm^2. P_4VPy-*co*-PS membranes have been used in the manufacturing of glucose biosensors with a sensitivity of 30 mA/(M cm^2) [867]. A biosensor based on MWCNT/P_4VPy/Prussian blue (PB) for the detection of hydrogen peroxide, formed from the enzymatic reaction of GOX with its substrate glucose, has also been reported [868]. This sensor was easily synthesized by immobilization of a composite film of MWCNT/P_4VPy on Au electrodes followed by electrochemical reaction of the obtained Au/MWCNT/P_4VPy matrix with an acidic solution of hexacyanoferrate(III), ferric chloride, and potassium chloride resulting in the deposition of PB on the electrode surface. The mechanism of action of this sensor is based on the capability of PB

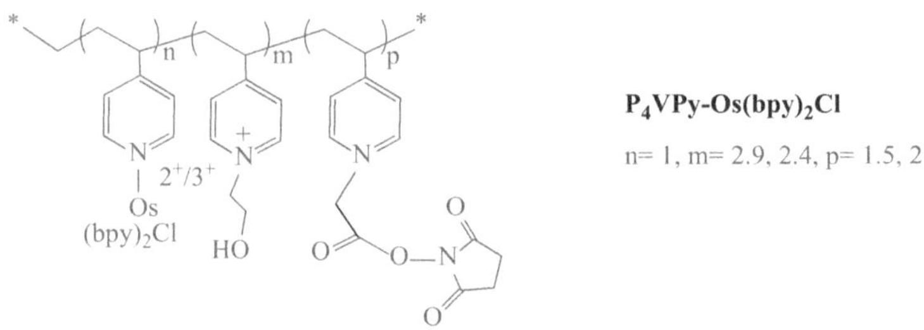

SCHEME 1.83 Structures of P_4VPy-Os(bpy)$_2$Cl used for manufacturing of glucose biosensors [865].

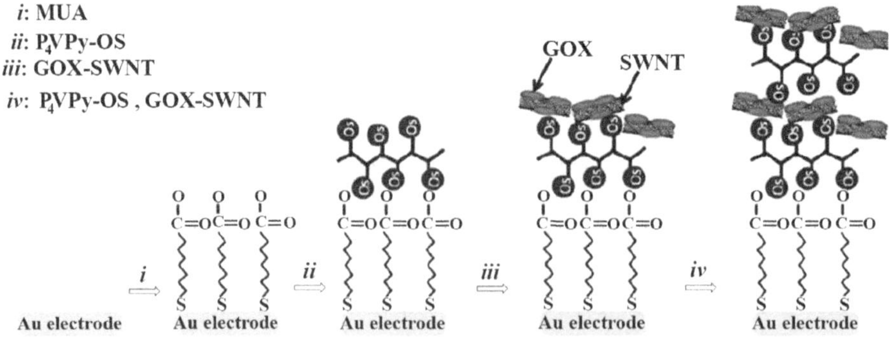

SCHEME 1.84 Functionalization of gold electrodes with P_4VPy-Os(bpy)$_2$Cl and GOX-SWCNTs by the LBL process [866]. (Reproduced with permission from Ref. [866]; Copyright 2009 American Chemical Society).

for the electroreduction of H_2O_2, which is aided by MWCNT with high surface area and high electrical conductivity. P_4VPy is bound to MWCNT surface through π-stacking interactions and helps to fix it on electrode. Au/MWCNT/P_4VPy/PB sensor exhibited a detection sensitivity of 1.3 μA/(μM cm^2) with a LOD of 2.5×10^{-8} M. The results showed that Au/MWCNT/PVPy/PB is superior over the PB–Au system in terms of more stability, lower detection limit, and faster response to H_2O_2. Interestingly, by addition of a buffer solution of GOX enzyme and Nafion on above electrode, a new glucose biosensor was produced, which upon addition to glucose solution was able to determine the H_2O_2 produced from the enzymatic reaction. The biosensor showed a LOD of 2 μM with a detection sensitivity of 38 μA/(mM cm^2). Very recently, a graphene-PVP-P_4VPy composite as a nonenzymatic glucose sensor with a detection sensitivity of 180 μA/(μM cm^2) has been reported [869].

One type of DNA damage is oxidation of the nucleobases by ROS, resulting from redox cycling of drug metabolites or certain pollutants. Voltammetry is an attractive method for rapid, inexpensive assays of detection of damaged DNA. This method is based on distinguishing intact double-stranded DNA from single-stranded DNA, and for this purpose, different biosensors have been designed [870,871]. P_4VPy-Os(5,6-dmphen)$_2$Cl (5,6-dmphen=5,6-dimethyl-1,10-phenanthroline) and P_4VPy-Ru(bpy)$_2$(ClO$_4$)$_2$ have been reported as a DNA biosensor for electrochemical detecting of DNA oxidation and damage (Scheme 1.85) [870,872–877]. They were coated on the surface of different electrodes such as pyrolytic graphite and Au disk electrodes. The simplest way to show the mechanism of action of these sensors, e.g., is based on the reactions shown in Scheme 1.86. The $P_4VPy\text{-}Ru(bpy)_2^{3+}$ generated in situ on the electrode surface (Scheme 1.86a) acts as the oxidizing agent for guanine in DNA (Scheme 1.86b) and the increase in catalytic oxidation current is measured by cyclic voltammetry [873].

Iridium complexes have attracted much attention in organic light-emitting diodes (OLEDs) and electrogenerated chemiluminescence (ECL), due to their unique luminescent characteristics and enhanced quantum yields [878]. Nevertheless, their sensing applications have often been restricted due to inadequate solubility and the

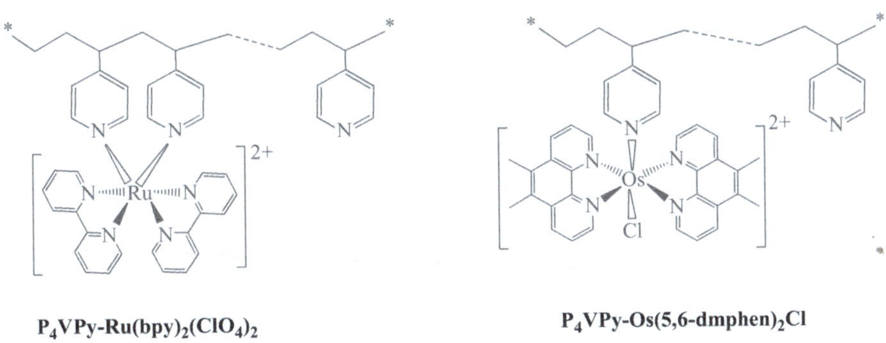

SCHEME 1.85 Structures of P_4VPy bearing Ru and Os complexes used in the construction of sensors for DNA detection [872,873].

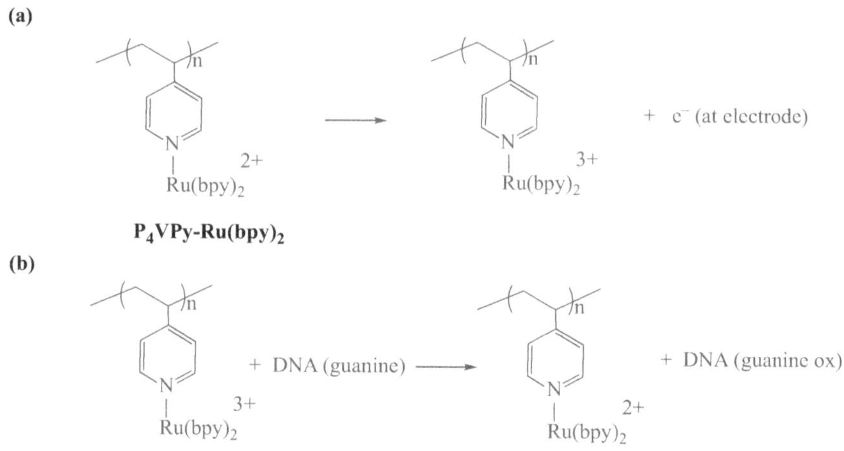

SCHEME 1.86 The catalytic oxidation pathway for the oxidation of guanine catalyzed by $P_4VPy\text{-}Ru(bpy)_2(ClO_4)_2$ [870,874].

quenching effects of H_2O and O_2 on their emission. One strategy to mitigate the above-mentioned problems is immobilization of iridium complexes on organic and inorganic carrier materials. In this context, immobilization of these complexes on PVPy and their use in ECL emitters have gained much importance in recent years [879]. For example, a new biosensor based on electrochemiluminescence of cyclometallated iridium(III)-[P_4VPy-*co*-PS] NPs (CIPNPs) for the detection of troponin I (TI) in human serum has been reported [880]. Blood troponin levels can be used to diagnose myocardial infarction. The method of preparation and operation of this sensor are shown in Scheme 1.87. In the first step, electrochemical pretreated carbon screen-printed electrode (CSPE) was coated with nitrogen-doped graphene (NG), in order to boost both the loading capacity and conductivity of the aptasensor. In the second step, Au NPs prepared from the chloroauric acid ($HAuCl_4$) and trisodium citrate were treated with the aptamer (Apt) to produce Au NPs-Apt. The resultant Au NPs-Apt was added to cyclometallated iridium(III) complex (CIPNPs) immobilized on commercially available P_4VPy-*co*-PS in 1-octyl-3-methylimidazolium hexafluorophosphate ionic liquid (IL), then the resulting suspension was accumulated on CSPE/NG electrode, and finally the electrode was rinsed with phosphate buffer saline solution and incubated in bovine serum albumin (BSA). When the sensor is exposed to TI, it attaches to the aptamer and consequently results in desorption of the aptamer from Au NPs-Apt which eventually leads to an increasing ECL emission depending on the amount of TI. The results showed that sensors constructed from CIPNPs exhibited more sensitivity than those made from benchmark $Ru(bpy)_3^{2+}$, because unlike $Ru(bpy)_3^{2+}$-containing sensors, those with CIPNPs segments are more adaptable to aqueous media and aggregation of P_4VPy-*co*-PS units protects the emitting centers from the quenching effect of H_2O and O_2 on the iridium complex-based ECL. The present ECL aptasensor is superior to the previous reported conventional ones (e.g., antibody bio-receptors) for the determination of TI with respect to easier operation and a wide dynamic range (0.1 pM to 10 nM) with a detection limit of 20 fM.

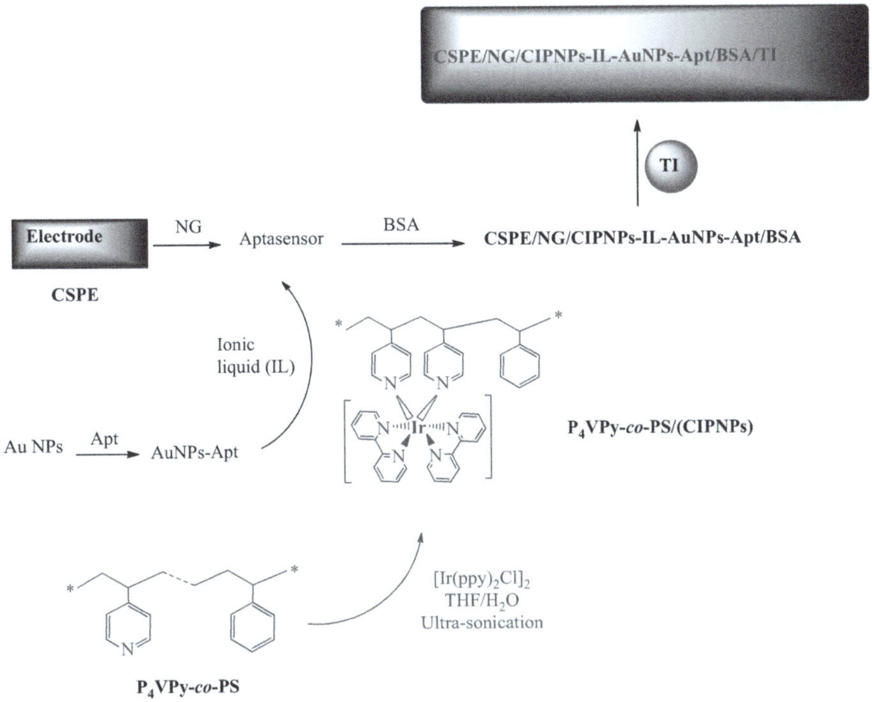

SCHEME 1.87 Fabrication and operation of an aptasensor based on [P_4VPy-*co*-PS]-Ir(bpy)$_2$ NPs for early diagnosis of myocardial infarction via detecting troponin I [880].

Optical sensors based on stimuli-responsive polymers have found extensive applications across diverse areas such as medical diagnostics and environmental monitoring [881]. A colorimetric pH sensor based on graphene oxide (GO) has been developed using two distinct pH-responsive polymers: PAA and P_2VPy [882]. These polymers serve as linkers between the GO and quantum dots (QDs) [882]. The PAA and P_2VPy chains were affixed to the surfaces of blue CdS/ZnS QDs (BQDs) and orange CdSe/ZnS QDs (OQDs), respectively, and then the resulting PAA-BQDs and P_2VPy-OQDs were simultaneously integrated onto a GO sheet to produce the desired sensor. In the obtained sensor, PAA and P_2VPy respond to the pH changes of the medium at ~pH 4.5 and ~pH 3, respectively, due to the ionization of carboxylic acid groups in PAA and the protonation of pyridine groups in P_2VPy. Consequently, the PL emissions of the two QDs can be independently regulated by adjusting pH. The color of the sensors changes from orange to white and then to blue as the pH of the medium rises from 1 to 7. Likewise, PyMMP-*b*-P_2VPy (PyMMP: (1-pyrene)methyl-2-methyl-2-propenoate)-coated CdSe/ZnS QDs have been introduced as a ratiometric pH sensor in which the color of QDs changes from blue to red upon alterations of pH of the medium [883].

In addition to the aforementioned applications of PVPy-based electrodes as humidity sensors, chemosensors for detecting chemicals, and biosensors, they have been used for various oxidation reduction processes and electrochemical studies and have been well reviewed [101]. In this regard, some of them are tabulated in Table 1.7.

TABLE 1.7
PVPy-based Electrodes for Electrochemical Processes Reported in the Literature

Entry	Components of Electrode	Application
1	Platinum microparticles incorporated into P_4VPy and Nafion films-coated carbon electrode	Formic acid and methanol oxidation [884]
2	Carbon cloth electrodes modified by glycerol dehydrogenase immobilized on P_4VPy	Oxidation of reduced nicotinamide adenine dinucleotide (NADH) [885]
3	Modified GCE with a cobalt phthalocyanine-doped P_4VPy films	Hydroquinone oxidation [886]
4	P_4VPy-bound $Ru(bpy)_2Cl^+$/oxalate system on indium tin oxide electrodes	Oxalate oxidation [887]
5	Indium tin oxide electrodes modified with poly(4-diazosulfonate styrene-*co*-4-VPy)	Photoelectron conversion materials [888]
6	Cu electrode coated with a P_4VPy film	Robust electroreduction of CO_2 [889]
7	Phthalocyanine cobalt incorporated into P_4VPy-*co*-PS film coated on graphite electrode	Electrocatalytic proton reduction [890]
8	Electrodes modified with cobalt phthalocyanine immobilized in P_4VPy film	Selective CO_2 reduction [891]
9	P_4VPy films with electrodeposited Pd NPs	Oxidation of oxalic acid [892]
10	P_4VPy-modified copper(I) oxide semiconductor	Selective reduction of CO_2 [893]

1.4.4 As a Polymeric Adsorbent

PVPy has been used as an efficient adsorbent for the removal of different organic compounds, gases, hormones, and drugs (Table 1.8). In comparison with PS resins without ion exchange functional group, PVPy exhibits higher adsorption capacities. Given that PVPy is a less potent base compared to primary, secondary, and tertiary aliphatic amines, it is simpler to separate the adsorbed species from PVPy than from stronger bases. Additionally, PVPy does not absorb counteranions like chloride and sulfate ions, which is a common problem faced when utilizing standard anion-exchange resins. In addition to the items listed in Table 1.8 (chemical compounds, gases, hormones, and drugs), PVPy derivatives have also been employed for the removal or adsorption of other species such as dyes and proteins from their relevant media which will be discussed below.

P_4VPy/treated clay composite [928], cross-linked poly(4-VPy-*b*-crotonic acid) and poly(4-VPy *N*-oxide-*b*-crotonic acid) [929], and poly(4-VPy-*g*-bromodecane) [930] have been used for the removal of brilliant orange, methylene blue, and textile blue basic dyes from aqueous solutions, respectively. Also, CS-grafted P_4VPy with a high uptake for acidic dyes [931], cross-linked PS-*b*-P_4VPy with dibromododecane as an organogel utilized for the removal of anionic dyes from their aqueous solutions [932], and poly(ethylene terephthalate) (PET) fibers grafted with 4-VPy and 2-methylpropenoic acid for the removal of methylene blue and rhodamine B from aqueous media [933] was introduced. Recently, the Fe_3O_4@P_4VPy composites

TABLE 1.8
Representative Examples of Applications of PVPy-based Sorbents for the Removal of Different Species

Entry	Polymeric Adsorbent	Sorbed Species
1	Cross-linked P_2VPy and P_4VPy	Phenol [894–896]
2	Cross-linked P_4VPy	Carboxylic acids [897]
3	Cross-linked P_4VPy	Lactic acid [898–900]
4	Cross-linked P_2VPy	Diclofenac [901]
5	Cross-linked P_4VPy	Amino acids [902,903]
6	P_4VPy	Sugars [904]
7	Cross-linked P_4VPy	[*N*-(Phosphonomethyl)imino]diacetic acid [905]
8	Cross-linked P_4VPy	SO_2 [906]
9	Quaternized P_4VPy gel coated on silica	Organic sulfonates [907,908]
10	Quaternized and cross-linked P_4VPy	Sulfide [909]
11	P_4VPy supported on bentonite[a]	Toluene [910]
12	Hybrid polymer (P_4VPy/SiO_2/APTMS[b]), cross-linked P_4VPy, and Fe_3O_4-P_2VPy	Bisphenol A [911–913]
13	P_4VPy-modified silica	Oil [914]
14	Cross-linked P_4VPy	Swage (peptone, urea, NaCl, KCl, etc). [915]
15	Poly(4-VPy-*co*-ethylene dimethacrylate) monolith	Steroid sex hormones [916] and non-steroidal anti-inflammatory drugs [917]
16	P_4VPy copolymer containing β-cyclodextrin (β-CD)	Substituted phenols [918]
17	Cross-linked P_4VPy	Phenolic species (*p*-aminophenol, *p*-cresol, *p*-chlorophenol, and *p*-nitrophenol) [919]
18	Poly(4-VPy-*g*-ethylene terephthalate) (P_4VPy-*g*-PET)	Selenite ions (SeO_3^{2-}) [920]
19	Poly(4-VPy-*co*-trimethylolpropane trimethacrylate) and poly(4-VPy-*co*-1,4-dimethacryloiloxybenzene) microspheres	Ibuprofen and ketoprofen [921]
20	A copolymer of 1-vinylimidazole/1,10-dibromode cane/4-VPy	Nitrate [922]
21	Magnetic $CoFe_2O_4@SiO_2$ decorated with P_4VPy	Bisphenol A [923]
22	P_4VPy	Hexagonal boron nitride (h-BN) sheets as the main impurity consist of boron nitride nanotubes (BNNTs) [924,925]
23	Poly[(4-styrenesulfonic acid)-*co*-(4-VPy)] cross-linked with DVB	(±)-Epinephrine [926]
24	Cross-linked PAA-*co*-P_4VPy	(±)-3,4-Methylenedioxymethamphetamine [927]

[a] Bleaching clay; [b] 3-aminopropyltrimethoxysilane.

$$FeCl_3{\cdot}6H_2O \xrightarrow[\text{2) 4-VPy, 190 °C}]{\text{1) Ethylene glycol, NaAc, r.t.}} \text{Fe}_3\text{O}_4@\text{P}_4\text{VPy nanospheres}$$

SCHEME 1.88 Synthetic pathway for Fe_3O_4@P_4VPy nanospheres [514].

were reported as an efficient adsorbent for the removal of Congo red from water, and it could be separated magnetically and reused (Scheme 1.88) [514]. Spin-coating of P_4VPy-TiO_2 and P_4VPy-ZnO on glass substrate produced composites which can be used for the removal of methyl orange and benzoic acid [934]. A composite of P_4VPy-*co*-poly(maleic anhydride)/Fe_3O_4 has been employed for the removal of Coomassie Brilliant Blue (CBB R250) dye in aqueous medium [935]. Very recently, carboxymethyl xanthan gum-*g*-P_4VPy copolymer and poly(*n*-hexadecyl-4-vinylpyridinium bromide) were studied for the removal of acid orange 10 dye [936] and the acid blue113 textile dye [937], respectively.

Controlling protein adsorption/desorption is a pressing issue in many biotechnology fields. pH-dependent swelling/deswelling behavior of P_4VPy makes its copolymer systems as promising candidates for controlling protein adsorption [793,938–941]. As previously stated, when the pH is below the IEP of P_4VPy [942], the P_4VPy chains expand as a result of the electrostatic repulsion among the positively charged repeating pyridine units. Conversely, when the pH exceeds the IEP, the polymer chains collapse due to the neutralization of pyridine groups, along with hydrophobic interactions between the vinyl groups present in the polymer chains. Poly(*N*-ethyl-4-vinyl-pyridinium)bromide-*b*-polymethylacrylate (P_4VPy-EtBr-*co*-PMA) was prepared via RAFT polymerization method (Scheme 1.89a) and used as a macroemulsifier for the emulsion polymerization of styrene using APS and DVB as initiator and cross-linking agent, respectively (Scheme 1.89b) [571]. The resultant polymeric microspheres with cationic P_4VPy-EtBr segments on their surface act as an efficient adsorbent for BSA with a structure of negative charge (at pH 7.4) via electrostatic interactions. The adsorption capacity of microspheres was found to be 900 mg/g, significantly more than other reported PS microspheres [943]. The desorption of protein from microspheres was accomplished by incubation at 25°C in the presence of phosphate-buffered saline and acetic buffer. Also, PDA/poly(2-methyl-2-oxazoline) (PMOXA)/P_4VPy as mixed polymer brushes was fabricated to investigate the controlled or switchable protein adsorption [944]. The results showed that by adjusting the fraction of PMOXA or P_4VPy, this system could adsorb a high amount of pepsin at pH 3, and achieve a desorption efficiency of over 92% at pH 7. It has been reported that the pH-sensitive chitosan-*g*-PVP-*co*-P_4VPy hydrogel cross-linked by MBAAM can act as enzyme (trypsin) carrier which improved both the enzyme stability and activity [945].

P_4VPy, P_2VPy, and P_3VPy-bonded silica with high durability have been used as stationary phases in supercritical fluid chromatography (SFC) [946–948], due to their ability to interact with analytes. The findings indicated that the patterns of interaction terms for the three polymer-based stationary phases were similar to those of the 2-ethylpyridine-bonded silica and the 2-picolylamine-bonded silica phases, which are the most well-known stationary phases used in SFC. The P_3VPy and P_4VPy phases generally provided improved selectivity and more retention

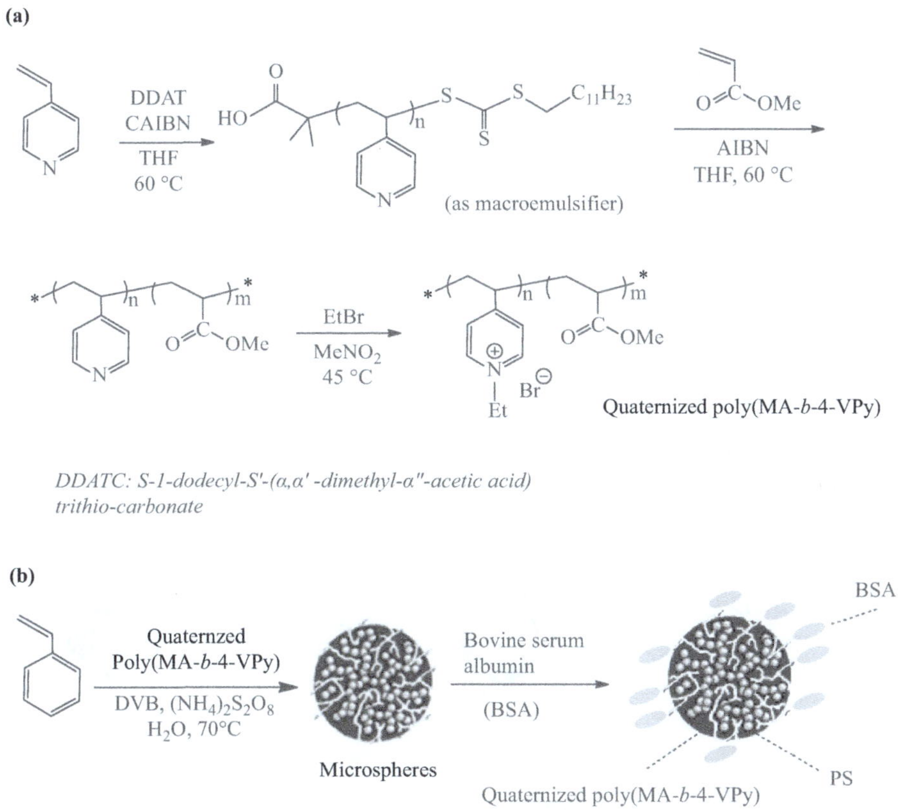

SCHEME 1.89 Schematic illustration of the synthesis of (a) P_4VPy-EtBr-*co*-PMA and (b) polystyrene microspheres containing P_4VPy-EtBr-*co*-PMA with BSA adsorbed on their surface [571]. (Reproduced with permission from Ref. [571]; Copyright 2012 Elsevier).

over P_2VPy for all the categories of analytes. A set of 140 drugs were analyzed on the above-mentioned stationary phases [946] and in this respect, P_4VPy showed better selectivity in complex mixtures. Solid-phase microextraction coupled with liquid chromatography based on a poly(4-VPy-*co*-ethylene dimethacrylate), P_4VPy-*co*-PEDMA, have been reported to be used for the determination of non-steroidal anti-inflammatory drugs [917,949] and poly(acrylamide-4-vinylpyridine -*N,N'*-methylene bisacrylamide), poly(AAM-4-VPy- MBAAM) monolithic capillaries for non-peptide angiotensin II receptor antagonists and estrogens [950,951]. Analyses were done in human plasma and water samples. SFC has been used for the separation of 15 co-formulated binary anti-hypertensive drug combinations using DCPak P_4VPy column [952]. Very recently, NVP, VIM, and 4-VPy combined with 1-octadecene were *co*-grafted on thiolpropyl silica (Sil@SH) to produce three new stationary phases [953], which showed high column performance in the separation of the examined organic analytes.

1.4.5 As a Polymeric Sorbent for the Removal of Metals

The application of combination of inorganic metals and PVPy polymers as an organic ligand can be traced back to 50 years ago [954–957]. Since that time, the removal of metals using PVPy as a polymer adsorbent has developed very quickly [71,74,958–980]. For example, resin Reillex™ HPQ (Scheme 1.11) in the sulfate form and in the nitrate form has been used in uranium [961] and plutonium [981] recovery. Fe_3O_4-PVP-SiO_2-P_4VPy with a core–shell structure has been investigated for the extraction of Cu(II), Pb(II), Cd(II), and Ni(II) [974]. P_4VPy was incorporated into Fe_3O_4-PVP-SiO_2 core–shell structures by ATRP method, using 2-bromoisobutyryl bromide-coated magnetic nanogels (Scheme 1.90, I) as a macroinitiator and CuBr-*N,N,N',N'',N''*-pentamethyldiethylenetriamine (PMDETA) as a catalyst–ligand complex. In resultant core–shell nanogels, P_4VPy acts as a pH-sensitive unit. At pH values < 6, the pyridine groups became protonated, consequently, leading to the expansion of nanogels and adsorption of metal ions on their surface. Conversely, pH levels around 6 cause the system to collapse and trap

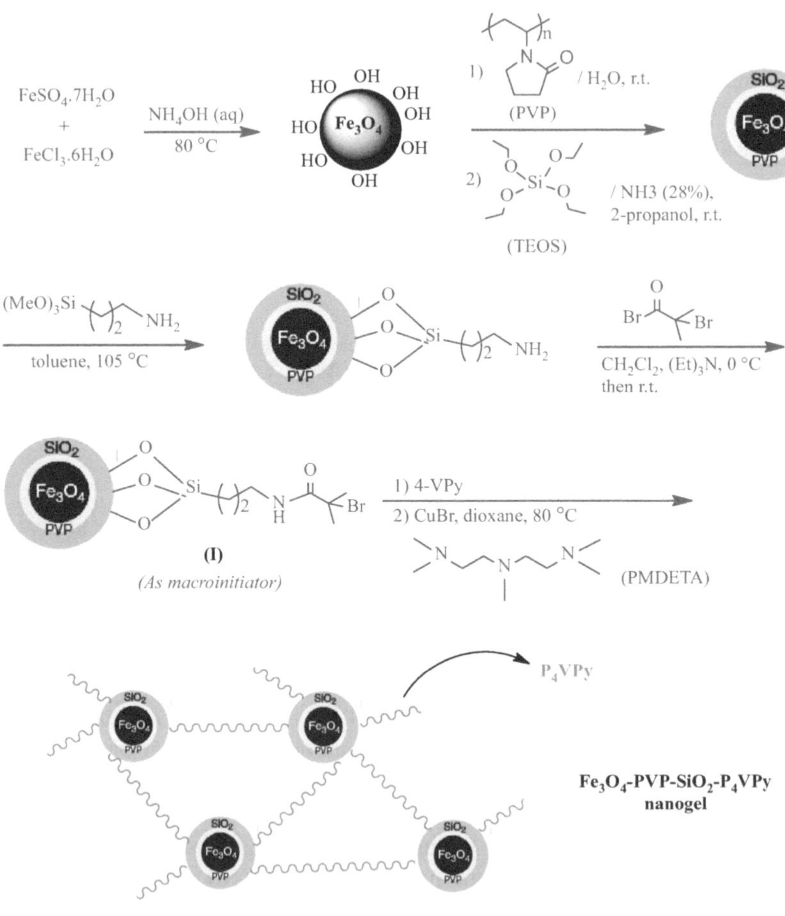

SCHEME 1.90 Synthetic pathway for Fe_3O_4-PVP-SiO_2-P_4VPy nanogels [974].

the target ions. At pH values above 6.0 the polymer aggregates because its structure is highly hydrophobic; at this stage the polymer, together with the trapped metals and ionic species, can be separated using an external magnetic field. A key feature of this work is high selectivity toward the aforementioned ions (even in ultra-trace amounts) over other ions.

PAAM-*co*-P_4VPy has been used for the removal of heavy metals [543,982]. Cross-linked P_4VPy resin bearing PAAM has shown high Hg-uptake capability (3.36 mmol/g) [982]. Scheme 1.91 shows the synthetic routes for this polymer and its mechanistic pathway for the removal of mercury. The mechanism includes redox initiation through carboxyl groups with cerium ammonium nitrate (CAN, Ce (IV)), followed by graft copolymerization of AAM monomers. Table 1.9 lists representative examples where PVPy complexing sorbents are employed for binding different metal ions.

1.4.6 PVPy-Stabilized Gold NPs

PVPy has been receiving much attention for immobilization and adsorption of Au NPs [39,347,995–1006]. Through *N*-ligand chelation, PVPy is an ideal surface modifier for supporting Au NPs onto various surfaces, e.g., metal glass, quartz, films, nanorods, GO, and alumina, as well as plastic surfaces such as polycarbonate, PS, polyethylene, and poly(dimethylsiloxane) [42,995–998,1007]. Colloidal golds are highly desirable as coloring and labeling agents in histochemistry and cytochemistry; in electronics; in medicine such as drug delivery, diagnostics, therapeutics, and bio-imaging; and in catalysis due to their optical properties, opportunity for surface modification, high electron density, atomic number, thermal and electrical conductivity of gold, and lower toxicity [39,999,1008,1009]. They may appear red, violet, or blue based on factors such as particle size, shape, and agglomeration. The colloidal stability of Au NPs is indeed a crucial parameter that significantly affects the physicochemical properties of NPs [1010]. Very recently, fundamentals and developments of polymer-grafted gold colloids have been reviewed [1011]. The stabilization of gold colloids by homo- and random copolymers

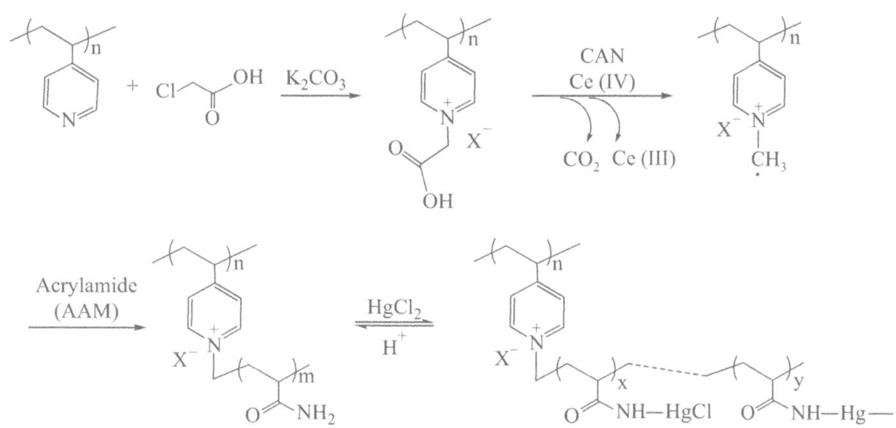

SCHEME 1.91 Synthetic routes for PAAM grafted onto cross-linked P_4VPy resin, which is used for the removal of mercury [982].

TABLE 1.9
PVPy Complexing Sorbents for Binding Different Metal Species

Entry	Sorbent	Metal Species
1	SiO_2/cross-linked quaternized P_4VPy-Cu(II)	Cu(II) [960]
2	Reillex™ HPQ	Cr(VI) [74]
3	Activated carbon-cross-linked quaternized P_4VPy	Cr(VI) [963]
4	SiO_2-quaternized P_4VPy	Cr(VI) [964]
5	Poly(4-VPy-*co*-2-hydroxyethylmethacrylate)/ *N*,*N*-bis(3-triethoxysilylpropyl)thiourea	Pt(IV) [966]
6	Cross-linked poly(4-VPy-*co*-protoporphyrin IX)	Mn^{2+} [967]
7	Cross-linked P_4VPy-dithizone	Au(III), Pd(II), Pt(IV) [968,975]
8	Cross-linked P_2VPy	Hg^{2+} [969]
9	P_4VPy-*co*-PAAM	Cu(II) [971]
10	Ion exchanger AN-251M bearing P_4VPy	$PdCl_4^{2-}$ [972]
11	SiO_2-poly(ST-*co*-4-VPy)	Cu(II), Ni(II), Fe(III), Pb(II) [973]
12	Fe_3O_4-PVP-SiO_2-P_4VPy	Cd(II), Pb(II), Cu(II), Ni(II) [974]
13	P_4VPy-grafted polyHIPE (high internal-phase emulsion) foams	Pu [983]
14	P_4VPy/SiO_2/APTMS (3-aminopropyltrimethoxysilane)	Cd(II) [911]
15	Cross-linked poly(4-VPy-*co*-protoporphyrin)	Co^{2+} [984]
16	P_4VPy-grafted reduced graphene oxide (RGO)	Pd(II) [985]
17	P_4VPy-grafted cellulose	Cu^{2+}, Fe^{2+} [727]
18	Nanocomposite based on MWCNTs and P_4VPy	Cd(II) [986]
19	Poly(4-VPy-*co*-vinylbenzyl chloride) and poly(4-VPy-*co*-*p*-xylylene dichloride)	Ru(IV), Rh(IV), Pd(II), Os(IV), Ir(IV), Pt(II), Au(III) [978]
20	Poly(protoporphyrin-*co*-4-VPy)	Cu^{2+} [987]
21	Silica gel/P_4VPy composite	Pb^{2+}, Fe^{3+}, Cu^{2+} [988]
22	Cross-linked poly(4-VPy-*co*-MAA)	Cu(II) [989]
23	Composite polymer monoliths of PAN and P_4VPy	As(V), Cr(VI) Ru(III), Rh(III), Pd(II) [990]
24	Cellulose-*g*-P_4VPy	Pb^{2+}, Fe^{3+} [991]
25	P_2VPy/MCM-41 composite	Cr(VI) [992]
26	P_4VPy resin	Th^{4+} [993]
27	Chitosan-*g*-4-VPy/thiol-amine-silica HZSM-5 zeolite nanocomposite	Cu^{2+}, Zn^{2+} [994]

of P_2VPy and P_4VPy has been investigated [39,998–1000,1012]. The pyridine moiety has strong affinity for metallic NPs, and thus, restrains the aggregation of metal NPs through steric stabilization [1001]. Usually, the method involves the reduction of tetrachloroauric acid ($HAuCl_4$), using a reducing agent or UV irradiation, in the presence of the corresponding polymers [1011]. PCL-*b*-P_4VPy [1013], PS-*b*-P_2VPy [1014,1015], and PS-*b*-P_4VPy [1016,1017] as well as star BCPs of PS-*b*-P_2VPy [1018] have been used for the stabilization of Au NPs. Self-assembly of PS-*b*-$P_2VPy/HAuCl_4$ at the liquid–liquid interface [1019] and also drug loading efficiency of pentablock terpolymers of PCL-*b*-PEO-*b*-P_2VPy-*b*-PEO-*b*-PCL and P_2VPy-stabilised Au NPs have been surveyed [1001,1020]. Very recently, stabilization of Au NPs by P_2VPy-*b*-PMMA-modified glassy carbon electrode for the detection of nicotine has been reported [851].

1.4.7 Optics and Electrochemistry

Polymers have found widespread applications in optics and electrochemistry due to several advantages, such as high quantum efficiencies, more convenient manufacturing process, low operating voltages, light weight, mechanical flexibility, and inexpensive [317,1021–1031]. The importance of using polymer compounds with electronic and optical properties was such that the 2000 Nobel Prize in Chemistry was awarded to this field of research. In this regard, PVPy polymers are attractive candidates for various electrooptic applications such as NLO, dye-sensitized solar cells (DSSCs), OLED, mechanochromic opal films, as polymer templates or ruthenium phthalocyanine dyes used in semiconductors for organic electronics, and PL [35,576,1021,1032–1038]. Recent advancements on materials such as fluorescent oligomers and PVPy derivatives for solution-processed OLEDs and perovskite LEDs [1039] and the optical and plasmonic response of the silver particulate thin films on polymer blends of PS/P_2VPy, PS/P_4VPy, and PVP/P_4VPy have been reviewed [1040]. PS-*b*-PVPy systems have been used for the development of nanoelectronic devices [221,274,1041–1045]. In the following discussion, I intend to introduce some reported examples of optical and electrochemical properties of PVPy-based materials which were used in electrical or optical devices or have the potential to be used in such devices. Representative examples of PVPy derivatives and their electrooptic ability are listed in Table 1.10.

It has been shown that P_4VPy dissolved in pyridine emits three distinct colors: blue, green, and red when photo-excited at various wavelengths [1077]. Nonetheless, these films exhibited moisture, density, and inhomogeneity, which hindered the production and assessment of any EL device. To address this issue, a protonation process was employed to modify the polymer, imparting a positive charge and making it suitable for the formation of self-assembled films. By this technique, earlier reports demonstrated the fabrication of LED devices utilizing self-assembled films of P_4VPy with additional transport layers. For example, self-assembly of P_4VPy with poly(*N*-vinyl-carbazole), PNVCz, and 2-(4-biphenylyl)-5-(4-*tert*-butyphenyl)-1,3,4-oxadiazole, PBD, as transport layers was introduced to be a useful method for the fabrication of an OLED device [1057,1078]. Also, a LED device through the use of self-assembled films of poly(4-VPy-*co*-vinylpyridinium-chloride) with poly(styrene-4-sulfonate) as a transport layer was introduced [1057]. Recently, a metallopolymer thin film as a LED based on grafting ligand-zinc(II) coordination fragments onto P_4VPy was prepared, which was miscible in ionic liquids (ILs) and exhibited red, green and blue emission colors [1058].

DSSCs have garnered significant interest for their application in solar energy, primarily due to their impressive energy conversion efficiency and affordability [1079]. However, a notable drawback of DSSCs is the leakage and evaporation of the liquid electrolytes employed. To address this issue, the use of solvent-free polymer electrolytes,

TABLE 1.10
Representative Examples of PVPy Derivatives and Their Electrooptic Applications

Entry	Polymer Composition	Electrooptic Ability
1	P_2VPy, P_2VPy-*b*-PNIPAM, nitroxide-terminated P_4VPy	Matrix-assisted laser desorption/ ionization mass spectrometry [1046–1049]
2	PS-*b*-P_4VPy and PS-*b*-P_2VPy	Lithography [237,313,1050–1054]
3	Printed PS-*b*-P_4VPy and PS-*b*-$P_4VPy/FeCl_3$ thin films	Photoelectrodes [1055]
4	P_4VPy-encapsulated TiO_2	Electrophoretic inks [1056]
5	Self-assembled P_4VPy with PNVCz	OLED [1057]
6	P_4VPy/ligand-zinc(II)	LED [1058].
7	Quaternized forms of P_4VPy-*co*-PAN and P_4VPy	DSSCs [1059–1061]
8	Polycyclopentadithiophene-benzothiadiazole-*b*-P_4VPy	Solar cells [1062]
9	PVPy-I_2	Li battery [1063]
10	P_4VPy/tungsten pentacarbonyl ($W(CO)_5$)	As NLO chromophore [102]
11	P_4VPy/pyridine	Memory devices [1038,1064]
12	Doping P_4VPy and graphene quantum dots made of chitin	Perovskite LEDs [1065]
13	Nanocomposite of P_4VPy and aluminum nitride	LEDs [1066]
14	P_4VPy bearing molybdenum disulfide (MoS_2)	Field-effect transistors [1067]
15	$CsPbBr_3$ perovskite quantum dots embedded in PS-*co*-P_2VPy	Robust and light-tunable memristors [1068]
16	Quaternized P_4VPy with (di)halomethanes	High-refractive index polymer thin films [1069]
17	Poly(4-vinylbenzoic acid) and P_4VPy	DSSCs [1070]
18	PS-*co*-P_4VPy	DSSCs [1071]
19	PS-*co*-P_4VPy bearing Cu(II) and Pd(II) chromophoric complexes	NLO [1072]
20	Multifunctional fullerenol/PS-*co*-P_4VPy composite	NLO [1073]
21	P_4VPy-nickel(II) chromophore	NLO [1074]
22	P_4VPy-CdSe	Ultralow-power memristors [1075]
23	P_2VPy/CdSe/CdS	Anti-counterfeiting [1076]

such as PVPy polymers, presents a viable solution [1080]. P_4VPy-*co*-(PAN: polyacrylonitrile) was employed for the production of polymer electrolytes in DSSCs, via chemically cross-linking solidification. The obtained DSSCs exhibited higher conversion efficiencies than the initial liquid DSSCs [1059]. The solidification of ionic-liquid electrolytes using cross-linked P_4VPy for DSSCs application has been investigated [1081,1082]. The effect of changing cross-linking agents (alky diiodide, alkyl dichloride, or alkyl dibromide) on efficiency of solar cells was surveyed [1081]. The findings indicated that the efficiency of the solar cells may be influenced by the length of dihalides and the type of halogen atoms (Cl, Br, I). Extended alkyl chains and alkyl diiodides had a more beneficial impact on the performance of solar cells. Quaternized P_4VPy with alkyl iodides-doped iodine and *N*-methylpyridine iodide was used for the fabrication of DSSCs with a conductivity of 6.41 mS/cm and an efficiency of 5.64% [1060,1083].

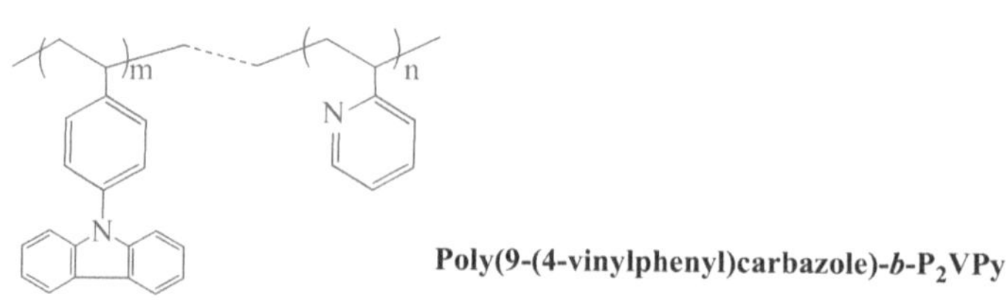

Also, the use of polycyclopentadithiophene-benzothiadiazole-*b*-P_4VPy as nanostructuring compatibilizer in solar cells has been reported [1062]. P_4VPy has been used in lead halide perovskite solar cells with the improved moisture stability and voltage [1084]. Very recently, it has been reported that introducing a thin layer of P_4VPy as an interlayer between the perovskite solar cells leads to an open-circuit voltage larger than 1.17 V, and a 30 mV increase in average compared to unmodified samples [1085].

Another reported application of PVPy in the field of optics and electronics is in second-order NLO applications [1086,1087]. NLO is a field within photonics that includes technologies for the acquisition, storage, processing, and transmission of photons instead of electrons in signal processing. Materials exhibiting NLO characteristics hold potential uses in optical devices, which has led to significant research efforts in this area. In this context, polymers have gained recognition as promising candidates for NLO applications. Such systems include either dispersion of NLO chromophores into polymer matrices, i.e., guest–host systems, or covalent bonding between the polymeric backbone and the chromophore [1088,1089]. Among the polymers used in this field, PVPy has also received much attention. For example, P_4VPy-supported tungsten pentacarbonyl ($W(CO)_5$) as a NLO chromophore has been reported [102]. Scheme 1.92 shows the preparation reaction of P_4VPy-$W(CO)_5$ from its stating materials. The obtained polymer complex showed modest second harmonic generation properties. The results revealed that P_4VPy has the potential to be used as a probe to attain the sign of the β value of chromophores (hyperpolarizability coefficient). In another study, P_4VPy-supported metal-containing chromophores have been investigated for NLO applications [1090]. The synthetic procedures for these metal–polymer complexes are depicted in Scheme 1.93. Binuclear complexes of Cu and Pd ($[ML]_2$) were synthesized from the reaction of bisbenzonitrile palladium(II) dichloride and copper(II) acetate monohydrate with ([4-(*N*,*N*-diethylamino)-2-hydroxyphenyl]methylene)hydrazide of 4-nitrobenzoic acid (L) as chromophore. The advantages of the aforementioned polymer include a significant concentration of the grafted chromophore within the polymers, a high value of the first hyperpolarizability coefficient of the chromophores, elevated T_g, excellent thermal stability in air (approximately 280°C), and good optical transparency of the films. NLO properties of the DMF solutions of different MWCNTs/polymer composites which are constructed from MWCNTs and poly(4-vinylphenol), P_4VPy, P_2VPy, and PEG have been reported [1091].

An additional use of PVPy derivatives is found in resistive random-access memory devices, which can switch their resistance between a low-resistance state and

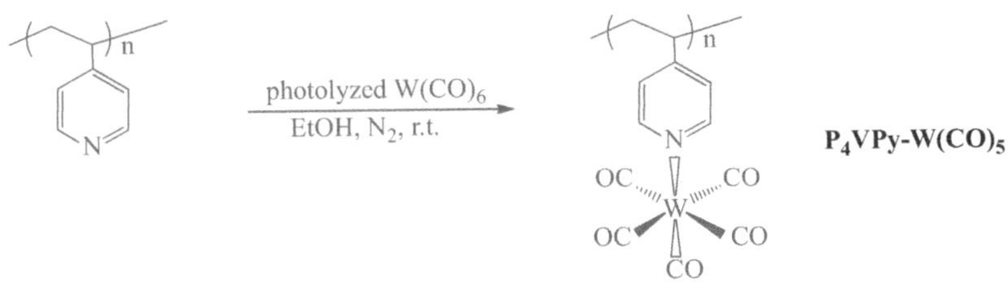

SCHEME 1.92 Preparation of P_4VPy-supported tungsten pentacarbonyl as a NLO chromophore [102].

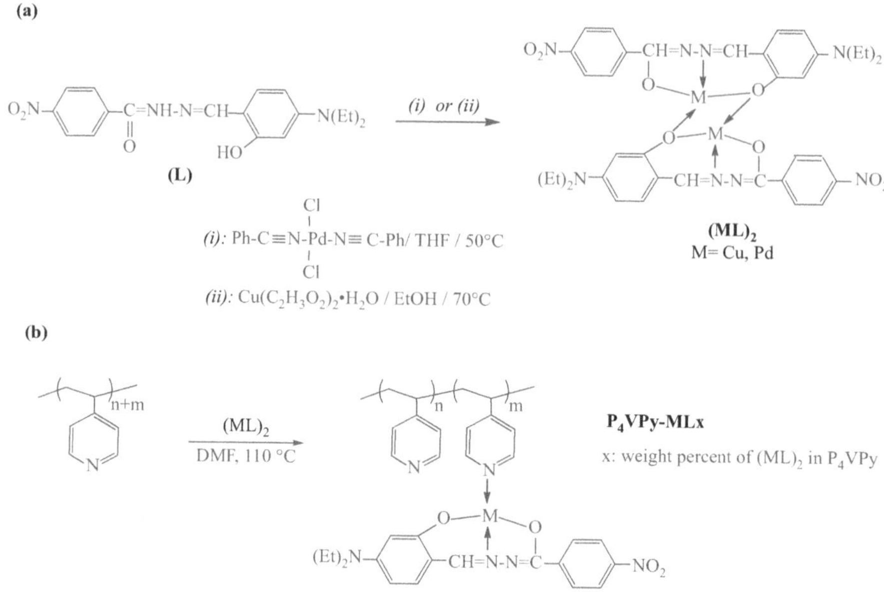

SCHEME 1.93 Preparation of (a) binuclear complexes of Cu and Pd ($[ML]_2$) and (b) P_4VPy-organometallic chromophores with NLO properties [1090].

a high-resistance state through the application of an external stimulus [1092]. For example, a kind of PS-*b*-P_4VPy micelles consisting of Au NPs in core has been prepared, which represents an effective approach for polymer memory fabrication as well as for integration into organic thin-film transistor-based circuits [1093,1094]. Also, BCPs bearing carbazole and pyridine moieties have been synthesized (Scheme 1.94) [1095]. The building blocks of pyridine and carbazole act as electron acceptor/donor centers. The fabricated devices from such systems with lamellar structure showed outstanding memory characteristics, retention characteristics ($>10^4$ s), a high ON/OFF ratio ($>10^4$), and good endurance cycling (>190 cycles). Memory devices fabricated from 5-mercapto-1-methyl tetrazole (MMT) and P_4VPy matrix [1096] and a composite of the colloidal monodispersed CdSe QDs and P_4VPy matrix [1097] with a maximum on-off current ratio of 10^5 were introduced. Very recently, $CsPbBr_3$ QDs and PS-*b*-P_2VPy as a memristor exhibits both negative differential resistance and resistive switching memory behavior [1068].

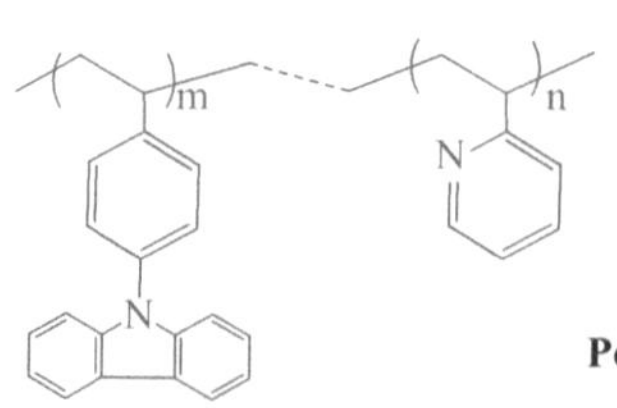

Poly(9-(4-vinylphenyl)carbazole)-*b*-P_2VPy

SCHEME 1.94 The molecular structure of block copolymers containing pyridine and carbazole moieties [1095].

Under UV irradiation, a viscous solution of P_4VPy/pyridine turns to photoactive gel (a smart gel) [63,64,1077,1098–1103]. The gel demonstrates distinctive tunable emission characteristics that can be adjusted quickly and reversibly as its conductivity increases. This gel is amorphous and sensitive to light. The excitation at wavelengths of 395, 455, and 557 nm emits blue, green, and red PL, respectively [1077]. A proton transfer with the formation of a pyridinium ion moieties was suggested to be responsible for the unique optical characteristics and photoinduced conductivity of P_4VPy gels. The photoinduced changes can be stored in the material over months [63,64,1103–1105]. The effect can be useful in memory devices or optical switching devices [1038,1064]. For example, it has been observed that the irradiation of P_4VPy gels at wavelengths of 254 or 312 nm results in the reversible formation of 5-amino-2,4-pentadienals via pyridine ring-opening reactions, which are subsequently followed by self-condensation to produce oligomers that can absorb light within the visible spectrum and extend into the near-IR range (Scheme 1.95) [1104]. The resulting polyacetylene analogs exhibit a conjugated molecular wire behavior, presenting potential applications in sensor devices and thermal energy harvesting.

The great need for electrically conductive polymers with potential applications in electronics has increasingly attracted attention. It is well known that the electrical characteristics of the majority of conducting polymers arise from the charge-transfer (CT) interactions occurring between the conjugated polymer backbone and the electron acceptors or donors, in addition to its improved electron transport capabilities. In this regard, PVPy as an electroactive material has been receiving much attention due to high tendency of its pyridyl ring to formation of significant CT interactions. The d.c. electrical conductivities of P_2VPy and P_4VPy charge transfer complexed with halobenzoquinones and tetracyanoethylene have been measured [1106,1107]. The results showed that the complex exhibits an interesting semiconductive behavior. Electrical and optical properties of poly(tetralincyclobutylhydroxy ethylmethacrylate-*co*-4-VPy) have been investigated [1108]. The electrical conductivities and bandgaps of the obtained samples having different copolymer compositions were found to be in the range of 10^{-9}–10^{-8} S/cm and 1.79–3.27 eV, respectively. P_4VPy-*b*-poly(dimethylsiloxane) doped with tetracyanoquinodimethane (TCNQ) has a conductivity of 10^{-2} S/cm [1109]. Scheme 1.96 shows the synthetic pathways for this polymer complex. Complexation of PVPy polymers with TCNQ [1110], which produce the corresponding conductive polymers, has been reported. Recently, a novel

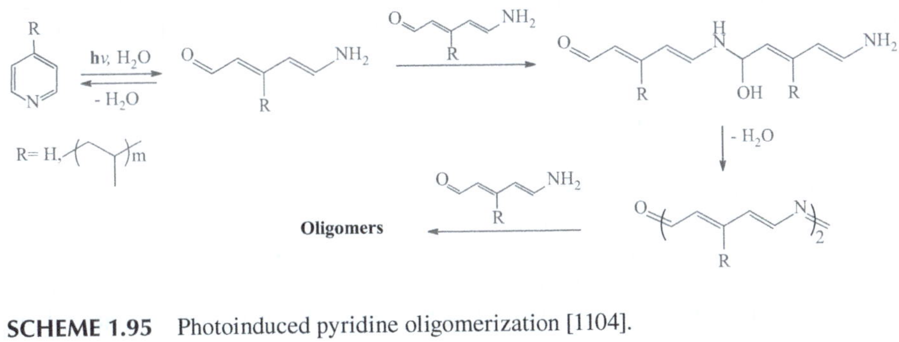

SCHEME 1.95 Photoinduced pyridine oligomerization [1104].

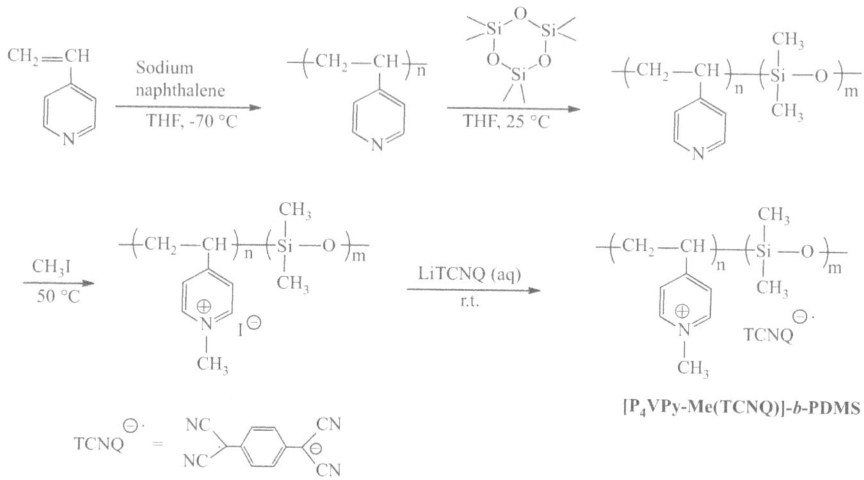

SCHEME 1.96 P_4VPy-*b*-poly(dimethylsiloxane) doped with TCNQ which includes the following steps: (1) anionic polymerization using sodium naphthalene as initiator, (2) ring-opening polymerization, (3) quaternization, and (4) the reaction of quaternized polymer with the lithium salt of TCNQ [1109].

n-type copolymer dopant PS–poly(4-vinyl-*N*-hexylpyridinium fluoride) exhibited an outstanding thermal stability with a high electrical conductivity of 58 S/cm at 88°C [1111]. Also, organic transistors of thin films of this polymeric fluoride dopant showed high electron mobility and Hall mobility of 0.86 and 1.70 $cm^2/(V.s)$, respectively. These results make them promising candidates for high-performance n-type all-polymer thermoelectric materials.

The optical and electrical properties of charge-transfer complexes of PVPy-iodine [320,1112–1120] have been extensively investigated. For example, the conductivity of complex of P_4VPy with I_2 was measured [1118]. The results showed that while pure P_4VPy has a resistivity of more than 10^{15} Ω cm, and I_2 itself is a semiconductor with a resistivity of 10^{11} Ω cm and a bandgap of 2 eV, the P_4VPy-I_2 (I_2/P_4VPy unit molar ratio is 2:1) exhibited a resistivity of 10^4 Ω cm at room temperature. P_4VPy-I_2 was stable to 147°C, whereupon it melts. In another study, UV–Vis spectroscopic studies showed that P_4VPy-I_2 and P_4VPy-I_2-PTSA form charge-transfer complex with lower transition energies than P_4VPy, I_2, and PTSA alone (Scheme 1.97) [1121]. While P_4VPy is an insulator with a bandgap of about 4.32 eV, both P_4VPy-I_2 and P_4VPy-I_2-PTSA are medium-bandgap semiconductors with energies of about 3.64 and 3.90 eV, respectively. The d.c. conductivities of P_4VPy, P_4VPy-PTSA (5% PTSA), P_4VPy-I_2 (1:1.5 molar ratio), and P_4VPy-PTSA-I_2 (1:3:15 molar ratio) at 300 K were found to be 10^{-14}, 3.89×10^{-10}, 10^{-3}, and 1.43×10^{-6} S/cm, respectively. Raising the temperature improved the conductivity of P_4VPy-PTSA-I_2 film, indicating semiconducting behavior of this compound.

PVPy-I_2 has been used in the industrial manufacturing of lithium batteries. The cathode and anode of this battery are made of PVPy-I_2 and lithium, respectively [1122]. In these batteries, PVPy-I_2 acts as a conductive polymer material and

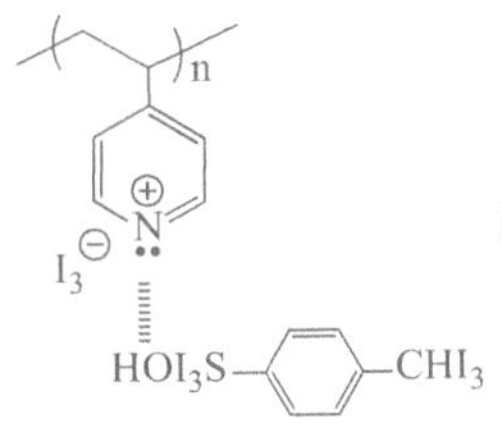

P_4VPy-I_2-PTSA

SCHEME 1.97 P_4VPy-iodine-*p*-toluene sulfonic acid complex [1121].

its performance depends on PVPy/I_2 ratio, and the peak value of conductivity is observed when this ratio reaches 1:8. The manufacture of battery is easily accomplished by addition of the molten PVPy-I_2 to lithium anode and eventually formation of electrolyte layer (LiI crystals) between two electrodes. The capacity of Li/PVPy-I_2 batteries can be varied through a wide range depending on their size, e.g., 3.5 Ah (ampere-hour) for size 14 mm × 45 mm × 52 mm. Furthermore, other PVPy polymers and their derivatives have been used in lithium-ion batteries [1123–1125]. A series of P_4VPy-based gel polymer electrolytes (viz., PEG-cross-linked P_4VPy) showing a high ionic conductivity of over 2.94×10^{-4} S/cm at 80°C and 1.34×10^{-5} S/cm at 25°C with potential application in lithium-ion batteries have been introduced (Scheme 1.98) [1126,1127]. They exhibited a high capacity (135 mAh/g) and retention of 94.4% of the initial capacity at the 30th cycle [1127]. It has been shown that P_4VPy can also play a hydrogen fluoride scavenger role in lithium-ion

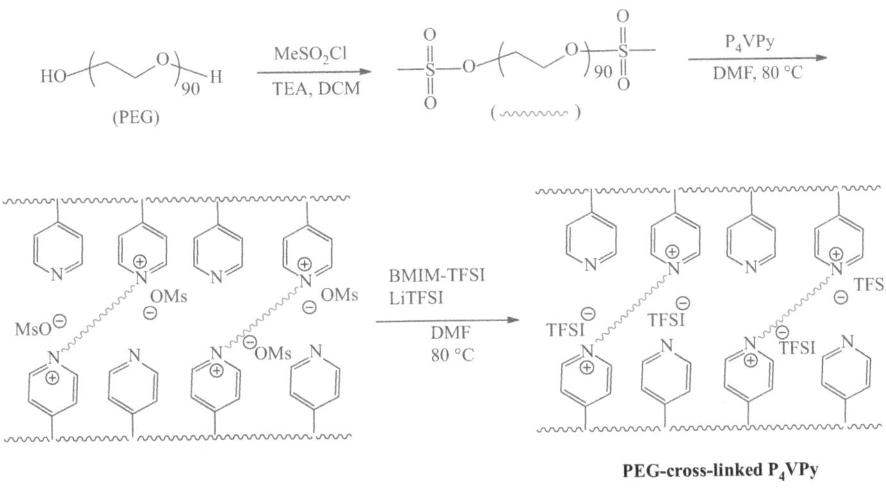

TEA: Triethylamine, DCM: dichloromethane, BMIM-TFSI: 1-Butyl-3-methylimidazolium bis(trifluoromethylsulfonyl)imide

LiTFSI: Bis(trifluoromethane)sulfonimide lithium salt

SCHEME 1.98 Synthesis of PEG-cross-linked P_4VPy -based gel polymer electrolytes which includes the following steps: (1) reacting PEG with methanesulfonyl chloride and (2) reaction of the obtained PEG-dimesylate with P_4VPy, followed by reaction with BMIM-TFSI ionic liquid in the presence of LiTFSI [1126].

batteries to improve capacity retention [1128]. Recently, an organodisulfide cathode (2,5-dimercapto-1,3,4-thiadiazole, DMcT) incorporated into P_4VPy with potential application in lithium batteries was introduced [1129].

Finally, P_4VPy-ICl, P_4VPy-IBr, and P_4VPy-I_2 thin films coated on quartz substrates were prepared, and it was revealed that the refractive index of these stable charge-transfer complexes increases from 1.58 to 2.08 for a 587.6 nm thick film [1130]. It was indicated that by controlling the reaction conditions as well as the degree of halogenation, the refractive index can be carefully tuned. These characteristics make them desired candidates as optical coating materials.

Newly, new golden metallopolymers based on P_4VPy, showing bright PL emission, have been reported [1131]. They are prepared according to the instructions shown in Scheme 1.99a. The reaction is rapid and a bright green luminescence immediately appears. It should be mentioned that both [Au(C_6X_5)THT] (THT = tetrahydrothiophene, X = Cl, F) and P_4VPy are nonluminescent materials. PL property of these polymers is ascribed to aurophilic Au(I)···Au(I) interactions (either intra- or interchain) as shown in Scheme 1.99b [878,1131]. Also, it has been reported that gold and zinc complexes of P_4VPy upon excitation with UV light show PL emission and will have potential applications in the optical devices including color display and lighting [1132] (Scheme 1.100). Very recently, it has been reported that a nonconjugated luminescent complex consisting of (−)-CSA and P_2VPy exhibited excellent cluster luminescence (CL) performance [32]. The CL performance of the complex improved through changing the stereoregularity from atactic to isotactic. Also, PL of poly[(methylenelactide)-*co*-(2-VPy)] having intense blue emissions with high quantum yields (>20%) has been investigated [1034].

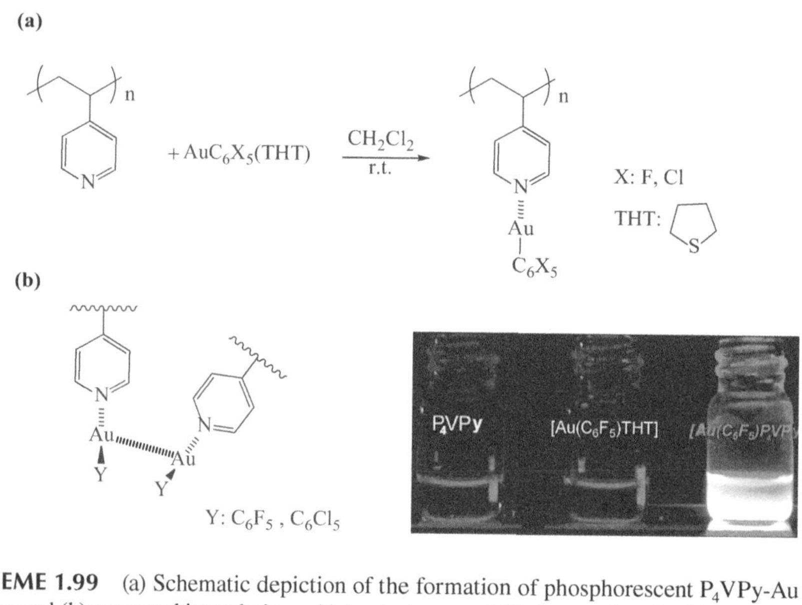

SCHEME 1.99 (a) Schematic depiction of the formation of phosphorescent P_4VPy-Au complexes and (b) proposed intrachain and interchain aurophilic interactions in this system [1131]. (Reproduced with permission from Ref. [1131]; Copyright 2009 American Chemical Society).

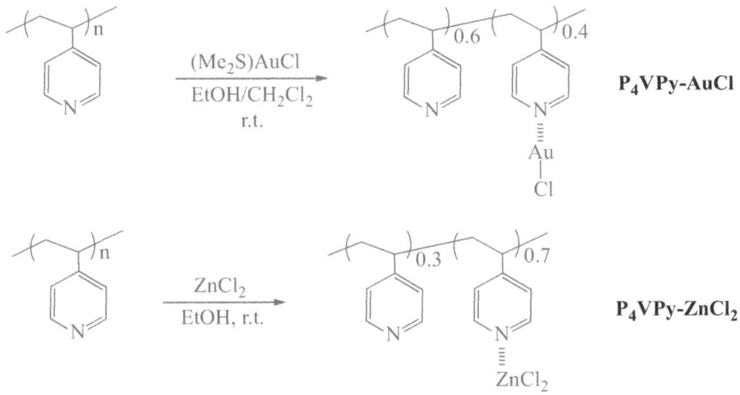

SCHEME 1.100 Synthesis of P_4VPy-AuCl and P_4VPy-ZnCl_2 [1132].

Very recently, a series of liquid crystalline polymers containing P_4VPy and cyanostilbene side chains with photothermoelectric properties was introduced [1133]. As the molar ratio of cyanostilbene increases, the lamellar structure of the system transforms into a hexagonal columnar structure. Upon irradiation of UV lamp for 10 s, the temperature of these liquid crystalline polymers can rise to 130°C, and 25 mA current can be detected in the circuit coupled with the thermoelectric module. Indeed, there was a conversion of optical energy into heat energy, which was then transformed into electric energy. This research holds promise for the potential development of novel liquid crystalline polymers exhibiting photothermoelectric properties [1134,1135].

CdSe QDs are semiconductor nanocrystals which have received increasing attention because of their optical properties [1136]. PS-*b*-P_4VPy diblock copolymers can be self-assembled into spherical micelles containing CdSe QDs [1037,1137–1139]. Generally, micelles are constituted of PS-*b*-P_4VPy, poly(3-hexylthiophene) (P_3HT) in the core, and CdSe-trioctylphosphine oxide (TOPO) ligands in the corona (Scheme 1.101). P_4VPy helps to the movement of QDs to polar media. P_3HT acts as a conductive polymer which leads to charge transfer across the core–corona (heterojunction) interface [1138].

1.4.8 As Corrosion Inhibitors for Some Metals

In the domain of corrosion inhibitor research, especially regarding the corrosion of metals in acidic conditions, it is widely recognized that the majority of acid inhibitors contain surface functionalities such as nitrogen, sulfur, or oxygen, which enable these substances to decrease the corrosion rate. The electronic factor is generally the main factor influencing the inhibiting effect. These inhibitors can impede or decelerate the electrochemical processes taking place in metallic substances. Within this context, various inhibitors, including acyclic and cyclic polymers that incorporate heteroatoms (e.g., N and O) with polar characteristics such as PVP, polyvinylimidazoles (PVIM), and polyethyleneimine, have been examined and reviewed in the literature [1140,1141]. The corrosion inhibition efficiency of polymers can be due to their low cost, stability in acid media, and complexation with metal ions, and

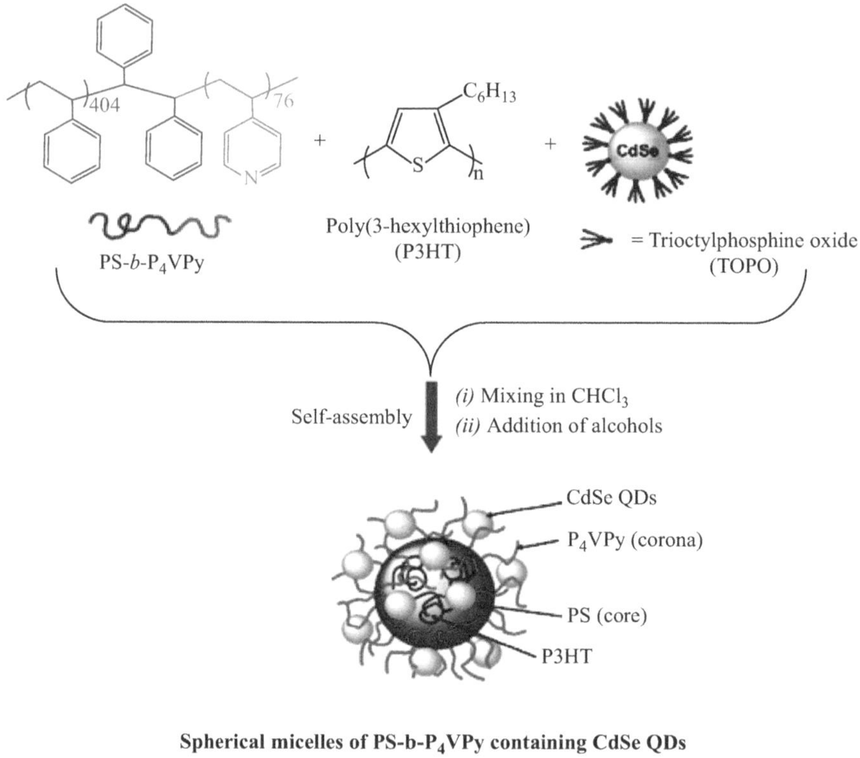

SCHEME 1.101 Polymeric micelles formed from diblock copolymers of PS-*b*-P_4VPy [1138]. (Reproduced with permission from Ref. [1138]; Copyright 2008 American Chemical Society).

consequently blanketing (protecting) the metal surface with these complexes from corrosive agents [1142,1143]. Along this line, PVPy, its quaternized forms, and its copolymer with other hydrophilic polymers have been employed as corrosion inhibitors for some metals such as Fe, Cu, Zn, and their alloys in hydrochloric acid, nitric acid, or sulfuric acid solutions [541,1140,1143–1154]. Table 1.11 shows representative examples of PVPy derivatives as corrosion inhibitors for metals.

Very early studies showed that 4-ethylpiperidine, 4-ethylpyridine, and low-molecular weight P_4VPy decrease the corrosion rate of steel in HCl solution by 80% [1144]. The results revealed that all three of the above compounds are chemisorbed on a steel surface to form a monolayer. Interestingly, P_4VPy showed greater extent of adsorption than non-polymeric ones, resulting in the greatly increased inhibition of acid corrosion. Following the initial reported works; further research into the anticorrosion application of PVPy polymers was quickly conducted. The effect of PVP, P_4VPy, and P_2VPy as corrosion inhibitors of zinc metal in 1.0 M H_2SO_4 solution has been reported [1148], and the formation of a protective film on the metal surface with the protection efficiency of ~87% was observed. Also, the inhibitory effect of P_4VPy and poly(*N*-poly-3-oxide ethylene pyridinium)bromide (P_4VPy-POEBr) on the corrosion of Cu60-Zn40 (brass) in 0.5 M nitric acid solution was examined [1145]. A schematic

TABLE 1.11
Representative Examples of PVPy-based Corrosion Inhibitors for Metals

Entry	Polymer Composition	Metal
1	P_4VPy (very low molecular weight)	Steel in aq HCl [1144]
2	Poly(4-VPy-poly-3-oxide ethylene)	Cu60-Zn40 in aq HNO_3 [1145]
3	Poly(AAM-*co*-4-VPy)	Mild steel in aq H_2SO_4 [1146]
4	P_4VPy	Mild steel in aq HCl [1147]
5	P_4VPy, P_2VPy, and PVP	Zn in aq H_2SO_4 [1148]
6	Poly(*N*-isopentyl-4-vinylpyridinium bromide)	Fe in aq H_2SO_4 [1150]
7	Poly(*N*-hexadecyl-4-vinylpyridinium bromide)	Mild steel in aq HCl [1151]
8	Quaternized P_4VPy-*g*-PEO	Fe in aq HCl [1152]
9	Poly (4-vinylpyridine-poly (3-oxide-ethylene) tosyle	Fe in aq H_2SO_4 [1153]
10	RGO and poly(4-VPy-*co*-butyl methacrylate) (P_4VPy-*co*-PBuMA)	Cu in 5% NaCl solution [1155]
11	P_4VPy (2% DVB)	Steel-37 in aq H_2SO_4 [1156]
12	4-VPy and 2-VPy oligomers	Mild steel in CO_2-saturated brine solution [1157–1160]
13	P_4VPy	Armco Fe in aq H_2SO_4 [1161]
14	P_4VPy and PVP	Low-carbon steel in aq H_2SO_4 [1162]

illustration of the fabrication of P_4VPy-POEBr is presented in Scheme 1.102. Weight loss measurements indicated that P_4VPy-POEBr and P_4VPy, at a concentration of 2.5×10^{-6M} in 0.5M HNO_3 solution, demonstrated inhibition efficiencies of 97% and 66%, respectively. The corrosion rate of brass in 0.5M HNO_3 was measured at 8.03 (mg/cm^2.h), while in the presence of P_4VPy-POEBr and P_4VPy inhibitors (10^{-6} M), the rates were significantly reduced to 3.60 and 2.72 (mg/cm^2.h), respectively. A significant feature of P_4VPy-POEBr is its efficiency that remains unaffected by temperature. Without the presence of polymer inhibitors, the corrosion rate of brass in 0.5M HNO_3 surged dramatically from 8.03 to 51.57 (mg/cm^2.h) across the temperature spectrum of 30°C–80°C. In contrast, when P_4VPy-POEBr (10^{-5} M) was introduced, the corrosion of brass was kept to a minimum, reaching a maximum of only 0.48 (mg/cm^2.h) at 70°C.

Another example of polymeric inhibitors based on PVPy for the protection of metal materials against corrosive acidic solutions is P_4VPy-*co*-PAAM [1146].

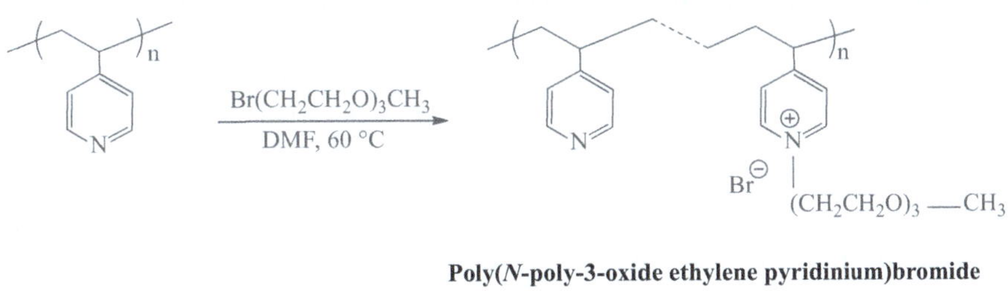

SCHEME 1.102 Preparation of the quaternized form of P_4VPy as an inhibitor for the corrosion of brass in acidic solutions [1145].

Weight loss measurements of mild steel in sulfuric acid solution (1 M) at 18°C in the presence of P_4VPy-*co*-PAAM (100 mg/L) showed that the corrosion rate of steel was 0.0671 (mg/cm^2.h) with an inhibition efficiency of 60.2%. It is thought that the adsorption of the P_4VPy-*co*-PAAM inhibitor onto the metal surface is linked to the interaction between the lone pairs of electrons present in the oxygen and nitrogen atoms of the AAM and VPy segments, respectively and the vacant d-orbitals of iron atoms. These unshared electron pairs, together with the π electrons from double bonds, are expected to contribute to the inhibitory effect of P_4VPy-*co*-PAAM, as they play a crucial role in enhancing the electron density on the metal surface, thereby decreasing the corrosion rate. Synergistic effect of P_4VPy-*co*-PAAM and KI on corrosion inhibition of mild steel in 1 M H_2SO_4 showed that inhibition efficiency of P_4VPy-*co*-PAAM (10 mg/L) and corrosion rate of steel in 1 M H_2SO_4 in the presence of KI (0.008%) were 88.6% and 0.0192 (mg/cm^2.h), respectively [1146]. Under similar conditions but in the absence of KI, inhibition efficiency and corrosion rate were found to be 51.3% and 0.0821 (mg/cm^2.h), respectively. A simple explanation for the above observations is that in the presence of KI chemisorption of P_4VPy-*co*-PAAM on the metal surface is enhanced, resulting in an increase in the values of the charge transfer resistance and eventually the enhancement of inhibition efficiencies. Very recently, P_4VPy and PVP were introduced as corrosion inhibitors for low carbon steel in H_2SO_4 (1 M) [1162]. Protection efficacies of 93.13% and 94.19% for P_4VPy and PVP, respectively, were found.

1.4.9 PVPy-Supported Reagents and Catalysts

These applications of PVPy polymers will be discussed in detail in the following chapters.

1.4.10 Miscellaneous Applications

Beyond the aforementioned categorized applications of PVPy polymers, the following represent additional miscellaneous uses of the polymer.

The dispersing of pigments in powder coatings plays a vital role in the industrial manufacturing of coatings, as it greatly influences the characteristics of the coating. Nevertheless, a significant limitation is the inadequate dispersion that occurs under typical extrusion conditions [1163]. To solve this problem, several attempts have been made, including the substitution of synthetic strategies of coatings as well as the utilization of pigment masterbatches [1164]. In this regard, BCP dispersants have also shown promising capability [1165–1167]. For example, poly(4-VPy-*b*-sodium methacrylate) has also been reported as a pigment dispersant for aqueous dispersions of alumina-coated TiO_2 [1168]. Also, P_2VPy-*b*-PCL has been used as TiO_2 pigment dispersing agent in polyester powder coatings [1163,1169–1171]. P_2VPy-*b*-PCL was easily synthesized from the corresponding monomers through anionic polymerization using *sec*-butyllithium as initiator (Scheme 1.103) [1163,1172]. The stabilization of pigments using P_2VPy-*b*-PCL is achieved through the attachment of pigments to the P_2VPy segments, which serve as the "anchor block", followed by their dispersion in polyester facilitated by the PCL units acting as the "buoy block". It is unequivocal

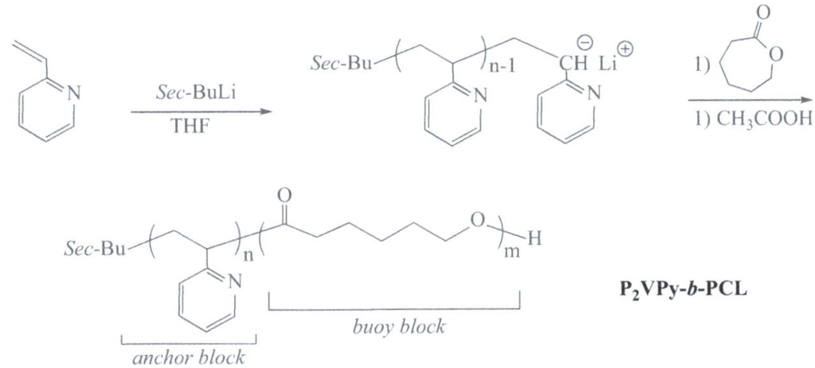

SCHEME 1.103 Synthesis of P_2VPy-*b*-PCL as TiO_2 pigment dispersing agent in polyester powder coatings [1172].

that PCL blocks containing hydrophobic alkyl chains will significantly enhance the miscibility of P_2VPy-*b*-PCL within the polyester resin. P_2VPy-*b*-PCL exhibited a high affinity for TiO_2 with reasonable adsorptions of 1–2 mg/m^2. The effect of P_2VPy-*b*-PCL on the properties of powder coatings such as modulus, haze, and gloss has been extensively studied [1173].

PET is one of the most utilized polymers in packaging (e.g., food packaging) and engineering applications as well as fabrication of fibers for clothing [1174,1175]. Although PET offers advantages such as resistance to acidic solutions, oxidizing agents, and sunlight, it encounters difficulties due to a lack of active chemical groups that can attract dyes and its inadequate moisture compatibility. A practical solution to the challenges faced by PET fibers is the grafting of hydrophilic monomers onto the fibers. This technique can greatly improve water absorbency and dye affinity, both of which are essential properties. In this context, the grafting of vinylpyridine monomers onto PET fibers has been explored [1176,1177]. The results showed that water absorption capacity, T_g, and viscosity of PET fibers were improved after grafting process. PET fibers were also grafted with a binary mixture of 2-hydroxyethylmethacrylate (HEMA) and 4-VPy via radical polymerization [1178]. The experimentations revealed that while P_4VPy/HEMA-grafted PET (at a graft yield of 318%) could be dyed well with Sudan IV and methylene blue dyes (about 1.57 mg of dye per gram of fibers), interaction of ungrafted PET fibers with these dyes was negligible, due to the lack of necessary functional groups. Undoubtedly, the incorporation of ester, pyridyl, and hydroxyl groups as hydrophilic entities into grafted fibers significantly contributes to enhancing dyeability.

In the tire manufacturing sector, a latex terpolymer made up of 2-VPy, ST, and butadiene (referred to as VPR, serving as a tire-cord binder) is incorporated with the rubber during the vulcanization process via sulfur cross-linking to improve the adhesion between rubber and reinforcing fibers [3,1179–1182]. A rubber composite constructed by mixing VPR and styrene-butadiene rubber (SBR) latexes with aqueous dispersions of GO, followed by reaction with H_2SO_4 to provide the coagulated GO/SBR/VPR composite, has been reported [1183]. When pH of the mixture was adjusted to 4, the demulsified VPR molecules reacted with acid via 2-VPy blocks and eventually generated pyridinium ions, leading to the adsorption of VPR on GO layers through electrostatic interactions. As the pH values decrease to approximately 3, the SBR undergoes gradual demulsification and subsequently co-coagulates with GO to form a GO/SBR composite. The main functions of VPR are to prevent the aggregation of GO layers and to act as a link between SBR and GO. Notably, in the absence of VPR, SBR could not coagulate with GO. The results demonstrated that the addition of GO into the rubber matrix improved the mechanical properties of the resulting GO/SBR/VPR composite, as shown by an increase in tensile strength, tear strength, and storage modulus, along with a decrease in tan delta. GO/SBR/VPR composite was also mixed with silica/SBR composite (used in the fabrication of passenger tires), and significant improvements in mechanical characteristics were observed (e.g., improved wear resistance, lower tan delta). Likewise, in another investigation, GO has been investigated as a promising reinforcing agent in the fabrication of synthetic rubbers containing VPR [1184]. Molecular dynamics and rheological behavior of VPR vulcanizates with reversible Zn^{2+}–pyridine coordination bond have been investigated [1185]. P_2VPy has been employed to build binary polymer brushes (usually with PS), to control and tune adhesion and friction behavior of solid surfaces [1186]. The curing of 2-VPy with butadiene and 1,4-butanedisulfonyl chloride has also produced rubber-like polymers [9].

One of the challenges associated with household laundry is the tendency for dyes to leach from colored fabrics. It is well known that certain polymers have the ability to draw in fugitive dyes due to robust electrostatic interactions and the formation of complexes with these dyes [1187]. Quaternized water-soluble vinylpyridine carboxylate polymers and P_4VPyNO has been used in the manufacturing of laundry detergents as dye transfer inhibitor compounds to prevent leaching of dyes in colored fabrics, during laundry [1187–1193]. Indeed, they would prevent the redeposition of the washed-off dyestuffs onto fabrics; in other words, they function as a dye scavenger, maintaining the washed-off dyestuffs in solution, which leads to the avoidance of redeposition on the fabrics. The charged characteristics of these polymers render them suitable for complexing with the charged dyes, ultimately limiting the "bleeding" of dyes from colored items during the laundering process. The dye scavenger capability of P_4VPyNO in the absence and presence of a certain surfactant (SDS) has been surveyed [1191–1193]. The results showed that there is a strong polymer–SDS interaction and formation of the "mixed" micelle structure having a core of hydrocarbon chains of SDS surrounded by the polar pyridine *N*-oxide and SDS sulfate groups. In fact, P_4VPyNO as a hydrophilic polymer is accumulated in the large available surface of SDS which results in the decrease of surface charge density of micelles and consequently decreasing negatively charged electrostatic repulsion between the SDS and the dye, and eventually leading to an improvement in the dye adsorption compared to micelles per se. This means that P_4VPyNO/SDS system is more effective at reducing dye deposition with respect to the SDS alone [1192,1193]. The following Equations 1.3 and 1.4 show the displacement of dye from the P_4VPyNO by SDS and relocation of dye in P_4VPyNO/SDS complexes, respectively. Finally, quaternized P_4VPy and its copolymer with AAM have been used as fabric softening compositions (in the washing machine) for depositing fabric softening actives on fabrics [3], due to the ability of the polymers to increase the electrostatic repulsion of materials.

$$P_4VPyNO/Dye + SDS \leftrightarrow P_4VPyNO/SDS + Dye \quad (1.3)$$

$$P_4VPyNO/SDS + Dye \leftrightarrow P_4VPyNO/SDS/Dye \quad (1.4)$$

It has been known that a pH-responsive material can act as a pH-responsive foam stabilizer [1194,1195]. The hydrophobicity–hydrophilicity balance of such materials plays the actual role in the bubble stabilization in these types of foams [1196]. As it was discussed before, PVPy is a pH-sensitive polymeric material, and therefore it seems suitable for stabilization of gas bubbles. For example, PS particles bearing P_4VPy colloids have been used as a pH-responsive particulate bubble stabilizer [1197]. As shown in Scheme 1.104, at $pH \geq 4$, P_4VPy segments are non-protonated, and bubbles can be stabilized through adsorption of PS-P_4VPy particles with hydrophobic character at the air–water interface of the bubbles. When the pH of the solution falls below 3, P_4VPy units become protonated and dissolve in water, leading to the absence of bubble formation. In other terms, the desorption of the PS-P_4VPy particles from the air–water interface occurs at low pH, which causes the disruption of the bubbles. Also, P_2VPy latexes have been used as effective foam stabilizers [1198]. With the presence of P_2VPy, significant swelling occurs, because at low pH values, latex migrates to microgel and leads to desorption of microgel from the air–water interface. Sterically stabilized P_2VPy latexes can be used in the preparation of Pickering-type emulsifiers [1199,1200] and millimeter-sized "liquid marbles" [1201]. In all cases, pH-responsive behavior was observed upon addition of sufficient acid.

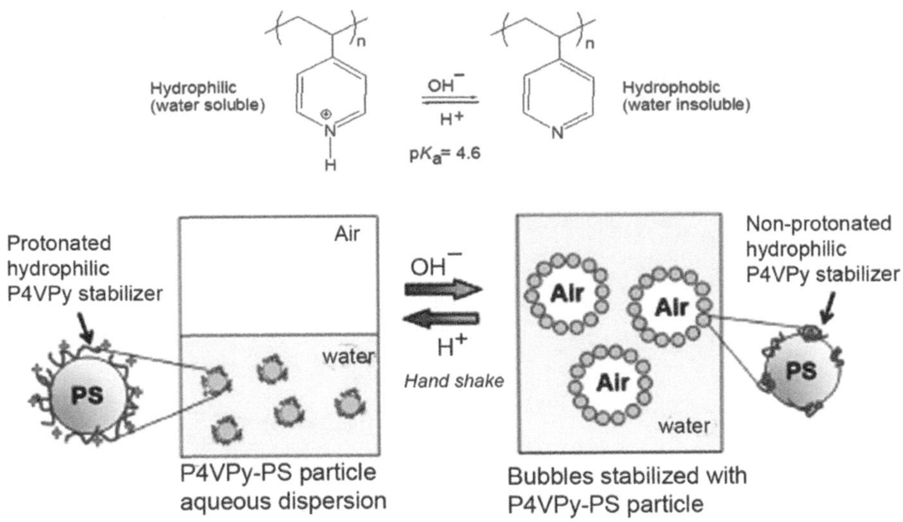

SCHEME 1.104 Adsorption and desorption of PS-P_4VPy particles on/from the air–water interface of the bubbles [1197]. (Reproduced with permission from Ref. [1197]; Copyright 2018 Frontiers in Chemistry).

Very recently, inverse vulcanization of sulfur and 4-VPy has been surveyed [1202]. The reaction occurs through the ring-opening polymerization of elemental sulfur, achieved by heating at 180°C in the presence of 4-VPy monomers. This process results in the formation of polysulfide chains, which are reddish brown in color and contain 4-VPy units distributed randomly throughout their backbone (Scheme 1.105). The resultant poly(sulfur-*r*-4-VPy) (with 50 wt.% 4-VPy) was subjected to the reaction with 1-bromododecane to get quaternized 4-VPy in polymer backbone in which the attack of the nitrogen atoms in pyridine rings to alkyl halide led to the formation of quaternized pyridine segments in polymer structure. The sulfur content of the copolymers could reach 95 wt.%. Polysulfide materials have exhibited various properties and applications [1203,1204], and it is expected that the polysulfides containing quaternized 4-VPy units will also show remarkable applications in different fields.

Various synthetic techniques have been reported for preparing metallic nanostructures [1131,1205–1220]. Among these, solid-state methods have increasingly gained attention recently due to its economical, eco-friendly, and industrially viable nature. In this regard, synthesis of metal and metal oxide NPs (M, M_xO_y) from solid-state pyrolysis of the macromolecular complex PS-*co*-P_4VPy·MX_n (where X denotes anion accompanying the cationic metal in the metal salt) alone or together with CS·MX_n as precursors has been reported and reviewed (Scheme 1.106) [1205]. Table 1.12 shows, typically, solid-state synthesis of some metal and metal oxide NPs from the pyrolysis of the corresponding macromolecular complex precursors.

Finally, electrospinnability of P_4VPy through the complexation with 3-pentadecylphenol (PDP) [1221] and a biphenol [1222] has been reported. The establishment of efficient hydrogen bonds between the phenolic OH groups and the pyridine group of P_4VPy, facilitated by the DMF solvent, has been proposed to enhance the electrospinnability of P_4VPy, leading to the development of a cross-linked network structure. Moreover, it was also demonstrated that by selective removing of phenol molecules from the electrospun nanofibers of P_4VP/PDP, porous nanofibers could be achieved [1221].

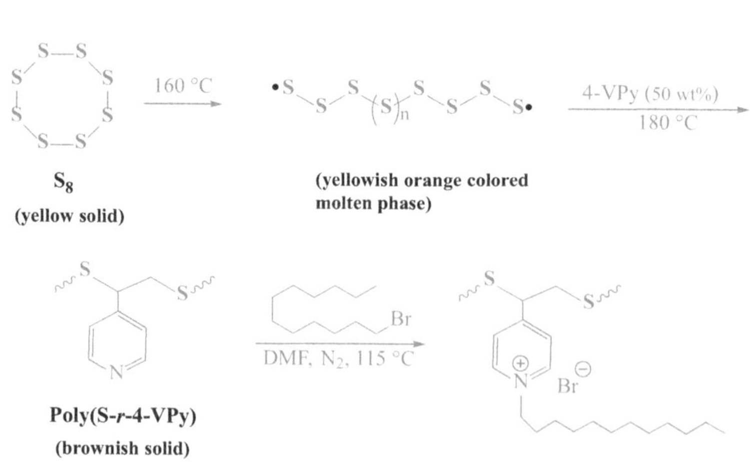

SCHEME 1.105 Inverse vulcanization of sulfur and 4-VPy monomers and quaternization of poly(sulfur-*r*-4-VPy) [1202].

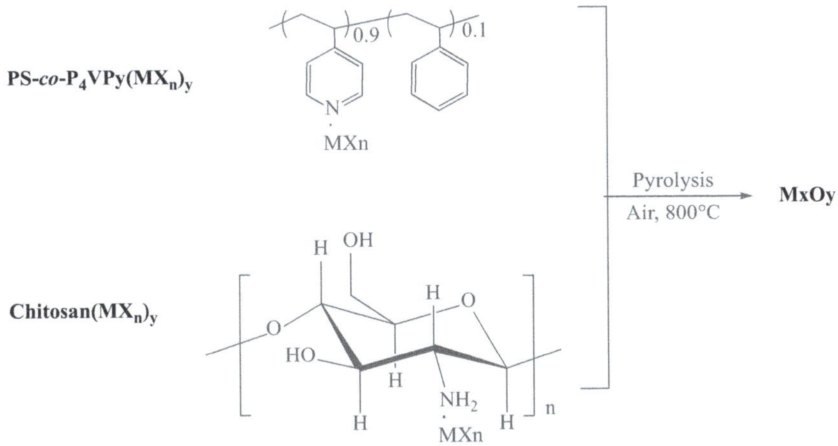

SCHEME 1.106 Solid-state preparation of metal and metal oxide NPs [1205].

TABLE 1.12
Solid-State Synthesis of Some Previously Reported Metal and Metal Oxide NPs from the Pyrolysis of the Corresponding Macromolecular Complex Precursors

Entry	Precursor	Pyrolytic Product
1	Chitosan·$(VCl_3)_n$ and PS-*co*-P_4VPy·$(VCl_3)_n$	V_2O_5 [1207]
2	Chitosan·$(CrCl_3)_n$ and PS-*co*-P_4VPy·$(CrCl_3)_n$	Cr_2O_3 [1208]
3	Chitosan·$(MoCl_4)_n$ and PS-*co*-P_4VPy·$(MoCl_3)_n$	MoO_3 [1208]
4	Chitosan·$(WCl_4)_n$ and PS-*co*-P_4VPy·$(WCl_4)_n$	WO_3 [1208]
5	Chitosan·$(MnCl_2)_n$ and PS-*co*-P_4VPy·$(MnCl_2)_n$	Mn_2O_3 [1205]
6	Chitosan·$(FeCl_2$ or $FeCl_3)_n$ and PS-*co*-P_4VPy·$(FeCl_2$ or $FeCl_3)_n$	Fe_2O_3 [1205]
7	Chitosan·$(NiCl_2.6H_2O)_n$ and PS-*co*-P_4VPy·$(NiCl_2.6H_2O)_n$	NiO [1216]
8	Chitosan·$(PtCl_2)_n$ and PS-*co*-P_4VPy·$(PtCl_2)_n$	Pt [1217]
9	Chitosan·$(AuCl_3)_n$·$(SiO_2)_m$ and PS-*co*-P_4VPy·$(AuCl_3)_n$·$(SiO_2)_m$	Au/SiO_2 composite [1218]
10	Chitosan·$(IrCl_3)_n$ and PS-*co*-P_4VPy·$(IrCl_3)_n$	IrO_2 [1219]
11	PS-*co*-P_4VPy·$(SnCl_2)_n$	SnO_2 [1220]
12	PS-*co*-P_4VPy·$(ZnCl_2)_n$	ZnO [1220]
13	PS-*co*-P_4VPy·$(Ce(NO_3)_3)_n$	CeO_2 [1220]

REFERENCES

1. Sahu D, Sreekanth PR, Behera PK, Pradhan MK, Patnaik A, Salunkhe S, Cep R. *European Journal of Medicinal Chemistry Reports*. 2024;12:100210.
2. Varshney S, Mishra N. Pyridine-based polymers and derivatives: Synthesis and applications. In: Singh P (ed.). *Recent Developments in the Synthesis and Applications of Pyridines*: Elsevier, Amsterdam, 2023, pp. 43–69.
3. Fink JK. *Handbook of Engineering and Specialty Thermoplastics: Water Soluble Polymers*: Scrivener Publishing LLC, New York, 2011, vol. 2.
4. Nuyken O. *Polystyrenes and Other Aromatic Poly(Vinyl Compound)s: Handbook of Polymer Synthesis*: Marcel Dekker, New York, 2005, pp. 85–162.

5. Brandrup J, Immergut EH, Grulke EA, Abe A, Bloch DR. *Polymer Handbook*: Wiley, *New York*, 1999.
6. Mishra M, Yagci Y. *Handbook of Vinyl Polymers: Radical Polymerization, Process, and Technology*: CRC Press, Boca Raton, 2016.
7. Shimizu S, Watanabe N, Kataoka T, Shoji T, Abe N, Morishita S, Ichimura H. *Pyridine and Pyridine Derivatives*: Ullmann's Encyclopedia of Industrial Chemistry: Wiley-VCH Verlag GmbH & Co. KGaA, Weinheim.
8. Jin R, Xu H, Easa J, Chapero-Planell A, O'Brien CP. *The Journal of Physical Chemistry C*. 2023;127(3):1441–54.
9. Takemoto K. *Polymer Reviews*. 1970;5(1):29–102.
10. Iddles H, Lang E, Gregg D. *Journal of the American Chemical Society*. 1937;59(10):1945–6.
11. Doering WE, Weil RAN. *Journal of the American Chemical Society*. 1947;69(10):2461–6.
12. Ishizone T, Kosaka Y, Goseki R. *Anionic Polymerization: Principles, Practice, Strength, Consequences and Applications*: Springer, Tokyo, 2015, pp. 127–89.
13. Nieswandt K, Georgopanos P, Abetz C, Filiz V, Abetz V. *Materials*. 2019;12(19):3145.
14. Bachman GB, Micucci DD. *Journal of the American Chemical Society*. 1948;70(7):2381–4.
15. Ko JH. *Polymer Data Handbook*: Oxford University Press, Oxford, 1999, p. 961.
16. Sanno N, Murakami I, Yamamura H. *Polymer Journal*. 1976;8(3):231–8.
17. Wu K, Wang Y, Hwu W. *Polymer Degradation and Stability*. 2003;79(2):195–200.
18. Elmaci A, Hacaloglu J. *Polymer Degradation and Stability*. 2009;94(4):738–43.
19. Batich CD, Yan J, Bucaria Jr C, Elsabee M. *Macromolecules*. 1993;26(17):4675–80.
20. Tantavichet N, Pritzker MD, Burns CM. *Journal of Applied Polymer Science*. 2001;81(6):1493–7.
21. Chanda M, O'Driscoll K, Rempel G. *Journal of Chemical Technology and Biotechnology Chemical Technology*. 1983;33(2):97–108.
22. Akin H, Hasirci N. *Journal of Applied Polymer Science*. 1992;46(7):1307–10.
23. Berkowitz JB, Yamin M, Fuoss RM. *Journal of Polymer Science*. 1958;28(116):69–82.
24. Han SH, Lee DH, Kim JK. *Macromolecules*. 2007;40:7416–9.
25. Alberda van Ekenstein G, Meyboom R, Ten Brinke G, Ikkala O. *Macromolecules*. 2000;33(10):3752–6.
26. Zha W, Han CD, Lee DH, Han SH, Kim JK, Kang JH, Park C. *Macromolecules*. 2007;40(6):2109–19.
27. Belfiore LA, McCurdie MP, Das PK. *Polymer*. 2001;42(25):09995–10006.
28. Yamamoto T, Lee B-L, Hayashi H, Saito N, Maruyama T. *Polymer*. 1997;38(16):4233–9.
29. Angelopoulou PP, Kearney LT, Keum JK, Collins L, Kumar R, Sakellariou G et al. *Journal of Materials Chemistry A*. 2023;11(18):9846–58.
30. Tamami B, Goudarzian N. *Polymer Bulletin*. 1990;23:295–8
31. Holt P, Tamami B. *Polymer*. 1970;11(10):553–60.
32. Du H, Zhao W, Xia Y, Xie S, Tao Y, Gao YQ, Zhang J, Wan X. *Aggregate*. 2023;4(2):e276.
33. Chen L-C, Mao Y-C, Lin S-C, Li M-C, Ho R-M, Tsai J-C. *Chemical Communications*. 2012;48(30):3668–70.
34. Tardi M, Rougé D, Sigwalt P. *European Polymer Journal*. 1967;3(1):85–99.
35. Caputo S, De Nicola A, Donati G, David A, Raos G, Milano G. *Journal of The Electrochemical Society*. 2019;166(9):B3309.
36. Du B, Wang ZL, He XH. *Advanced Materials Research*. 2012;562:123–8.
37. Soldera A. *Molecular Simulation*. 2012;38(8-9):762–71.
38. Kennemur JG. *Macromolecules*. 2019;52(4):1354–70.
39. Mayer AB, Mark JE. *European Polymer Journal*. 1998;34(1):103–8.
40. Gupta S, Henson A. *Composites Part B: Engineering*. 2024;272:111207.
41. Lee J, Kwak J, Choi C, Han SH, Kim JK. *Macromolecules*. 2017;50(23):9373–9.

42. Malynych S, Luzinov I, Chumanov G. *The Journal of Physical Chemistry B*. 2002;106(6):1280–5.
43. Ding Q, Zhao W, Li F, He J, Zhang Y. *Science China Chemistry*. 2024;67(8):2737–46.
44. Price CC, Greene CE. *Journal of Polymer Science*. 1951;6(1):111–5.
45. Peppas NA, Hoffman AS. *Hydrogels: Biomaterials Science*: Elsevier, Amsterdam, 2020, pp. 153–66.
46. Chirani N, Yahia L, Gritsch L, Motta FL, Chirani S, Farè S. *Journal of Biomedical Sciences*. 2015;4(02):1–23.
47. Pavlova L, Pavlov M, Davankov V. *Doklady Chemistry*, 2006;406:6–8.
48. Luz CL, Coutinho F. *Polymer*. 2001;42(11):4931–8.
49. Fontanals N, Marcé RM, Galià M, Borrull F. *Journal of Polymer Science Part A: Polymer Chemistry*. 2003;41(13):1927–33.
50. Li S, Han G, Zhang W. *Polymer Chemistry*. 2020;11(29):4681–92.
51. Ishizu K, Ikemoto T, Ichimura A. *Polymer*. 1999;40(11):3147–51.
52. Ishizu K, Ikemoto T, Ichimure A. *Polymer*. 1998;39(2):449–54.
53. Tsubaki K, Ishizu K. *Polymer*. 2001;42(20):8387–93.
54. He G, Bennett TM, Alias K, Jiang L, Schwab ST, Alauhdin M, Howdle SM. *Polymer Chemistry*. 2019;10(29):3960–72.
55. Könemann T, Savage NJ, Huffman JA, Pöhlker C. *Atmospheric Measurement Techniques*. 2018;11(7):3987–4003.
56. Cáceres P, Piérola IF, Horta A. *Die Makromolekulare Chemie, Rapid Communications*. 1987;8(11):573–8.
57. Nishiyama S, Tajima M, Yoshida Y. *Journal of Photopolymer Science and Technology*. 2004;17(5):727–34.
58. Mavronasou K, Zamboulis A, Klonos P, Kyritsis A, Bikiaris DN, Papadakis R, Deligkiozi I. *Polymers*. 2022;14(4):804.
59. Pierola IF, Turro NJ, Kuo PL. *Macromolecules*. 1985;18(3):508–11.
60. Pierola I, Caceres M, Caceres P, Castellanos M, Nunez J. *European Polymer Journal*. 1988;24(9):895–8.
61. Utena Y, Yajima H, Ishii T, Handa T. *European Polymer Journal*. 1987;23(8):643–7.
62. Utena Y, Yajima H, Ishii T, Handa T. *European Polymer Journal*. 1988;24(1):71–80.
63. Vaganova E, Meshulam G, Kotler Z, Rozenberg M, Yitzchaik S. *Journal of Fluorescence*. 2000;10(2):81–8.
64. Vaganova E, Yitzchaik S. *Macromolecular Symposia*. 2004;207(1):95–104.
65. Satoko N, Masahiro T, Yasuhiko Y. *Journal of Photopolymer Science and Technology*. 2005;18(1):109–16.
66. Nishiyama S, Tajima M, Yoshida Y. *Thin Solid Films*. 2006;509(1-2):177–81.
67. Vyprachtický D, Sung K, Okamoto Y. *Journal of Polymer Science Part A: Polymer Chemistry*. 1999;37(9):1341–5.
68. Bucatariu F, Teodosiu C, Morosanu I, Fighir D, Ciobanu R, Petrila L-M, Mihai M. *Polymers*. 2021;13(22):3963.
69. Bunten KA, Kakkar AK. *Journal of Materials Chemistry*. 1995;5(11):2041–3.
70. Ma GH, Fukutomi T. *Journal of Applied Polymer Science*. 1991;43(8):1451–7.
71. Sonmez HB, Bicak N. Reactive and Functional *Polymers*. 2002;51(1):55–60.
72. Sahiner N. *Turkish Journal of Chemistry*. 2009;33(1):23–31.
73. Beauvais RA, Alexandratos SD. *Reactive and Functional Polymers*. 1998;36(2):113–23.
74. Chanda M, Rempel G. *Reactive Polymers*. 1993;21(1-2):77–88.
75. Tamami B, Parvanak BK. *Iranian Polym Journal*. 2009;18:191–206.
76. Boroujeni KP, Taheri S, Seyfipour G. *Synthesis and Reactivity in Inorganic, Metal-Organic, and Nano-Metal Chemistry*. 2014;44(1):84–8.
77. Boroujeni KP, Shojaei P. *Turkish Journal of Chemistry*. 2013;37(5):756–64.
78. Jaeger W, Bohrisch J, Laschewsky A. *Progress in Polymer Science*. 2010;35(5):511–77.

79. Izumrudov VA, Zhiryakova MV, Melik-Nubarov NS. *European Polymer Journal.* 2015;69:121–31.
80. Laschewsky A. *Current Opinion in Colloid & Interface Science.* 2012;17(2):56–63.
81. Nakano H, Goto T, Yamamoto Y. *Journal of Polymer Science: Polymer Letters Edition.* 1980;18(5):381–3.
82. Vishalakshi B, Ghosh S, Kalpagam V. *Polymer.* 1993;34(15):3270–5.
83. Bicak N, Gazi M. *Journal of Macromolecular Science, Part A.* 2003;40(6):585–91.
84. Chovino C, Gramain P. *Polymer.* 1999;40(17):4805–10.
85. Salamone J, Snider B, Fitch W. *Macromolecules.* 1970;3(5):707–9.
86. Salamone JC, Snider B, Fitch W. *Journal of Polymer Science Part B: Polymer Letters.* 1971;9(1):13–7.
87. Moore J, Goldstein J. *Journal of Polymer Science Part A-1: Polymer Chemistry.* 1972;10(7):2103–13.
88. Mondal P, Saha S, Chowdhury P. *Journal of Applied Polymer Science.* 2013;127(6):5045–50.
89. Chovino C, Gramain P. *Macromolecules.* 1998;31(20):7111–4.
90. Morcellet-Sauvage J, Loucheux C. *Die Makromolekulare Chemie: Macromolecular Chemistry and Physics.* 1975;176(2):315–31.
91. Frere Y, Gramain P. *Macromolecules.* 1992;25(12):3184–9.
92. Crooks III WJ, Kyser III EA, Walter SR. *Solvent Extraction and Ion Exchange.* 2002;20(4-5):543–59.
93. Boucher EA, Mollett CC. *Journal of the Chemical Society, Faraday Transactions 1: Physical Chemistry in Condensed Phases.* 1982;78(1):75–88.
94. Li Y, Yang M, She Y. *Sensors and Actuators B: Chemical.* 2005;107(1):252–7.
95. D'Aprano A, Fuoss RM. *Journal of Polymer Science Part A-2: Polymer Physics.* 1969;7(6):1101–9.
96. Mane S, Ponrathnam S, Chavan N. *Canadian Chemical Transactions.* 2015;3(4):473–85.
97. Briones X, Tapia RA, Campodónico PR, Urzúa M, Leiva Á, Contreras R, González-Navarrete J. *Reactive and Functional Polymers.* 2018;124:64–71.
98. Toral MI, González-Navarrete J, Leiva A, Ríos HE, Urzúa MD. *European Polymer Journal.* 2009;45(3):730–7.
99. Silva RA, Urzúa MD, Petri DF. *Journal of Colloid and Interface Science.* 2009;330(2):310–6.
100. Zhai L, Li H. *Molecules.* 2019;24(19):3425.
101. Behzadi pour G, Nazarpour fard H, Fekri aval L, Esmaili P. *Ionics.* 2020;26:549–63.
102. Lacroix P, Lin W, Wong G. *Chemistry of Materials.* 1995;7(7):1293–8.
103. Zander N, Orlicki JA, Rawlett AM. *Thermal and FTIR Characterization of Poly (4-Vinylpyridine) Crosslinked with Metal Salts*: Army Research Laboratory Research Triangle Park, NC, 2010.
104. Tennouga L, Bensalah W, Mansri A. e-*Polymers.* 2018;18(6):551–8.
105. Ladenheim H, Morawetz H. *Journal of Polymer Science.* 1957;26(113):251–4.
106. Bohrisch J, Wendler U, Jaeger W. *Macromolecular Rapid Communications.* 1997;18(11):975–82.
107. Soto VM, Galin J. *Polymer.* 1984;25(1):121–8.
108. Lu CH. Developer Compositions Containing Quaternized Vinylpyridine Polymers, and Copolymers. US Patent 4397934, 1983.
109. Kirsh YE, Komarova O, Lukovkin G. *European Polymer Journal.* 1973;9(12):1405–15.
110. Roach JD, Lane RF, Hussain Y. *Water Research.* 2011;45(3):1387–93.
111. Yoshida M. *European Polymer Journal.* 1997;33(6):943–8.
112. Satoh M, Yoda E, Hayashi T, Komiyama J. *Macromolecules.* 1989;22(4):1808–12.
113. Roach JD, Bondaruk MM, Al-Abdulghani A, Shahrori Z. *Advances in Materials Physics and Chemistry.* 2016;6(09):249.

114. Varshney SK, Zhong XF, Eisenberg A. *Macromolecules.* 1993;26(4):701–6.
115. Tsai S-C, Lin Y-C, Lin E-L, Chiang Y-W, Kuo S-W. *Polymer Chemistry.* 2016;7(13):2395–409.
116. Shaw M. *Journal of Applied Polymer Science.* 1974;18(2):449–72.
117. Jordan D, Mathieson A, Porter M. *Journal of Polymer Science.* 1956;21(99):473–82.
118. Arichi S, Matsuura H, Tanimoto Y, Murata H. *Bulletin of the Chemical Society of Japan.* 1966;39(3):434–9.
119. Shirai M, Saito T, Tsunooka M, Tanaka M, Nishijima K. *Journal of Applied Polymer Science.* 1990;41(9-10):2527–32.
120. Su M, Wei Y, Qi L, Pang Y, Guo Y, Li P, Li J, Li HP. *Express Polymer Letters.* 2011;5(12):1102–12.
121. Liu Y, Wei Y, Qin R, Zhang L, Yang L, Li H, Li X. *Express Polymer Letters.* 2011;5(1):60–72.
122. Lou X, Detrembleur C, Pagnoulle C, Jérôme R, Bocharova V, Kiriy A, Stamm M. *Advanced Materials.* 2004;16(23-24):2123–7.
123. Yoshida M, Sakamoto N, Ikemi K, Arichi S. *Bulletin of the Chemical Society of Japan.* 1992;65(11):3108–11.
124. Yoshida M, Sakamoto N, Ikemi K, Arichi S. *Bulletin of the Chemical Society of Japan.* 1993;66(6):1598–602.
125. Schmidt M. *Die Makromolekulare Chemie, Rapid Communications.* 1989;10(2):89–96.
126. Yamaguchi M, Yamaguchi Y, Matsushita Y, Noda I. *Polymer Journal.* 1990;22(12):1077–83.
127. Shen H, Zhang L, Eisenberg A. *Journal of the American Chemical Society.* 1999;121(12):2728–40.
128. Martin TJ, Prochazka K, Munk P, Webber S. *Macromolecules.* 1996;29(18):6071–3.
129. Zhang W, Shi L, Ma R, An Y, Xu Y, Wu K. *Macromolecules.* 2005;38(21):8850–2.
130. Iwaku M, Haseba T, Tatsuma T, Oyama N. *Journal of Electroanalytical Chemistry.* 1998;442(1-2):27–33.
131. Loxley A, Vincent B. *Colloid and Polymer Science.* 1997;275:1108–14.
132. Boucher E. *Progress in Polymer Science.* 1978;6(2):63–122.
133. Strauss UP, Fuoss RM. *Journal of Polymer Science.* 1949;4(4):457–72.
134. Fuoss RM, Strauss UP. *Journal of Polymer Science.* 1948;3(2):246–63.
135. Fuoss RM, Cathers GI. *Journal of Polymer Science.* 1949;4(2):97–120.
136. Boucher EA, Khosravi-Babadi E, Mollett CC. *Journal of the Chemical Society, Faraday Transactions 1: Physical Chemistry in Condensed Phases.* 1979;75:1728–35.
137. Plotnikova A, Merekalov A, Derikov Y, Shandryuk G, Kudryavtsev Y. *Polymer Science, Series C.* 2024:1–6.
138. Brannon-Peppas L, Peppas NA. *Chemical Engineering Science.* 1991;46(3):715–22.
139. Chang R. *Physical Chemistry for the Biosciences*: McMillian, New York, 1981.
140. Shieh JJ, Chung TS. *Journal of Polymer Science Part B: Polymer Physics.* 1999; 37(20):2851–61.
141. Chung TS, Kafchinski ER, Kohn RS, Foley P, Straff RS. *Journal of Applied Polymer Science.* 1994;53(5):701–8.
142. Khairou KS. *Polymer Degradation and Stability.* 1994;46(3):315–8.
143. Leza ML, Cortazar M, Casinos I, Guzmán GM. *Die Angewandte Makromolekulare Chemie: Applied Macromolecular Chemistry and Physics.* 1989;168(1):195–203.
144. Azhari SJ, Diab M. *Polymer Degradation and Stability.* 1998;60(2-3):253–6.
145. Jandrey A, De Aguiar A, De Aguiar M, de Santa Maria L. *Thermochimica Acta.* 2004;424(1-2):63–8.
146. Zhang BQ, Chen GD, Pan CY, Luan B, Hong CY. *Journal of Applied Polymer Science.* 2006;102(2):1950–8.
147. Cesteros LC, Velada J, Katime I. *Polymer.* 1995;36(16):3183–9.

148. Allan J, Beaumont P, Milburn G, Wood I. *Thermochimica Acta*. 1993;224:119–25.
149. Li X-G. *Reactive and Functional Polymers*. 1999;42(1):53–8.
150. Khairou K, Diab M. *Polymer Degradation and Stability*. 1994;44(1):17–20.
151. Awad MK, Khairou KS, Diab MA. *Polymer Degradation and Stability*. 1994;46(2):165–70.
152. Harden F, Kargar M, Handge UA. *Advances in Polymer Technology*. 2022;2022(1):5602902.
153. Winslow FH, Matreyek W. *Journal of Polymer Science*. 1956;22(101):315–24.
154. Joshi R. *Journal of Polymer Science*. 1962;56(164):313–38.
155. Haas HC, Moreau RD. *Journal of Polymer Science: Polymer Chemistry Edition*. 1977;15(5):1225–38.
156. Sarma NS, Dass N. *Materials Science and Engineering: B*. 2001;79(1):78–82.
157. Rao RV, Shridhar M, Ashokan P. *Materials Science and Engineering: A*. 2000;292(1):125–9.
158. Gatica N, Medina F, Moraga C, Vergara L. *Journal of the Chilean Chemical Society*. 2019;64(3):4575–81.
159. Hasegawa D, Teramoto Y, Nishio Y. *Journal of Wood Science*. 2008;54:143–52.
160. McCurdie MP, Belfiore LA. *Polymer*. 1999;40(11):2889–902.
161. Rafique U, Mazhar M, Ali S, Khwaja FA. *Synthetic Metals*. 1996;78(1):73–7.
162. Forster RJ, Vos JG. *Macromolecules*. 1990;23(20):4372–7.
163. Bekturov EA, Bimendina LA. *Journal of Macromolecular Science, Part C: Polymer Reviews*. 1997;37(3):501–18.
164. Santana A, Noda L, Pires A, Bertolino J. *Polymer Testing*. 2004;23(7):839–45.
165. Atorngitjawat P, Runt J. *The Journal of Physical Chemistry B*. 2007;111(48):13483–90.
166. Kuo S-W, Wu C-H, Chang F-C. *Macromolecules*. 2004;37(1):192–200.
167. Rodrigues Jd S, Gonçalves D, Mangrich A, Soldi V, Bertolino J, Pires A. *Advances in Polymer Technology: Journal of the Polymer Processing Institute*. 2000;19(2):113–9.
168. Belfiore LA, McCurdie MP, Ueda E. *Macromolecules*. 1993;26(25):6908–17.
169. Khan AK, Rashid R, Hussain M, Yunus U, Akhtar Z, Ali S et al. *Arabian Journal for Science and Engineering*. 2016;41:105–9.
170. Lee JY, Moskala EJ, Painter PC, Coleman MM. *Applied Spectroscopy*. 1986;40(7):991–4.
171. de Meftahi MV, Frechet JM. *Polymer*. 1988;29(3):477–82.
172. Huglin MB, Rego JM. *Polymer*. 1990;31(7):1269–76.
173. Winkler R, Beena Unni A, Tu W, Chat K, Adrjanowicz K. *The Journal of Physical Chemistry B*. 2021;125(22):5991–6003.
174. Serghei A, Huth H, Schick C, Kremer F. *Macromolecules*. 2008;41(10):3636–9.
175. Serghei A, Chen D, Lee D, Russell TP. *Soft Matter*. 2010;6(6):1111–3.
176. Tress M, Mapesa EU, Kossack W, Kipnusu WK, Reiche M, Kremer F. Molecular dynamics of condensed (semi-) isolated *polymer* chains. In: Kremer F (ed.). *Dynamics in Geometrical Confinement: Advances in Dielectrics*: Springer, Cham, 2014, pp. 61–93.
177. Efremov MY, Olson EA, Zhang M, Zhang Z, Allen LH. *Physical Review Letters*. 2003;91(8):085703.
178. Madkour S, Yin H, Füllbrandt M, Schönhals A. *Soft Matter*. 2015;11(40):7942–52.
179. Napolitano S, Lupascu V, Wübbenhorst M. *Macromolecules*. 2008;41(4):1061–3.
180. Zanten JHv, Wallace WE, Wu W-L. *Physical Review-Section E-Statistical Physics Plasma Fluids Related Interdiscpl Topics*. 1996;53(3):R2053.
181. Park CH, Kim JH, Ree M, Sohn B-H, Jung JC, Zin W-C. *Polymer*. 2004;45(13):4507–13.
182. Roth CB, McNerny KL, Jager WF, Torkelson JM. *Macromolecules*. 2007;40(7):2568–74.
183. Glor EC, Angrand GV, Fakhraai Z. *The Journal of Chemical Physics*. 2017;146(20).
184. Rahman MS, Samal S, Lee J-S. *Macromolecules*. 2007;40(26):9279–83.
185. Changez M, Kang N-G, Kim DW, Lee J-S. *Nanoscale*. 2013;5(23):11554–60.
186. Berman D, Sha Y, Shevchenko EV. *Molecules*. 2021;26(3):679.

187. Feng H, Lu X, Wang W, Kang N-G, Mays JW. *Polymers.* 2017;9(10):494.
188. Hahn J, Filiz V, Rangou S, Lademann B, Buhr K, Clodt JI, Jung A, Abetz C, Abetz A. *Macromolecular Materials and Engineering.* 2013;298(12):1315–21.
189. Quirk RP, Corona-Galvan S. *Macromolecules.* 2001;34(5):1192–7.
190. Schieder M, Lunkenbein T, Martin T, Milius W, Auffermann G, Breu J. *Journal of Materials Chemistry A.* 2013;1(2):381–7.
191. Regenbrecht M, Akari S, Förster S, Möhwald H. *Surface and Interface Analysis.* 1999;27(5-6):418–21.
192. Walther A, Goldmann AS, Yelamanchili RS, Drechsler M, Schmalz H, Eisenberg A, Müller AH. *Macromolecules.* 2008;41(9):3254–60.
193. Noro A, Matsushima S, He X, Hayashi M, Matsushita Y. *Macromolecules.* 2013;46(20):8304–10.
194. Noro A, Matsushita Y, Lodge TP. *Macromolecules.* 2008;41(15):5839–44.
195. Noro A, Matsushita Y, Lodge TP. *Macromolecules.* 2009;42(15):5802–10.
196. Mendrek S, Mendrek A, Adler HJ, Dworak A, Kuckling D. *Journal of Polymer Science Part A: Polymer Chemistry.* 2009;47(7):1782–94.
197. Chung T-M, Wang H-F, Lin T, Chiang Y-W, Chen Y-C, Ko B-T, Ho R-M. *Macromolecules.* 2012;45(24):9727–33.
198. Szwarc M. *Nature.* 1956;178(4543):1168–9.
199. Hoheisel TN, Hur K, Wiesner UB. *Progress in Polymer Science.* 2015;40:3–32.
200. Xiao L-L, Zhou X, Yue K, Guo Z-H. *Polymers.* 2020;13(1):110.
201. Park C, Yoon J, Thomas EL. *Polymer.* 2003;44(22):6725–60.
202. Mäkinen R, Ruokolainen J, Ikkala O, de Moel K, ten Brinke G, De Odorico W, Stamm M. *Macromolecules.* 2000;33(9):3441–6.
203. Yu YY, Chien WC, Chen ST. *Polymer International.* 2008;57(12):1369–76.
204. Ribbe AE, Okumura A, Matsushige K, Hashimoto T. *Macromolecules.* 2001;34(23):8239–45.
205. Rahman MS, Samal S, Lee J-S. *Macromolecules.* 2006;39(15):5009–14.
206. Tiitu M, Torkkeli M, Serimaa R, Mäkelä T, Ikkala OT. *Solid State Ionics.* 2005;176(13-14):1291–9.
207. Xu F, Zheng S-Z, Luo Y-L, Chen T-T. *Macromolecular Research.* 2013;21:977–86.
208. Kim SJ, Kim DJ, Heo SY, Ahn H, Ryu DY, Kim JH. *Electrochimica Acta.* 2013;105:15–22.
209. Kang M, Chung K, Baeg K-J, Kim DH, Kim C. *Applied Physics Letters.* 2015;106(4).
210. Fu X, Song L, Liu J, Li X, Zhang X, Jia Y. *Macromolecular Chemistry and Physics.* 2012;213(16):1663–8.
211. Kim EJ, Shin JJ, Lee GS, Kim S, Park S, Park J, Choe D, Lee D, Choi J, Bang J, Chim YH, Li S. *Macromolecules.* 2022;55(5):1590–9.
212. Choi JW, Li Z, Black CT, Sweat DP, Wang X, Gopalan P. *Nanoscale.* 2016; 8(22):11595–601.
213. Bugakov M, Boiko N, Chernikova E, Abramchuk S, Shibaev V. *Polymer Science, Series C.* 2018;60:3–13.
214. Rider AN. *Materials Science and Materials Engineering.* 2018;6:352–79.
215. Zhou J, Thapar V, Chen Y, Wu B-X, Craig GS, Nealey PF, Hur SM, Chang TH, Xiong S. *ACS Applied Materials & Interfaces.* 2021;13(34):41190–9.
216. Blanazs A, Armes SP, Ryan AJ. *Macromolecular rapid Communications.* 2009;30(4-5):267–77.
217. Alexandridis P, Tsianou M. *European Polymer Journal.* 2011;47(4):569–83.
218. Jones MR, Osberg KD, Macfarlane RJ, Langille MR, Mirkin CA. *Chemical Reviews.* 2011;111(6):3736–827.
219. Mai Y, Eisenberg A. *Accounts of Chemical Research.* 2012;45(10):1657–66.
220. Li W, Müller M. *Progress in Polymer Science.* 2016;54:47–75.

221. Ji S, Wan L, Liu C-C, Nealey PF. *Progress in Polymer Science*. 2016;54:76–127.
222. Zhang Q, Lin J, Wang L, Xu Z. *Progress in Polymer Science*. 2017;75:1–30.
223. Zhang S, Pelligra CI, Feng X, Osuji CO. *Advanced Materials*. 2018;30(18):1705794.
224. Kennemur J, Yao, L., Bates, FS, Hillmyer, MA. *Macromolecules*. 2014;47(4):1411–8.
225. Maria S, Susha AS, Sommer M, Talapin DV, Rogach AL, Thelakkat M. *Macromolecules*. 2008;41(16):6081–8.
226. Hofman AH, van Ekenstein GOA, Woortman AJ, ten Brinke G, Loos K. *Polymer Chemistry*. 2015;6(39):7015–26.
227. Ju J, Hayward R. *APS March Meeting Abstracts*. 2022;2022:W18–012.
228. Lee D, Kim J, Ku KH, Li S, Shin JJ, Kim BJ. *Polymer Chemistry*. 2022;13(18):2570–88.
229. Biswas S, Das A. *ChemNanoMat*. 2021;7(7):748–72.
230. Dai KH, Kramer EJ. *Polymer*. 1994;35(1):157–61.
231. Shull KR, Kramer EJ, Hadziioannou G, Tang W. *Macromolecules*. 1990;23(22):4780–7.
232. Peng H, Chen D, Jiang M. *Langmuir*. 2003;19(26):10989–92.
233. Nguyen HT, Tran TTT, Nguyen-Thai NU. *Journal of Nanostructure in Chemistry*. 2018;8:61–9.
234. Wang Y, Xu X, Xu P, Feng X, Zhang Y, Fu F, Liu X. *Polymer International*. 2018;67(6):619–26.
235. Liu R, Rong Z, Han G, Yang X, Zhang W. *Polymer*. 2021;215:123431.
236. He Q, Zhang Y, Zhang Z, Qian H-J, Chen Q. *Frontiers in Soft Matter*. 2022;2:980210.
237. Cummins C, Borah D, Rasappa S, Chaudhari A, Ghoshal T, O'Driscoll B, Carolan P, Petkov N, Holmes JD, Morris MA. *Journal of Materials Chemistry C*. 2013;1(47):7941–51.
238. Glass R, Möller M, Spatz JP. *Nanotechnology*. 2003;14(10):1153.
239. Tang C, Lennon EM, Fredrickson GH, Kramer EJ, Hawker CJ. *Science*. 2008;322(5900):429–32.
240. Nabeela K, Thomas RT, Nair JB, Maiti KK, Warrier KGK, Pillai S. ACS *Applied Materials & Interfaces*. 2016;8(43):29242–51.
241. Yu C-H, Chuang Y-H, Tung S-H. *Polymer*. 2011;52(18):3994–4000.
242. Nunes SP. *Macromolecules*. 2016;49(8):2905–16.
243. Peinemann K-V, Abetz V, Simon PF. *Nature Materials*. 2007;6(12):992–6.
244. Kosonen H, Valkama S, Ruokolainen J, Torkkeli M, Serimaa R, Ten Brinke G, Ikkala O. *The European Physical Journal E*. 2003;10:69–75.
245. Connal LA, Lynd NA, Robb MJ, See KA, Jang SG, Spruell JM, Hawker CJ. *Chemistry of Materials*. 2012;24(21):4036–42.
246. Huang C, Tan J, Xu Q, He J, Li X, Liu D, Zhang L. *RSC Advances*. 2017;7(73):46069–81.
247. He X, Qu Y, Gao C, Zhang W. *Polymer Chemistry*. 2015;6(35):6386–93.
248. Shi P, Zhou H, Gao C, Wang S, Sun P, Zhang W. *Polymer Chemistry*. 2015;6(27):4911–20.
249. Ren Z, Guo X, Lu Z, Guo X, Ma D. *Polymer Bulletin*. 2007;59:371–80.
250. Dai C-A, Yen W-C, Lee Y-H, Ho C-C, Su W-F. *Journal of the American Chemical Society*. 2007;129(36):11036–8.
251. Kim J-S, Kim Y, Kim H-J, Kim HJ, Yang H, Jung YS, Stein GE, Kim BJ. *Macromolecules*. 2017;50(5):1902–8.
252. Mukundan T. *Macromolecules*. 2012;45:3070.
253. Kim J-S, Han J, Kim Y, Park H, Coote JP, Stein GE, Kim BJ. *Macromolecules*. 2018;51(11):4077–84.
254. Sary N, Rubatat L, Brochon C, Hadziioannou G, Ruokolainen J, Mezzenga R. *Macromolecules*. 2007;40(19):6990-7.\
255. Phillip WA, Dorin RM, Werner Jr, Hoek EM, Wiesner U, Elimelech M. *Nano Letters*. 2011;11(7):2892–900.
256. Faber M, Hofman AH, Polushkin E, van Ekenstein GA, Seitsonen J, Ruokolainen J et al. *Macromolecules*. 2013;46(2):500–17.

257. Jung A, Filiz V, Rangou S, Buhr K, Merten P, Hahn J, Clodt J, Abetz C, Abetz V. *Macromolecular Rapid Communications.* 2013;34(7):610–5.
258. Han SH, Kim JK. *Reactive and Functional Polymers.* 2009;69(7):493–7.
259. Dyakonova M, Stavrouli N, Popescu M, Kyriakos K, Grillo I, Philipp M, Jaksch S, Tsitsillianis C, Papadakis CM. *Macromolecules.* 2014;47(21):7561–72.
260. Stavrouli N, Aubry T, Tsitsilianis C. *Polymer.* 2008;49(5):1249–56.
261. Li X, Deng H. *ACS Applied Polymer Materials.* 2020;2(8):3601–11.
262. Stewart-Sloan CR, Wang R, Sing MK, Olsen BD. *Journal of Polymer Science Part B: Polymer Physics.* 2017;55(15):1181–90.
263. Stewart-Sloan CR, Olsen BD. *ACS Macro Letters.* 2014;3(5):410–4.
264. Sinturel C, Bates FS, Hillmyer MA. *ACS Macro Letters.* 2015;4(9):1044–1050.
265. Bates FS, Fredrickson GH. *Annual Review of Physical Chemistry.* 1990;41(1):525–57.
266. Cho Y-H, Yang J-E, Lee J-S. *Materials Science and Engineering: C.* 2004;24(1-2):293–5.
267. Azuma K, Sun J, Choo Y, Rokhlenko Y, Dwyer JH, Schweitzer B, Hayakawa T, Osuji CO, Gopalan P. *Macromolecules.* 2018;51(16):6460–7.
268. Faber M, Hofman AH, Loos K, Brinke GT. *Macromolecular Rapid Communications.* 2016;37(11):911–9.
269. Gobius du Sart G, Vukovic I, Alberda van Ekenstein G, Polushkin E, Loos K, Ten Brinke G. *Macromolecules.* 2010;43(6):2970–80.
270. Ding S-P, Zhang Z-K, Ye Z, Du B-Y, Xu J-T. *ACS Macro Letters.* 2021;10(10):1321–5.
271. Kong Y, Kawahara H, Kikuchi R, Ohta N, Mufundirwa A, Hirai T, Hatakeyama-Sato K, Nabae Y, Hayakawa T. *Polymer.* 2025:128098.
272. Lundy R, Flynn SP, Cummins C, Kelleher SM, Collins MN, Dalton E, Daniels S, Morris MA, Enright R. *Physical Chemistry Chemical Physics.* 2017;19(4):2805–15.
273. Sinturel C, Vayer M, Morris M, Hillmyer MA. *Macromolecules.* 2013;46(14):5399–415.
274. Kim H-C, Park S-M, Hinsberg WD. *Chemical Reviews.* 2010;110(1):146–77.
275. Lehn J-M. *Supramolecular Chemistry*: VCH, Weinheim, New York, 1995.
276. Karayianni M, Pispas S. *Journal of Polymer Science.* 2021;59(17):1874–98.
277. Mansky P, Haikin P, Thomas E. *Journal of Materials Science.* 1995;30:1987–92.
278. Lodge TP, Pudil B, Hanley KJ. *Macromolecules.* 2002;35(12):4707–17.
279. Nandan B, Vyas MK, Böhme M, Stamm M. *Macromolecules.* 2010;43(5):2463–73.
280. Gowd EB, Böhme M, Stamm M. *IOP Conference Series: Materials Science and Engineering.* 2010;14(1):012015.
281. Zhang X, Harris KD, Wu NL, Murphy JN, Buriak JM. *ACS Nano.* 2010;4(11):7021–9.
282. Xiong S, Wan L, Ishida Y, Chapuis Y-A, Craig GS, Ruiz R, Nealey PF. *ACS Nano.* 2016;10(8):7855–65.
283. Knoll A, Horvat A, Lyakhova K, Krausch G, Sevink G, Zvelindovsky A, Magerle R. *Physical Review Letters.* 2002;89(3):035501.
284. Sun Z, Chen Z, Zhang W, Choi J, Huang C, Jeong G, Coughlin EB, Hsu Y, Yang X, Lee KY, Kuo DS. *Advanced Materials.* 2015;27(29):4364–70.
285. Chaudhari A, Ghoshal T, Shaw MT, Cummins C, Borah D, Holmes JD, Morris MA, *Advances in Patterning Materials and Processes XXXI.* 2014: 905110.
286. Ku KH, Shin JM, Kim MP, Lee C-H, Seo M-K, Yi G-R, Jang S-G, Kim B-J. *Journal of the American Chemical Society.* 2014;136(28):9982–9.
287. Qiu H, Hudson ZM, Winnik MA, Manners I. *Science.* 2015;347(6228):1329–32.
288. Zhang Y, Pearce S, Eloi J-C, Harniman RL, Tian J, Cordoba C, Kang Y, Fukui T, Qiu H, Blackburn A, Richardson RM. *Journal of the American Chemical Society.* 2021;143(15):5805–14.
289. Ku KH, Kim Y, Yi G-R, Jung YS, Kim BJ. *ACS Nano.* 2015;9(11):11333–41.
290. Ku KH, Ryu JH, Kim J, Yun H, Nam C, Shin JM et al. *Chemistry of Materials.* 2018;30(23):8669–78.

291. Klinger D, Wang CX, Connal LA, Audus DJ, Jang SG, Kraemer S, Killops KL, Fredrickson GH, Kramer EJ, Hawker CJ. *Angewandte Chemie*. 2014;126(27):7138–42.
292. Schmidt BV, Wang CX, Kraemer S, Connal LA, Klinger D. *Polymer Chemistry*. 2018;9(13):1638–49.
293. Navarro L, Thünemann AF, Yokosawa T, Spiecker E, Klinger D. *Angewandte Chemie International Edition*. 2022;61(35):e202208084.
294. Yan N, Zhu Y, Jiang W. *Chemical Communications*. 2018;54(94):13183–95.
295. Wong CK, Qiang X, Müller AH, Gröschel AH. *Progress in Polymer Science*. 2020;102:101211.
296. Nghiem TL, Chakroun R, Janoszka N, Chen C, Klein K, Wong CK, Gröschel AH. *Macromolecular Rapid Communications*. 2020;41(18):2000301.
297. Qiu H, Russo G, Rupar PA, Chabanne L, Winnik MA, Manners I. *Angewandte Chemie International Edition*. 2012;51(47):11882–5.
298. Lee J, Ku KH, Kim J, Lee YJ, Jang SG, Kim BJ. *Journal of the American Chemical Society*. 2019;141(38):15348–55.
299. Ku KH, Shin JM, Yun H, Yi GR, Jang SG, Kim BJ. *Advanced Functional Materials*. 2018;28(42):1802961.
300. Navarro L, Thünemann AF, Klinger D. *ACS Macro Letters*. 2022;11(3):329–35.
301. Chen C, Xiao Z, Connal LA. *Australian Journal of Chemistry*. 2016;69(7):741–5.
302. Li Y, Wang X, Sun J. *Chemical Society Reviews*. 2012;41(18):5998–6009.
303. Deng Y, Wang T, Guo Y, Qiu X, Qian Y. *ACS Sustainable Chemistry & Engineering*. 2015;3(6):1215–20.
304. Zhang H, Fu Y, Wang D, Wang L, Wang Z, Zhang X. *Langmuir*. 2003;19(20):8497–502.
305. Yin Q, Beyou E, Portinha D, Charlot A. *European Polymer Journal*. 2019;117:188–99.
306. Fu Y, Bai S, Cui S, Qiu D, Wang Z, Zhang X. *Macromolecules*. 2002;35(25):9451–8.
307. Karayianni M, Pispas S. Self-assembly of amphiphilic block copolymers in selective solvents. In: Procházka K (ed.). *Fluorescence Studies of Polymer Containing Systems*: Springer, Cham, 2016, vol. 16.
308. Kwak J, Han SH, Moon HC, Kim JK, Pryamitsyn V, Ganesan V. *Macromolecules*. 2015;48(17):6347–52.
309. Kuo S-W, Lin C-L, Chang F-C. *Polymer*. 2002;43(14):3943–9.
310. Aizawa M, Buriak JM. *Journal of the American Chemical Society*. 2006;128(17):5877–86.
311. Soininen AJ, Tanionou I, Ten Brummelhuis N, Schlaad H, Hadjichristidis N, Ikkala O et al. *Macromolecules*. 2012;45(17):7091–7.
312. Korhonen JT, Verho T, Rannou P, Ikkala O. *Macromolecules*. 2010;43(3):1507–14.
313. Mun JH, Cha SK, Kim YC, Yun T, Choi YJ, Jin HM et al. *Small*. 2017;13(17):1603939.
314. Shin DO, Mun JH, Hwang G-T, Yoon JM, Kim JY, Yun JM et al. *ACS Nano*. 2013;7(10):8899–907.
315. Cha SK, Lee GY, Mun JH, Jin HM, Moon CY, Kim JS et al. *ACS Applied Materials & Interfaces*. 2017;9(18):15727–32.
316. Kim SY, Yoon E, Joo T, Park MJ. *Macromolecules*. 2011;44(13):5289–98.
317. Young WW, Tabuchi H, Iguchi R, Konishi T, Fukao K, Katsumata R. *Macromolecules*. 2022;55(15):6590–7.
318. Ruokolainen J, Makinen R, Torkkeli M, Makela T, Serimaa R, Brinke Gt, Ikkala O. *Science*. 1998;280(5363):557–60.
319. Ishizu K, Honda K, Kanbara T, Yamamoto T. *Polymer*. 1994;35(22):4901–6.
320. Stanković RI, Lenz RW, Karasz FE. *European Polymer Journal*. 1990;26(3):359–69.
321. Dai L, White JW. *Polymer*. 1997;38(4):775–83.
322. Zhang Y-D, Ping J, Wu Q-W, Pan H-B, Fan X-H, Shen Z, Zhou Q-F. *Polymer Chemistry*. 2017;8(10):1689–98.
323. Pelz A, Dörr TS, Zhang P, de Oliveira PW, Winter M, Wiemhöfer H-D, Kraus T. *Chemistry of Materials*. 2018;31(1):277–85.

324. Wang DR, Wujcik KH, Teran AA, Balsara NP. *Macromolecules.* 2015;48(14):4863–73.
325. Mäki-Ontto R, de Moel K, Polushkin E, Alberda van Ekenstein G, ten Brinke G, Ikkala O. *Advanced Materials.* 2002;14(5):357–61.
326. Kosonen H, Valkama S, Hartikainen J, Eerikäinen H, Torkkeli M, Jokela K et al. *Macromolecules.* 2002;35(27):10149–54.
327. Cho G, Park K-p, Jang J, Jung S, Moon J, Kim T. *Electrochemistry Communications.* 2002;4(4):336–9.
328. Schulze MW, McIntosh LD, Hillmyer MA, Lodge TP. *Nano Letters.* 2014;14(1):122–6.
329. Mäki-Ontto R, de Moel K, de Odorico W, Ruokolainen J, Stamm M, ten Brinke G, Ikkala O. *Advanced Materials.* 2001;13(2):117–21.
330. Ruokolainen J, Torkkeli M, Serimaa R, Komanschek B, Ikkala O, Ten Brinke G. *Physical Review E.* 1996;54(6):6646.
331. Ruokolainen J, Torkkeli M, Serimaa R, Komanschek E, Ten Brinke G, Ikkala O. *Macromolecules.* 1997;30(7):2002–7.
332. Ruotsalainen T, Torkkeli M, Serimaa R, Mäkelä T, Mäki-Ontto R, Ruokolainen J et al. *Macromolecules.* 2003;36(25):9437–42.
333. Lin E-L, Hsu W-L, Chiang Y-W. *ACS Nano.* 2018;12(1):485–93.
334. Kang Y, Walish JJ, Gorishnyy T, Thomas EL. *Nature Materials.* 2007;6(12):957–60.
335. Kang Y-J. *Bulletin of the Korean Chemical Society.* 2008;29(12):2329–30.
336. Kang C, Kim E, Baek H, Hwang K, Kwak D, Kang Y, Thomas EL. *Journal of the American Chemical Society.* 2009;131(22):7538–9.
337. Ayyub OB, Ibrahim MB, Kofinas P. *Polymer.* 2014;55(24):6227–31.
338. Terzic I, Meereboer NL, Acuautla Mn, Portale G, Loos K. *Macromolecules.* 2018;52(1):354–64.
339. Terzic I, Meereboer NL, Acuautla M, Portale G, Loos K. *Nature Communications.* 2019;10(1):601.
340. Förster S, Antonietti M. *Advanced Materials.* 1998;10(3):195–217.
341. Gaucher G, Dufresne M-H, Sant VP, Kang N, Maysinger D, Leroux J-C. *Journal of Controlled Release.* 2005;109(1-3):169–88.
342. Israelachvili JN, Mitchell DJ, Ninham BW. *Biochimica et Biophysica Acta (BBA)-Biomembranes.* 1977;470(2):185–201.
343. Doncom KE, Blackman LD, Wright DB, Gibson MI, O'Reilly RK. *Chemical Society Reviews.* 2017;46(14):4119–34.
344. Damas C, Brembilla A, Lochon P, Baros F, Viriot M-L. *Polymer.* 1995;36(10):2095–101.
345. Chu B. *Langmuir.* 1995;11(2):414–21.
346. Förster S, Abetz V, Müller AH. *Advance Polymer Science.* 2004;166:173–210.
347. Spatz JP, Mößmer S, Möller M. *Angewandte Chemie International Edition in English.* 1996;35(13-14):1510–2.
348. Zhang L, Eisenberg A. *Science.* 1995;268(5218):1728–31.
349. Dutta G, Guha N, Ghosh Dastidar D, Das A. Dutta, G, Guha, N, Ghosh Dastidar, D, Das, A. Applications of block copolymers as stimuli-responsive copolymers. In: Mishra N, Pandey V (eds.). *Block Co-Polymeric Nanocarriers: Design, Concept, and Therapeutic Applications*: Springer, Singapore, 2023.
350. Z, Ma C, Huang X, Lu G, Winnik MA, Feng C. *Macromolecules.* 2021;54(14):6705–17.
351. Bronstein LM, Sidorov SN, Valetsky PM, Hartmann J, Cölfen H, Antonietti M. *Langmuir.* 1999;15(19):6256–62.
352. Antonietti M, Wenz E, Bronstein L, Seregina M. *Advanced Materials.* 1995;7(12):1000–5.
353. Popescu M-T, Korogiannaki M, Marikou K, Tsitsilianis C. *Polymer.* 2014;55(13):2943–51.
354. Memic A, Alhadrami HA, Hussain MA, Aldhahri M, Al Nowaiser F, Al-Hazmi F et al. *Biomedical Materials.* 2015;11(1):014104.
355. T, Kishi S, Takahashi S, Utsumi I. *Journal of Pharmaceutical Sciences.* 1962;51(11):1061–8.

356. Brewer K, Bai F, Blencowe A. *Pharmaceutics.* 2023;15(3):977.
357. Mishra AK, Lim J, Lee J, Park S, Seo Y, Hwang H, Kim JK. *Polymer.* 2021;213:123329.
358. Poggi E, Guerlain C, Debuigne A, Detrembleur C, Gigmes D, Hoeppener S et al. *European Polymer Journal.* 2015;62:418–25.
359. Pippa N, Pispas S, Demetzos C. *Design and Development of New Nanocarriers.* 2018:57–88.
360. Natalello A, Tonhauser C, Berger-Nicoletti E, Frey H. *Macromolecules.* 2011;44(24):9887–90.
361. Bronstein LM, Sidorov SN, Zhirov V, Zhirov D, Kabachii YA, Kochev SY et al. *The Journal of Physical Chemistry B.* 2005;109(40):18786–98.
362. Rocha N, Mendes J, Duraes L, Maleki H, Portugal A, Geraldes CF et al. *Journal of Materials Chemistry B.* 2014;2(11):1565–75.
363. Gohy JF, Willet N, Varshney S, Zhang JX, Jérôme R. *Angewandte Chemie International Edition.* 2001;40(17):3214–6.
364. Azzam T, Bronstein L, Eisenberg A. *Langmuir.* 2008;24(13):6521–9.
365. Nghiem T-L, Löbling TI, Gröschel AH. *Polymer Chemistry.* 2018;9(13):1583–92.
366. Borchert U, Lipprandt U, Bilang M, Kimpfler A, Rank A, Peschka-Süss R et al. *Langmuir.* 2006;22(13):5843–7.
367. Gohy J-F, Willet N, Varshney SK, Zhang J-X, Jérôme R. e-*Polymers.* 2002;2(1):034.
368. Vamvakaki M, Papoutsakis L, Katsamanis V, Afchoudia T, Fragouli PG, Iatrou H et al. *Faraday Discussions.* 2005;128:129–47.
369. Nyrkova IA, Semenov AN. *Faraday Discussions.* 2005;128:113–27.
370. McLean SC, Scholes CA, Smith TA, Gee ML. *Australian Journal of Chemistry.* 2011;64(6):825–32.
371. Gohy J-F, Varshney SK, Antoun S, Jérôme R. *Macromolecules.* 2000;33(25):9298–305.
372. McLean SC, Gee ML. *Macromolecules.* 2007;40(13):4644–53.
373. Teng Y, Morrison M, Munk P, Webber S, Procházka K. *Macromolecules.* 1998;31(11):3578–87.
374. Albrecht K, Mourran A, Zhu X, Markkula T, Groll J, Beginn U et al. *Macromolecules.* 2008;41(5):1728–38.
375. Yeh C-L, Hou T, Chen H-L, Yeh L-Y, Chiu F-C, Müller AJ, Hadjichristidis N. *Macromolecules.* 2011;44(3):440–3.
376. Gohy J-F, Lohmeijer BG, Varshney SK, Décamps B, Leroy E, Boileau S, Schubert US. *Macromolecules.* 2002;35(26):9748–55.
377. de Jeu WH, Markkula T, Albrecht K, Ivanov DA, Möller M, Mourran A. *Macromolecules.* 2017;50(12):4754–8.
378. Atanase LI, Riess G. *Journal of Colloid and Interface Science.* 2013;395:190–7.
379. Abandansari HS, Nabid MR, Rezaei SJT, Niknejad H. *Polymer.* 2014;55(16):3579–90.
380. Haladjova E, Rangelov S. *Pharmaceutics* 2024; 16(5): 646.
381. Xu X-Y, Gao Y-T, Wang Y-Y, Zhou Y-S, Xiong B-J, Zhu J-T. *Chinese Journal of Polymer Science.* 2023;41(6):897–904.
382. Van Butsele K, Cajot S, Van Vlierberghe S, Dubruel P, Passirani C, Benoit JP et al. *Advanced Functional Materials.* 2009;19(9):1416–25.
383. Musteata V, Sutisna B, *Polymer* opoulos G, Avgeropoulos A, Meneau F, Peinemann K-V et al. *European Polymer Journal.* 2018;100:121–31.
384. Talingting MR, Munk P, Webber S, Tuzar Z. *Macromolecules.* 1999;32(5):1593–601.
385. Prochazka K, Martin TJ, Webber SE, Munk P. *Macromolecules.* 1996;29(20):6526–30.
386. Fujita N, Otsuka H, Takahara A, Shinkai S. *Chemistry Letters.* 2003;32(4):352–3.
387. Chen D, Peng H, Jiang M. *Macromolecules.* 2003;36(8):2576–8.
388. Han H, Lee J, Hong J, Shim SE. *Macromolecular Research.* 2009;17:469–75.
389. Calderara F, Riess G. *Macromolecular Chemistry and Physics.* 1996;197(7):2115–32.

390. Jang SG, Kramer EJ, Hawker CJ. *Journal of the American Chemical Society*. 2011;133(42):16986–96.
391. Aizawa M, Buriak JM. *Journal of the American Chemical Society*. 2005;127(25):8932–3.
392. Spatz JP, Chan VH, Mößmer S, Kamm FM, Plettl A, Ziemann P, Möller M. *Advanced Materials*. 2002;14(24):1827–32.
393. Hahn J, Webber S. *Langmuir*. 2004;20(10):4211–9.
394. Meli M-V, Badia A, Grütter P, Lennox RB. *Nano Letters*. 2002;2(2):131–5.
395. Park S-Y, Sul W-H, Chang Y-J. *Macromolecules*. 2007;40(10):3757–64.
396. Fujita N, Yamashita T, Asai M, Shinkai S. *Angewandte Chemie*. 2005;117(8):1283–7.
397. Laiho A, Ras RH, Valkama S, Ruokolainen J, Österbacka R, Ikkala O. *Macromolecules*. 2006;39(22):7648–53.
398. Koh HD, Lee JP, Lee JS. *Macromolecular Rapid Communications*. 2009;30(12):976–80.
399. Li M, Liu X, Zhao X, Yang F, Wang X, Li Y. *Single-Walled Carbon Nanotubes: Preparation, Properties and Applications*. 2019:25–67.
400. Fu Q, Huang S, Liu J. *The Journal of Physical Chemistry B*. 2004;108(20):6124–9.
401. Bhaviripudi S, Reina A, Qi J, Kong J, Belcher AM. *Nanotechnology*. 2006;17(20):5080.
402. Lu J, Yi SS, Kopley T, Qian C, Liu J, Gulari E. *The Journal of Physical Chemistry B*. 2006;110(13):6655–60.
403. Lu J, Kopley T, Dutton D, Liu J, Qian C, Son H et al. *The Journal of Physical Chemistry B*. 2006;110(22):10585–9.
404. Yabu H, Endo A, Koike K, Motoyoshi K, Higuchi T, Shimomura M. *Polymer Journal*. 2011;43(3):301–5.
405. Göbel C, Hils C, Drechsler M, Baabe D, Greiner A, Schmalz H, Weber B. *Angewandte Chemie International Edition*. 2020;59(14):5765–70.
406. Klimm O, Göbel C, Rosenfeldt S, Puchtler F, Miyajima N, Marquardt K et al. *Nanoscale*. 2016;8(45):19058–65.
407. Weber B. *Chemistry– A European Journal*. 2017;23(72):18093–100.
408. Göbel C, Klimm O, Puchtler F, Rosenfeldt S, Förster S, Weber B. *Beilstein Journal of Nanotechnology*. 2017;8(1):1318–27.
409. Zhang Q, Lu M, Wu H, Zhang L, Feng X, Jin Z. *Langmuir*. 2022;38(41):12441–9.
410. Sfika V, Tsitsilianis C. *Macromolecules*. 2003;36(13):4983–8.
411. Kamachi M, Kurihara M, Stille J. *Macromolecules*. 1972;5(2):161–7.
412. Schacher F, Walther A, Ruppel M, Drechsler M, Müller AH. *Macromolecules*. 2009;42(10):3540–8.
413. Schacher F, Walther A, Müller AH. *Langmuir*. 2009;25(18):10962–9.
414. Hanisch A, Gröschel AH, Förtsch M, Drechsler M, Jinnai H, Ruhland TM et al. *ACS Nano*. 2013;7(5):4030–41.
415. Xu W, Ledin PA, Iatridi Z, Tsitsilianis C, Tsukruk VV. *Macromolecules*. 2015;48(10):3344–53.
416. Xu W, Ledin PA, Iatridi Z, Tsitsilianis C, Tsukruk VV. *Angewandte Chemie International Edition*. 2016;55(16):4908–13.
417. Erwin AJ, Korolovych VF, Iatridi Z, Tsitsilianis C, Ankner JF, Tsukruk VV. *Macromolecules*. 2018;51(13):4800–12.
418. Du Sart GG, Rachmawati R, Voet V, van Ekenstein GA, Polushkin E, Ten Brinke G, Loos K. *Macromolecules*. 2008;41(17):6393–9.
419. Giebeler E, Stadler R. *Macromolecular Chemistry and Physics*. 1997;198(12):3815–25.
420. Iatridi Z, Tsitsilianis C. *Chemical Communications*. 2011;47(19):5560–2.
421. Stavrouli N, Kyriazis A, Tsitsilianis C. *Macromolecular Chemistry and Physics*. 2008;209(21):2241–7.

422. García MC. Stimuli responsive polymeric nanocarriers for drug delivery applications. In: Makhlouf ASH, Abu-Thabit NY (eds.), *Advanced Nanocarriers for Therapeutics: Woodhead Publishing Series in Biomaterials*: Woodhead, Oxford, 2019, vol. 2, pp. 345–92.
423. Shao G, Liu Y, Cao R, Han G, Yuan B, Zhang W. *Polymer Chemistry.* 2023;14(16):1863–80.
424. Balamurugan S, Mendez S, Balamurugan SS, O'Brien MJ, López GP. *Langmuir.* 2003;19(7):2545–9.
425. Kuckling D, Wohlrab S. *Polymer.* 2002;43(4):1533–6.
426. Wohlrab S, Kuckling D. *Journal of Polymer Science Part A: Polymer Chemistry.* 2001;39(21):3797–804.
427. Tarabukina E, Rozanova A, Filippov A, Constantin M, Harabagiu V, Fundueanu G. *Colloid and Polymer Science.* 2018;296:557–65.
428. Xu Y, Shi L, Ma R, Zhang W, An Y, Zhu X. *Polymer.* 2007;48(6):1711–7.
429. Wu C, Zhou S. *Macromolecules.* 1995;28(15):5388–90.
430. Kuckling D, Vo CD, Wohlrab SE. *Langmuir.* 2002;18(11):4263–9.
431. Atanase LI, Riess G. *Journal of Colloid and Interface Science.* 2019;549:171–8.
432. Pelton R. *Advances in Colloid and Interface Science.* 2000;85(1):1–33.
433. Nur H, Pinkrah V, Mitchell J, Benée L, Snowden M. *Advances in Colloid and Interface Science.* 2010;158(1-2):15–20.
434. Pinkrah V, Snowden M, Mitchell J, Seidel J, Chowdhry B, Fern G. *Langmuir.* 2003;19(3):585–90.
435. Shulkin A, Stöver HD. *Macromolecules.* 2003;36(26):9836–9.
436. Mu M, Ebara M. Smart polymers. In: Narain R (ed.). Polymer Science and Nanotechnology: Fundamentals and Applications: Elsevier, Amsterdam, 2020, pp. 257–279.
437. Liu X, Jiang M. *Angewandte Chemie.* 2006;118(23):3930–4.
438. Zheng L, Sun Z, Li C, Wei Z, Jain P, Wu K. *Polymer Degradation and Stability.* 2017;139:1–19.
439. Singh PK, Singh VK, Singh M. e-*Polymers.* 2007;7(1):030.
440. Alfrey Jr T, Morawetz H. *Journal of the American Chemical Society.* 1952;74(2):436–8.
441. Apen PG, Rasmussen PG. *Journal of Polymer Science Part A: Polymer Chemistry.* 1992;30(2):203–10.
442. Xue W, Huglin MB, Khoshdel E. *Polymer International.* 1999;48(1):8–14.
443. Lowe AB, McCormick CL. *Chemical Reviews.* 2002;102(11):4177–90.
444. Soto VM, Galin J. *Polymer.* 1984;25(2):254–62.
445. Billing M, Elter JK, Schacher FH. *Polymer.* 2016;104:40–8.
446. Lee S, Park S, Sylvianti N, Choi HK, Kim JH. *Synthetic Metals.* 2015;209:441–6.
447. Galin M, Marchal E, Mathis A, Meurer B, Soto YM, Galin J. *Polymer.* 1987;28(11):1937–44.
448. Xue W, Huglin MB, Liao B. *European Polymer Journal.* 2006;42(11):3015–23.
449. Malekzadeh A, Zahedi P, Abdouss M. *New Journal of Chemistry.* 2022;46(18):8698–707.
450. Gugoasa AI, Racovita S, Vasiliu S, Popa M. *Polymers.* 2023;15(3):490.
451. Xuan F, Liu J. *Polymer International.* 2009;58(12):1350–61.
452. Xue W, Huglin MB, Russell AT. *Macromolecular Rapid Communications.* 1999;20(4):239–43.
453. Hart R, Timmerman D. *Journal of Polymer Science.* 1958;28(118):638–40.
454. Chiu C-Y, Chang Y, Liu T-H, Chou Y-N, Yen T-J. *Journal of Materials Chemistry B.* 2021;9(40):8437–50.
455. Wu C, Zhou Y, Wang H, Hu J. *Nanomaterials.* 2019;9(5):706.
456. Gui Z, Qian J, An Q, Xu H, Zhao Q. *European Polymer Journal.* 2009;45(5):1403–11.
457. Xue W, Huglin MB, Liao B, Jones TG. *European Polymer Journal.* 2007;43(3):915–27.

458. Xue W, Huglin MB, Liao B. *European Polymer Journal*. 2007;43(10):4355–70.
459. Duarte-Peña L, Magaña H, Bucio E. *Pharmaceutics*. 2023;15(3):960.
460. Huang Y-Y, Hsu J-Y, Chen H-L, Hashimoto T. *Macromolecules*. 2007;40(10):3700–7.
461. Cesteros LC, Isasi JR, Katime I. *Macromolecules*. 1994;27(26):7887–92.
462. Wang J, Cheung MK, Mi Y. *Polymer*. 2001;42(7):3087–93.
463. Isasi JR, Cesteros LC, Katime I. *Macromolecular Rapid Communications*. 1994;15(11):903–8.
464. Yi J, Goh S. *Polymer*. 2001;42(22):9313–6.
465. Liu S, Pan Q, Xie J, Jiang M. *Polymer*. 2000;41(18):6919–29.
466. Lee JY, Painter PC, Coleman MM. *Macromolecules*. 1988;21(4):954–60.
467. de Ilarduya AM, Eguiburu JL, Espi E, Iruin JJ, Fernández-Berridi MJ. *Die Makromolekulare Chemie: Macromolecular Chemistry and Physics*. 1993;194(2):501–10.
468. Urakawa O, Yasue A. *Polymers*. 2019;11(7):1153.
469. Cesteros LC, Meaurio E, Katime I. *Macromolecules*. 1993;26(9):2323–30.
470. Lee L-T, Woo EM, Hou SS, Förster S. *Polymer*. 2006;47(25):8350–9.
471. Chen W-C, Kuo S-W, Lu C-H, Jeng U-S, Chang F-C. *Macromolecules*. 2009;42(10):3580–90.
472. Kuo SW. *Polymer International*. 2009;58(5):455–64.
473. Jiang M, Huang T, Xie J. *Macromolecular Chemistry and Physics*. 1995;196(3):803–13.
474. Kuo S-W, Tung P-H, Chang F-C. *European Polymer Journal*. 2009;45(7):1924–35.
475. Dobrosielska K, Wakao S, Takano A, Matsushita Y. *Macromolecules*. 2008;41(20):7695–8.
476. Dobrosielska K, Wakao S, Suzuki J, Noda K, Takano A, Matsushita Y. *Macromolecules*. 2009;42(18):7098–102.
477. Kosonen H, Ruokolainen J, Nyholm P, Ikkala O. *Polymer*. 2001;42(23):9481–6.
478. Hameed N, Guo Q. *Polymer*. 2008;49(24):5268–75.
479. Hameed N, Liu J, Guo Q. *Macromolecules*. 2008;41(20):7596–605.
480. Hameed N, Guo Q. *Polymer*. 2008;49(4):922–33.
481. Salim NV, Hameed N, Guo Q. *Journal of Polymer Science Part B: Polymer Physics*. 2009;47(19):1894–905.
482. Hameed N, Salim NV, Guo Q. *The Journal of Chemical Physics*. 2009;131(21).
483. Asari T, Matsuo S, Takano A, Matsushita Y. *Macromolecules*. 2005;38(21):8811–5.
484. Chen W-C, Kuo S-W, Chang F-C. *Polymer*. 2010;51(18):4176–84.
485. Kuo SW, Tung PH, Lai CL, Jeong KU, Chang FC. *Macromolecular Rapid Communications*. 2008;29(3):229–33.
486. Tseng T-C, Kuo S-W. *Macromolecules*. 2018;51(16):6451–9.
487. Tanaka H, Hasegawa H, Hashimoto T. *Macromolecules*. 1991;24(1):240–51.
488. Zhou Y, Firkowska-Boden I, Arras MM, Jandt KD. *Langmuir*. 2020;37(1):391–9.
489. Kwei T. *Journal of Polymer Science: Polymer Letters Edition*. 1984;22(6):307–13.
490. Makhija S, Pearce E, Kwei T, Liu F. *Polymer Engineering & Science*. 1990;30(13):798–801.
491. Pu H. *Polymer International*. 2003;52(9):1540–5.
492. Schult K, Paul D. *Journal of Polymer Science Part B: Polymer Physics*. 1997;35(6):993–1007.
493. Matsushita Y. *Macromolecules*. 2007;40(4):771–6.
494. Kapoor U, Kulshreshtha A, Jayaraman A. *Polymers*. 2020;12(11):2764.
495. Leibler L. *Macromolecules*. 1980;13(6):1602–17.
496. Pen X, Shen J. *Polymer International*. 1998;46(2):106–10.
497. Lu X, Weiss R. *Macromolecules*. 1991;24(21):5763–8.
498. Pan Y, Huang Y, Cheng M, Cong G, Leung LM. *European Polymer Journal*. 1998;34(Part I):213–7.

499. Pan Y, Huang Y, Liao B, Cong G, Leung LM. *European Polymer Journal.* 1998;34(Part II):207–11.
500. Wang Y, Qi G, Peng H, Yang S. *Polymer.* 2002;43(9):2811–8.
501. Xiang M, Jiang M, Zhang Y, Wu C, Feng L. *Macromolecules.* 1997;30(8):2313–9.
502. Xu S, Chen B, Tang T, Huang B. *Polymer.* 1999;40(12):3399–406.
503. Kerres J, Ullrich A, Meier F, Häring T. *Solid State Ionics.* 1999;125(1-4):243–9.
504. Guliyeva A, Vayer M, Warmont F, Faugère AM, Andreazza P, Takano A et al. *ACS Macro Letters.* 2018;7(7):789–94.
505. Asai Y, Takano A, Matsushita Y. *Macromolecules.* 2015;48(5):1538–42.
506. Asai Y, Yamada K, Yamada M, Takano A, Matsushita Y. *ACS Macro Letters.* 2014;3(2):166–9.
507. Mogi Y, Nomura M, Kotsuji H, Ohnishi K, Matsushita Y, Noda I. *Macromolecules.* 1994;27(23):6755–60.
508. Agnew NH. *Journal of Polymer Science: Polymer Chemistry Edition.* 1976;14(11):2819–30.
509. Lyons A, Vasile M, Pearce E, Waszczak J. *Macromolecules.* 1988;21(11):3125–34.
510. Ramos J, Millan A, Palacio F. *Polymer.* 2000;41(24):8461–4.
511. Millan A, Palacio F. *Applied Organometallic Chemistry.* 2001;15(5):396–400.
512. Bill E. *Practical Approaches to Biological Inorganic Chemistry*, 2nd edn. Elsevier, Amsterdam, 2020, 201–28.
513. Coey J, Meagher A, Kelly J, Vos J. *Journal of Polymer Science: Polymer Chemistry Edition.* 1984;22(2):303–18.
514. Zeng B YL, Zheng W, Zhu J, Ma X, Liu X, Dai, L. *Journal of Materials Science & Technology.* 2018;34(6):999–1007.
515. Wu L, Pang T, Wu L, Guan Y, Yin L. *Nanocomposites.* 2022;8(1):227–37.
516. Ren J, Wang C, Ding J, Li T, Ma Y. *ACS Applied Polymer Materials.* 2022;4(12):9449–62.
517. Wagner DR, Ament K, Mayr L, Martin T, Bloesser A, Schmalz H et al. *Environmental Science and Pollution Research.* 2021;28:6222–31.
518. Ochiai E. *The Journal of Organic Chemistry.* 1953;18(5):534–51.
519. Gregson L, Holt P. *Die Makromolekulare Chemie: Macromolecular Chemistry and Physics.* 1969;128(1):193–6.
520. Lee C-M, Pearce EM, Kwei T. *Polymer.* 1996;37(19):4283–8.
521. Okamoto Y, Kwei T, Vyprachtický D. *Macromolecules.* 1998;31(26):9201–5.
522. Raja V, Sarma A, Rao VN. *Materials Letters.* 2003;57(30):4678–83.
523. Holt P, Nasrallah E. *Journal of the Chemical Society B: Physical Organic.* 1968:233–7.
524. Holt P, Lindsay H. *Journal of the Chemical Society C: Organic.* 1969(6):1012–5.
525. Holt P, Tamami B. *Polymer.* 1973;14(12):645–8.
526. Tamikado T, Sakai T, Sagisaka K. *Die Makromolekulare Chemie: Macromolecular Chemistry and Physics.* 1961;50(1):244–52.
527. Natta G, Mazzanti G, Longi P, Dall'Asta G, Bernardini F. *Journal of Polymer Science.* 1961;51(156):487–504.
528. Geuskens G, Lubikulu J, David C. *Polymer.* 1966;7(1):63–5.
529. Alshaikh-Kadir K, Holt P. *Die Makromolekulare Chemie: Macromolecular Chemistry and Physics.* 1976;177(2):311–7.
530. Ghesquière D, Chachaty C. *Die Makromolekulare Chemie: Macromolecular Chemistry and Physics.* 1978;179(9):2305–14.
531. Alshaikh-Kadir K, Holt PF. *Die Makromolekulare Chemie: Macromolecular Chemistry and Physics.* 1977;178(2):329–35.
532. Holt P, Lindsay H. *Journal of the Chemical Society B: Physical Organic.* 1969:54–61.
533. Holt P, Tamami B. *Die Makromolekulare Chemie: Macromolecular Chemistry and Physics.* 1972;155(1):55–60.
534. Kabanov V, Aliev K, Kargin V. *Polymer Science USSR.* 1968;10(7):1873–92.

535. Iwatsuki S, Kokubo T, Motomatsu K, Tsuji M, Yamashita Y. *Die Makromolekulare Chemie: Macromolecular Chemistry and Physics*. 1968;120(1):154–60.
536. Laurin D, Parravano G. *Journal of Polymer Science Part A-1: Polymer Chemistry*. 1968;6(4):1047–9.
537. Laurin D, Parravano G. *Journal of Polymer Science Part B: Polymer Letters*. 1966;4(11):797–801.
538. Hong S, Kim M, Hong CK, Jung D, Shim SE. *Synthetic Metals*. 2008;158(21-24):900–7.
539. de Wit J, van Ekenstein GA, Polushkin E, Korhonen J, Ruokolainen J, Ten Brinke G. *Macromolecules*. 2009;42(6):2009–14.
540. Bouras B, Mansri A, Tennouga L, Grassl B. *Research on Chemical Intermediates*. 2015;41:5839–58.
541. Gao B, Lv Y, Jiu H. *Polymer international*. 2003;52(9):1468–73.
542. Medjahed K, Tennouga L, Mansri A, Baba O. *Research on Chemical Intermediates*. 2013;39:4099–107.
543. El-Hamshary H, El-Garawany M, Assubaie FN, Al-Eed M. *Journal of Applied Polymer Science*. 2003;89(9):2522–6.
544. Dannley RL, Schufle JA, Cohen I, Chambers JR. *Journal of Polymer Science*. 1956;19(92):285–96.
545. Bengough W, Henderson W. *Transactions of the Faraday Society*. 1965;61:141–9.
546. Fisher M, Szwarc M. *Macromolecules*. 1970;3(1):23–7.
547. Kandelhard F, Pashayev E, Schymura J, Georgopanos P. Industrial & *Engineering Chemistry Research*. 2023;62(22):8696–708.
548. Tsuchida E, Nishide H, Nishiyama T. *Journal of Polymer Science: Polymer Symposia*. 1974;47(1):35–46.
549. Mika A, Childs R, Dickson J, McCarry B, Gagnon D. *Journal of Membrane Science*. 1995;108(1-2):37–56.
550. Mika A, Childs R, Dickson J, McCarry B, Gagnon D. *Journal of Membrane Science*. 1997;135(1):81–92.
551. Stachera DM, Childs RF. *Journal of Membrane Science*. 2001;187(1-2):213–25.
552. Lee H-C, Fantin M, Antonietti M, Matyjaszewski K, Schmidt BV. *Chemistry of Materials*. 2017;29(21):9445–55.
553. Tate DP, Graves DF. Modified Rubber Compositions Containing Aromatic Six-Membered Heterocyclic Nitrogen-Containing Groups. Canada Patent 1257049A, 1990.
554. Meitzner EF, Oline JA, Reticular Crosslinked Monovinylidene N-Heterocyclic Copolymer. US Patent 4221871A, 1980.
555. Zhou X, Wang W, Yu H, Zhao J, Zhang Z, Zhu X. *Polymer Chemistry*. 2013;4(12):3575–81.
556. Fontanals N, Puig P, Galià M, Marcé RM, Borrull F. *Journal of Chromatography A*. 2004;1035(2):281–4.
557. Arichi S, Mitsuta S, Sakamoto N, Murata H. *Bulletin of the Chemical Society of Japan*. 1966;39(3):428–34.
558. Sambasivudu K, Maheedhar G, Srinivasa Rao V, Syam Kumar M, Shailaja D. *Journal of Applied Polymer Science*. 2006;102(1):192–7.
559. Liu X, Wu F, Au C, Tao Q, Pi M, Zhang W. *Journal of Applied Polymer Science*. 2019;136(3):46984.
560. Dayamoy B, Rao M, Joseph A, Ajay M, Inderjit S, Wattal P. *Desalination and Water Treatment*. 2012;38(1/3):254–8.
561. Shih JS, Yap ET, Hornby JC, Srinivas B, Process for Making Quaternized Vinylpyridine Carboxylate Polymers Using Suspension Polymerization. US Patent 6201046B1, 2001.
562. Li G, Shen J, Zhu Y. *Journal of Applied Polymer Science*. 2000;78(3):668–75.

563. Kumaresan R, Sabharwal K, Srinivasan T, Rao PV, Dhekane G. *Solvent Extraction and Ion Exchange*. 2007;25(4):515–28.
564. Coutinho F, Siqueira M, Barbosa C. European *Polymer Journal*. 1990;26(11):1189–93.
565. Maillard-Terrier M, Cazé C. *European Polymer Journal*. 1984;20(2):113–8.
566. Luz CL, Coutinho FM. *European Polymer Journal*. 2000;36(3):547–53.
567. Coutinho F, Luz CL. *European Polymer Journal*. 1993;29(8):1119–23.
568. Jeřábek K, Widdecke H, Fleischer B. *Reactive Polymers*. 1993;19(1-2):81–6.
569. Hamid MA, Naheed R, Fuzail M. *Polymer Journal*. 1996;28(12):1052–9.
570. de Santa Maria LC, de Aguiar AP, Aguiar MR, Jandrey AC, Guimarães PIC, Nascimento LG. *Materials Letters*. 2004;58(5):563–8.
571. Cao J, Pan X, Huang W, Wang Y, Hua D, Zhu X, Liang H. *Journal of Colloid and Interface Science*. 2012;381(1):137–42.
572. Pathak S, Greci MT, Kwong RC, Mercado K, Prakash GS, Olah GA, Thompson ME. *Chemistry of Materials*. 2000;12(7):1985–9.
573. Fan S, Luan Y, Wang J, Gao H, Zhang X, Wang G. *Journal of Molecular Catalysis A: Chemical*. 2015;404:186–92.
574. Dupin D, Fujii S, Armes SP, Reeve P, Baxter SM. *Langmuir*. 2006;22(7):3381–7.
575. Fitzgerald EB, Fuoss RM. *Industrial & Engineering Chemistry*. 1950;42(8):1603–6.
576. Siegwardt L, Glößner V, Boehm A, Schneider M, Gallei M. ACS *Applied Materials & Interfaces*. 2024;16(8):10722–35.
577. Yan N, Ning C, Liu X, Zhu Y. *Macromolecular Rapid Communications*. 2025:2401023.
578. Kim JW, Lee D, Shum HC, Weitz DA. *Advanced Materials*. 2008;20(17):3239–43.
579. Pickering SU. *Journal of Chemical Society, Transactions*. 1907;91:2001–21.
580. Atta AM, Ezzat AO, Al-Lohedan HA, Tawfeek AM, Alobaidi AA. *Nanomaterials*. 2019;9(12):1693.
581. Ni H, Ma G, Nagai M, Omi S. *Journal of Applied Polymer Science*. 2001;80(11):1988–2001.
582. Kumar S, Kim D-W, Lee H-J, Changez M, Yoon T-H, Lee J-S. *Macromolecules*. 2013;46(18):7166–72.
583. Takahashi K, Miyamori S, Uyama H, Kobayashi S. *Macromolecular Rapid Communications*. 1997;18(6):471–5.
584. Percy M, Michailidou V, Armes S, Perruchot C, Watts J, Greaves S. *Langmuir*. 2003;19(6):2072–9.
585. Ishizu K, Yasuda M, Tamura T. *Journal of Colloid and Interface Science*. 2003;267(2):320–5.
586. Han H, Hong CK, Hong J, Park DW, Shim SE. *Journal of Applied Polymer Science*. 2009;111(6):2900–7.
587. Charbonnier A, Brochon C, Cloutet E, Navarro C, Hadziioannou G. *Journal of Polymer Science Part A: Polymer Chemistry*. 2013;51(21):4608–17.
588. Matyjaszewski K. *Macromolecules*. 2012;45(10):4015–39.
589. Yang R, Wang Y, Wang X, He W, Pan C. *European Polymer Journal*. 2003;39(10):2029–33.
590. Davis KA, Matyjaszewski K. *Macromolecules*. 2001;34(7):2101–7.
591. Jaymand M. *Journal of Polymer Research*. 2011;18:1617–24.
592. Vidts KR, Du Prez FE. *European Polymer Journal*. 2006;42(1):43–50.
593. Xia J, Zhang X, Matyjaszewski K. *Macromolecules*. 1999;32(10):3531–3.
594. Huang C-F, Kuo S-W, Chen J-K, Chang F-C. *Journal of Polymer Research*. 2005;12:449–56.
595. Chavan V, Seo KS, Wesdemiotis C, Quirk RP. *Polymer Chemistry*. 2020;11(4):876–81.
596. Li L, Yan G, Cheng Z, Wu J, Yu X, Guo Q. *Surface and Interface Analysis*, 2009;41(1):69–74.
597. Chen H, Chen L, Hao Z, Fu X, Lu Z. *Journal of Macromolecular Science®, Part A: Pure and Applied Chemistry*. 2009;46(8):832–6.

598. Lewis GT, Nguyen V, Cohen Y. *Journal of Polymer Science Part A: Polymer Chemistry.* 2007;45(24):5748–58.
599. Wu H, Wan Y, Wang W, Wang Y, Zhou N, Zhang W et al. *Polymer Chemistry.* 2015;6(14):2620–5.
600. Tsarevsky NV, Braunecker WA, Brooks SJ, Matyjaszewski K. *Macromolecules.* 2006;39(20):6817–24.
601. Rutnakornpituk B, Wichai U, Vilaivan T, Rutnakornpituk M. *Journal of Nanoparticle Research.* 2011;13:6847–57.
602. Luo J, Li M, Xin M, Sun W, Xiao W. *Macromolecular Chemistry and Physics.* 2016;217(16):1777–84.
603. Carmean RN, Figg CA, Scheutz GM, Kubo T, Sumerlin BS. *ACS Macro Letters.* 2017;6(2):185–9.
604. Willcock H, O'Reilly RK. *Polymer Chemistry.* 2010;1(2):149–57.
605. Fultz BA, Beery D, Coia BM, Hanson K, Kennemur JG. *Polymer Chemistry.* 2020;11(37):5962–8.
606. Postma A, Davis TP, Evans RA, Li G, Moad G, O'Shea MS. *Macromolecules.* 2006;39(16):5293–306.
607. Chong YK, Moad G, Rizzardo E, Thang SH. *Macromolecules.* 2007;40(13):4446–55.
608. Mattson KM, Pester CW, Gutekunst WR, Hsueh AT, Discekici EH, Luo Y et al. *Macromolecules.* 2016;49(21):8162–6.
609. Discekici EH, Shankel SL, Anastasaki A, Oschmann B, Lee IH, Niu J et al. *Chemical Communications.* 2017;53(11):1888–91.
610. Lee JM, Lee BH, Choe S. *Polymer.* 2006;47(11):3838–44.
611. Nieswandt K, Georgopanos P, Abetz V. *Polymer Chemistry.* 2021;12(15):2210–21.
612. Öztürk T, Göktaş M, Savaş B, Işıklar M, Atalar MN, Hazer B. e-*Polymers.* 2014;14(1):27–34.
613. Yuan J, Ma R, Gao Q, Wang YF, Cheng SY, Feng LX et al. *Journal of Applied Polymer Science.* 2003;89(4):1017–25.
614. Jana S, Uchman M. *Progress in Polymer Science.* 2020;106:101252.
615. Convertine AJ, Sumerlin BS, Thomas DB, Lowe AB, McCormick CL. *Macromolecules.* 2003;36(13):4679–81.
616. Dan M, Huo F, Zhang X, Wang X, Zhang W. *Journal of Polymer Science Part A: Polymer Chemistry.* 2013;51(7):1573–84.
617. Božović-Vukić J, Mañon HT, Meuldijk J, Koning C, Klumperman B. *Macromolecules.* 2007;40(20):7132–9.
618. Resendiz-Lara DA, Wurm FR. *ACS Macro Letters.* 2021;10(10):1273–9.
619. Zamfir M, Patrickios CS, Montagne F, Abetz C, Abetz V, Oss-Ronen L, Talmon Y. *Journal of Polymer Science Part A: Polymer Chemistry.* 2012;50(8):1636–44.
620. Luo J, Li M, Xin M, Sun W. *Macromolecular Chemistry and Physics.* 2015;216(15):1646–52.
621. Zhu H, Mumtaz F, Zhang C, Tan L, Liu S, Zhang Y et al. *Applied Surface Science.* 2017;426:817–26.
622. Demirci S, Kinali-Demirci S, Caykara T. *Polymer.* 2013;54(20):5345–50.
623. Gilbert RG. *Emulsion Polymerization*: Academic Press, London, 1995.
624. Cao J, Li Y, Tan Y, Zhang L, *Tan Journal of Polymer Chemistry.* 2024;15(2):106–17.
625. Lansalot M, Rieger J. *Polymerization-Induced Self-Assembly.* 2019;40(2);1800885.
626. Canning SL, Smith GN, Armes SP. *Macromolecules.* 2016;49(6):1985–2001.
627. Liu C, Hong C-Y, Pan C-Y. *Polymer Chemistry.* 2020;11(22):3673–89.
628. Wen S-P, Fielding LA. *Soft Matter.* 2022;18(7):1385–94.
629. Lohmeijer BG, Schubert US. *Journal of Polymer Science Part A: Polymer Chemistry.* 2005;43(24):6331–44.
630. Fischer A, Brembilla A, Lochon P. *Macromolecules.* 1999;32(19):6069–72.

631. Baumann M, Schmidt-Naake G. *Macromolecular Chemistry and Physics.* 2000;201(18):2751–5.
632. Diaz T, Fischer A, Jonquieres A, Brembilla A, Lochon P. *Macromolecules.* 2003;36(7):2235–41.
633. Miura Y, Dote H. *Journal of Polymer Science Part A: Polymer Chemistry.* 2005;43(16):3689–700.
634. Montaudo M, Puglisi C, Battiato S, Zappia S, Destri S, Samperi F. *Journal of Applied Polymer Science.* 2019;136(3):46976.
635. Lessard BH, Marić M. e-*Polymers.* 2012;12(1):063.
636. Chen Z, Wang Y, Feng Y, Jiang X, Yang C, Wang M. *Journal of Applied Polymer Science.* 2004;91(3):1842–7.
637. Zaremski MY, Chen X, Orlova A, Blagodatskikh I, Nikonorova N. *Polymer Science Series B.* 2015;57:181–96.
638. Di Mauro AE, Toscanini M, Piovani D, Samperi F, Curri ML, Corricelli M et al. *European Polymer Journal.* 2014;60:222–34.
639. Günther M, Kuckling D, Corten C, Gerlach G, Sorber J, Suchaneck G, Arndt K-F. *Sensors and Actuators B: Chemical.* 2007;126(1):97–106.
640. Lokaj J, Holler P. *Journal of Applied Polymer Science.* 2001;80(11):2024–30.
641. Qiao X, Zhou Z, Pang X, Lansalot M, Bourgeat-Lami E. *Polymer.* 2019;172:330–8.
642. Chalari I, Pispas S, Hadjichristidis N. *Journal of Polymer Science Part A: Polymer Chemistry.* 2001;39(17):2889–95.
643. Ding XZ, Fischer A, Brembilla A, Lochon P. *Journal of Polymer Science Part A: Polymer Chemistry.* 2000;38(17):3067–73.
644. Ding XZ, Fischer A, Yang S, Wu P, Brembilla A, Lochon P. *European Polymer Journal.* 2001;37(8):1561–9.
645. Rahim NA, Audouin F, Twamley B, Vos JG, Heise A. *European Polymer Journal.* 2012;48(5):990–6.
646. Zhou YN, Chen ZC, Wei C, Luo ZH. *Macromolecular Chemistry and Physics.* 2015;216(3):329–33.
647. Shoji M, Eguchi M, Layman JM, Cashion MP, Long TE, Nishide H. *Macromolecular Chemistry and Physics.* 2009;210(7):579–84.
648. Hsieh H, Quirk RP. *Anionic Polymerization: Principles and Practical Applications*: CRC Press, Florida, 1996.
649. Creutz S, Teyssié P, Jérôme R. *Macromolecules.* 1997;30(1):1–5.
650. Lee J, Hogen-Esch TE. *Macromolecules.* 2001;34(9):2805–11.
651. Nugay N, Küçükyavuz Z, Küçükyavuz S. *Polymer International.* 1993;32(1):93–6.
652. Shunsuke T, Raita G, Takashi I, Akira H. *Macromolecules.* 2014;47:2333–9.
653. Tien C, Hogen-Esch T. *Journal of the American Chemical Society.* 1976;98(22):7109–10.
654. Tien C, Hogen-Esch T. *Macromolecules.* 1976;9(5):871–2.
655. Tien C, Hogen-Esch T. *Journal of Polymer Science: Polymer Letters Edition* 1978;16:297–302.
656. Shin YD, Han SH, Samal S, Lee JS. *Journal of Polymer Science Part A: Polymer Chemistry.* 2005;43(3):607–15.
657. Hiemenz PC, Lodge TP. Polymer *Chemistry*: CRC Press, Boca Raton, 2007.
658. Kim H, Kang B-G, Choi J, Sun Z, Yu DM, Mays J, Russell TP. *Macromolecules.* 2018;51(3):1181–8.
659. Masuda J, Takano A, Nagata Y, Noro A, Matsushita Y. *Physical Review Letters.* 2006;97(9):098301.
660. Natalello A, Morsbach J, Friedel A, Alkan A, Tonhauser C, Müller AH, Frey H. *Organic Process Research & Development.* 2014;18(11):1408–12.
661. Appold M, Rüttiger C, Kuttich B, Stühn B, Gallei M. *Macromolecular Chemistry and Physics.* 2018;219(1):1700187.

662. Manesi G-M, Moutsios I, Moschovas D, Papadopoulos G, Ntaras C, Rosenthal M et al. *Polymers*. 2023;15(21):4227.
663. Watanabe H, Amemiya T, Shimura T, Kotaka T. *Macromolecules*. 1994;27(8):2336–8.
664. Hadjichristidis N, Hirao A. *Anionic Polymerization: Principles, Practice, Strength, Consequences and Applications*: Springer, Japan, 2015.
665. Tsutsumi K, Funaki Y, Hirokawa Y, Hashimoto T. *Langmuir*. 1999;15(16):5200–3.
666. Champetier G, Fontanille M, Sigwalt P, Korn AC. *Journal of Polymer Science*. 1962;58(166):911–24.
667. Tardi M, Sigwalt P. *European Polymer Journal*. 1973;9(12):1369–79.
668. Atanase LI, Lerch JP, Caprarescu S, Iurciuc CE, Riess G. *Journal of Applied Polymer Science*. 2017;134(38):45313.
669. Sweat DP, Kim M, Larson SR, Choi JW, Choo Y, Osuji CO, Gopalan P. *Macromolecules*. 2014;47(19):6687–96.
670. Creutz S, Teyssié P, Jérôme R. *Macromolecules*. 1997;30(19):5596–601.
671. Rao PR, Masson P, Lutz P, Beinert G, Rempp P. *Polymer Bulletin*. 1984;11:115–20.
672. Yin R, Hogen-Esch TE. *Journal of Polymer Science Part A: Polymer Chemistry*. 1994;32(2):363–8.
673. Luxton A, Quig A, Delvaux M-J, Fetters L. *Polymer*. 1978;19(11):1320–4.
674. Krasnoselskaya IG, Erussalimsky BL. *Die Makromolekulare Chemie, Rapid Communications*. 1985;6(3):191–5.
675. Klein JW, Lamps J-P, Gnanou Y, Rempp P. *Polymer*. 1991;32(12):2278–82.
676. Bellas V, Iatrou H, Pitsinos EN, Hadjichristidis N. *Macromolecules*. 2001;34(16):5376–8.
677. Hückstädt H, Göpfert A, Abetz V. *Macromolecular Chemistry and Physics*. 2000;201(3):296–307.
678. Ludwigs S, Böker A, Abetz V, Müller AH, Krausch G. *Polymer*. 2003;44(22): 6815–23.
679. Rüttiger C, Gemmer L, Schöttner S, Kuttich B, Stühn B, Gallei M. *Journal of Organometallic Chemistry*. 2019;882:80–9.
680. Vanderark L, Januszewski E, Gwyther J, Manners I. *European Polymer Journal*. 2011;47(4):823–6.
681. Gentekos DT, Fors BP. *ACS Macro Letters*. 2018;7(6):677–82.
682. Ishizone T, Hirao A, Nakahama S. *Macromolecules*. 1993;26(25):6964–75.
683. Hirao A, Goseki R, Ishizone T. *Macromolecules*. 2014;47(6):1883–905.
684. Ishizone T, Kosaka Y, Goseki R. Anionic polymerization of polar vinyl monomers: Vinylpyridines, (meth)acrylates, (meth)acrylamides, (meth)acrylonitrile, phenyl vinyl sulfoxide, benzofulvene, and other monomers. In: Hadjichristidis N, Hirao A (eds.). *Anionic Polymerization*: Springer, Japan, 2015, pp. 127–89.
685. Morsbach J, Müller AH, Berger-Nicoletti E, Frey H. *Macromolecules*. 2016;49(14):5043–50.
686. Natta G, Mazzanti G, Dall'asta G, Longi P. *Die Makromolekulare Chemie: Macromolecular Chemistry and Physics*. 1960;37(1):160–2.
687. Soum A, Fontanille M. *Die Makromolekulare Chemie: Macromolecular Chemistry and Physics*. 1980;181(4):799–808.
688. Soum A, Fontanille M. *Die Makromolekulare Chemie: Macromolecular Chemistry and Physics*. 1982;183(5):1145–59.
689. Brigodiot M, Cheradame H, Fontanille M, Vairon J. *Polymer*. 1976;17(3):254–6.
690. Soum A, Fontanille M, Sigwalt P. *Journal of Polymer Science: Polymer Chemistry Edition*. 1977;15(3):659–73.
691. Matsuzaki K, Sugimoto T. *Journal of Polymer Science Part A-2: Polymer Physics*. 1967;5(6):1320–2.
692. Matsuzaki K, Kanai T, Matsubara T, Matsumoto S. *Journal of Polymer Science: Polymer Chemistry Edition*. 1976;14(6):1475–84.

693. Weill G, Hermann G. *Journal of Polymer Science Part A-2: Polymer Physics.* 1967;5(6):1293–5.
694. Natta G, Longi P, Nohdio U. *Die Makromolekulare Chemie.* 1965;83:161–6.
695. Matsuzaki K, Matsubara T, Kanai T. *Journal of Polymer Science: Polymer Chemistry Edition.* 1977;15(7):1573–83.
696. Zhang N, Salzinger S, Soller BS, Rieger B. *Journal of the American Chemical Society.* 2013;135(24):8810–3.
697. Altenbuchner PT, Soller BS, Kissling S, Bachmann T, Kronast A, Vagin SI, Rieger B. *Macromolecules.* 2014;47(22):7742–9.
698. Kaneko H, Nagae H, Tsurugi H, Mashima K. *Journal of the American Chemical Society.* 2011;133(49):19626–9.
699. Oishi M, Yoshimura R, Nomura N. *Inorganic Chemistry.* 2019;58(20):13755–60.
700. Yan C, Xu T-Q, Lu X-B. *Macromolecules.* 2018;51(6):2240–6.
701. Li M, Wang C, Chen J, Guo Z, Mou Z, Luo Y. *Dalton Ttransactions.* 2018;47(44):15967–76.
702. Mou Z, Zhuang Q, Xie H, Luo Y, Cui D. *Dalton Transactions.* 2018;47(42):14985–91.
703. Kronast A, Reiter D, Altenbuchner PT, Vagin SI, Rieger B. *Macromolecules.* 2016;49(17):6260–7.
704. Dimov DK, Hogen-Esch TE. *Macromolecules.* 1995;28(22):7394–400.
705. Wang C, Chen J, Xu W, Mou Z, Yao Y, Luo Y. *Inorganic Chemistry.* 2020;59(5):3132–41.
706. Wang Y, Jiang H, Wang H, Mou Z. *Journal of Rare Earths.* 2023;41(3):381–7.
707. Mou Z, Wang Y. *Russian Journal of Coordination Chemistry.* 2022;48(8):526–35.
708. Holt P, Lindsay H, Beck E. *British Journal of Pharmacology.* 1970;38(1):192–201.
709. Schlipköter H, Brockhaus A. *Klinische Wochenschrift.* 1961;39:1182–9.
710. Mao Y, Daniel LN, Knapton AD, Shi X, Saffiotti U. *Applied Occupational and Environmental Hygiene.* 1995;10(12):1132–7.
711. Ernst H, Rittinghausen S, Bartsch W, Creutzenberg O, Dasenbrock C, Görlitz B-D et al. *Experimental and Toxicologic Pathology.* 2002;54(2):109–26.
712. Knaapen AM, Albrecht C, Becker A, Höhr D, Winzer A, Haenen GR et al. *Carcinogenesis.* 2002;23(7):1111–20.
713. Prügger F, Mallner B, Schlipköter H. *Wien Klin Wochenschr.* 1984;96(23):848–53.
714. Zhao J, Liu J, Li G. *Zentralblatt fur Bakteriologie, Mikrobiologie und Hygiene 1 Abt Originale B, Hygiene.* 1983;178(3):259–62.
715. Gulumian M, Van Wyk A. *La Medicina del Lavoro.* 1987;78(2):124–8.
716. Chen F, Lu Y, Demers LM, Rojanasakul Y, Shi X, Vallyathan V, Castranova V. *Annals of Clinical & Laboratory Science.* 1998;28(1):1–13.
717. Goldstein S, Czapski G, Heller A. *Chemical Research in Toxicology.* 2006;19(1):86–91.
718. Liu Y-S, Della Rocca J, Schenck L, Koynov A, Sifri RJ, Winston MS, Frank DS. *Molecular Pharmaceutics.* 2024;21(3):1182–91.
719. Yao C, Li X, Neoh K, Shi Z, Kang E. *Journal of Membrane Science.* 2008;320(1-2):259–67.
720. Li G, Shen J. *Journal of Applied Polymer Science.* 2000;78(3):676–84.
721. Gao B, He S, Guo J, Wang R. *Materials Letters.* 2007;61(3):877–83.
722. Hu F, Neoh K, Cen L, Kang E. *Biotechnology and Bioengineering.* 2005;89(4):474–84.
723. Li G, Shen J, Zhu Y. *Journal of Applied Polymer Science.* 1998;67(10):1761–8.
724. Kawabata N, Kawato S. *Epidemiology & Infection.* 1998;121(2):349–56.
725. Kantouch A, El-Sayed AA. *International Journal of Biological Macromolecules.* 2008;43(5):451–5.
726. Wang C-c, Yang F-l, Liu L-F, Fu Z-m, Xue Y. *Journal of Membrane Science.* 2009;345(1-2):223–32.
727. Chauhan GS, Singh B, Dhiman SK. *Journal of Applied Polymer Science.* 2004;91(4):2454–64.
728. Sundar S, Vijayalakshmi N, Gupta S, Rajaram R, Radhakrishnan G. *Progress in Organic Coatings.* 2006;56(2–3):178–84.

729. Moor KJ, Osuji CO, Kim J-H. *ACS Applied Materials & Interfaces.* 2016;8(49):33583–91.
730. Xue Y, Xiao H. *Polymers.* 2015;7(11):2290–303.
731. Gokkaya D, Topuzogullari M, Arasoglu T, Trabzonlu K, Ozmen MM, Abdurrahmanoğlu S. *Polymer International.* 2021;70(6):829–36.
732. Nagasako T, Ogata T, Kurihara S, Nonaka T. *Journal of Applied Polymer Science.* 2010;116(5):2580–9.
733. Patel M, Patel R, Chi WS, Kim JH, Sung J-S. *Chinese Journal of Polymer Science.* 2015;33:265–74.
734. Kawabata N, Ujino I. *Reactive and Functional Polymers.* 1998;37(1-3):213–8.
735. Kawabata N, Yamazaki K, Otake T, Oishi I, Minekawa Y. *Epidemiology & Infection.* 1990;105(3):633–42.
736. Hernández-Orta M, Pérez E, Cruz-Barba LE, Sánchez-Castillo MA. *Journal of Materials Science.* 2018;53(12):8766–85.
737. Chen J, Wen X, Xu F, Xiang S, Zhao S, Fu F et al. *Journal of Applied Polymer Science.* 2023;140(19):e53836.
738. Ma X, Sun X, Li T, Wang R, He Y, Song P. *ACS Applied Nano Materials.* 2024;7(17):21083–93.
739. Gómez-Lázaro B, López-Saucedo F, Flores-Rojas GG, Camacho-Cruz A, Bucio E. *Journal of Polymer Research.* 2024;31(8):234.
740. Arakkal A, Rathinam P, Sirajunnisa P, Gopinathan H, Vengellur A, Bhat SG, Sailaja G. *Reactive and Functional Polymers.* 2022;172:105190.
741. Flores-Rojas G, Vázquez E, López-Saucedo F, Buendía-González L, Vera-Graziano R, Mendizabal E, Bucio E. *Cellulose.* 2023;30(6):3853–68.
742. Tasso M, Tuninetti JS, Pissinis DE, Moya S, Azzaroni O. *Journal of Polymer Science.* 2025;63(3):699–708.
743. Diana R, Gentile FS, Concilio S, Petrella A, Belvedere R, Schibeci M et al. *Polymers.* 2023;15(4):883.
744. Qu R, Wei W, Gu Y, Li G, Luan Y, Suo H et al. *Chemical Engineering Journal.* 2025:160467.
745. Zhang C, Maric M. *Journal of Polymer Science Part A: Polymer Chemistry.* 2012;50(20):4341–57.
746. Arizaga A, Ibarz G, Piñol R. *Journal of Colloid and Interface Science.* 2010;348(2):668–72.
747. Deka SR, Quarta A, Di Corato R, Falqui A, Manna L, Cingolani R, Pellegrino T. *Langmuir.* 2010;26(12):10315–24.
748. Hu X, Zhang Y, Xie Z, Jing X, Bellotti A, Gu Z. *Biomacromolecules.* 2017;18(3):649–73.
749. Gil ES, Hudson SM. *Progress in Polymer Science.* 2004;29(12):1173–222.
750. de Menezes RNL, Felisberti MI. *European Polymer Journal.* 2022;180:111598.
751. Iurciuc-Tincu C-E, Cretan MS, Purcar V, Popa M, Daraba OM, Atanase LI, Ochiuz L. *Polymers.* 2020;12(7):1450.
752. Brewer K, Blencowe A. *Pharmaceutics.* 2023;15(2):454.
753. Wen H, Park K. *Oral Controlled Release Formulation Design and Drug Delivery: Theory to Practice.* Wiley, New York, 2011.
754. Pawłowski Ł, Akhtar MA, Zieliński A, Boccaccini AR. *Materials Letters.* 2023;336:133885.
755. Onn S, Kamarudin NHN, Azali NS, Jaafar JA, Timmiati SN, Setiabudi HD. *Malaysian Journal of Catalysis.* 2020;4:13–18.
756. Raczkowska J, Stetsyshyn Y, Awsiuk K, Zemła J, Kostruba A, Harhay K et al. *RSC Advances.* 2016;6(90):87469–77.
757. Sidorov SN, Bronstein LM, Kabachii YA, Valetsky PM, Soo PL, Maysinger D, Eisenberg A. *Langmuir.* 2004;20(9):3543–50.

758. Liu C, Xu J, Chen H. *Journal of Inorganic and Organometallic Polymers and Materials.* 2015;25:153–8.
759. Lim Soo P, Sidorov SN, Mui J, Bronstein LM, Vali H, Eisenberg A, Maysinger D. *Langmuir.* 2007;23(9):4830–6.
760. Willott JD, Nielen WM, de Vos WM. *ACS Applied Polymer Materials.* 2019;2(2):659–67.
761. Mika AM, Childs RF, Dickson JM. *Journal of Membrane Science.* 1999;153(1):45–56.
762. Jackson EA, Hillmyer MA. *ACS Nano.* 2010;4(7):3548–53.
763. Ying L, Zhai G, Winata A, Kang E, Neoh K. *Journal of Colloid and Interface Science.* 2003;265(2):396–403.
764. Zhai G, Kang E, Neoh K. *Journal of Membrane Science.* 2003;217(1-2):243–59.
765. Zhai G, Ying L, Kang E, Neoh K. *Journal of Materials Chemistry.* 2002;12(12):3508–15.
766. Ghalehnooiy MR, Marjani A, Ghadiri M. *RSC Advances.* 2015;5(116):95994–6001.
767. Tokarev I, Orlov M, Minko S. *Advanced Materials.* 2006;18(18):2458–60.
768. Dorin RM, Phillip WA, Sai H, Werner J, Elimelech M, Wiesner U. *Polymer.* 2014;55(1):347–53.
769. Choudhury S, Ray SK. *Separation and Purification Technology.* 2021;254:117618.
770. Dessouki AM, Hegazy E-SA, El-Dessouky MM, El-Sawy NM. *Radiation Physics and Chemistry* (1977). 1985;26(2):157–63.
771. McGinty KM, Brittain WJ. *Polymer.* 2008;49(20):4350–7.
772. Neagu V, Untea I, Tudorache E, Luca C. *Reactive and Functional Polymers.* 2003;57(2-3):119–24.
773. Malik MA, Mukhtar R, Zaidi S, Ahmed S, Awan MA. *Reactive and Functional Polymers.* 2002;51(2-3):117–20.
774. Mika AM, Childs RF, West M, Lott JN. *Journal of Membrane Science.* 1997;136(1-2):221–32.
775. Luca C, Vlad CD, Bunia I. *Revue Roumaine de Chimie.* 2009;54(2):107–17.
776. Stachera DM, Childs RF, Mika AM, Dickson JM. *Journal of Membrane Science.* 1998;148(1):119–27.
777. Naidu S, Ahn H, Gong J, Kim B, Ryu DY. *Macromolecules.* 2011;44(15):6085–93.
778. Chung T-S, Shieh J-J, Lau WW, Srinivasan M, Paul D. *Journal of Membrane Science.* 1999;152(2):211–25.
779. Shieh J-J, Chung T-S, Paul D. *Chemical Engineering Science.* 1999;54(5):675–84.
780. Lai J, Wei S. *Journal of Applied Polymer Science.* 1986;32(7):5763–75.
781. Yoshikawa M, Ezaki T, Sanui K, Ogata N. *Journal of Applied Polymer Science.* 1988;35(1):145–54.
782. Kawakami M, Yamashita Y, Iwamoto M, Kagawa S. *Journal of Membrane Science.* 1984;19(3):249–58.
783. Goodwin A, Goodwin KM, Wang W, Yu Y-G, Lee J-S, Mahurin SM et al. *Macromolecules.* 2016;49(17):6213–25.
784. Liu Y, Cai L, Ma L, Li M, Yang J, Chen K, Yin P. *Nano Letters.* 2021;21(21):9021–9.
785. Shieh J-J, Chung T-S, Wang R, Srinivasan M, Paul DR. *Journal of Membrane Science.* 2001;182(1-2):111–23.
786. Jin H, An Q, Zhao Q, Qian J, Zhu M. *Journal of Membrane Science.* 2010; 347(1-2):183–92.
787. Liu T, An Q-F, Zhao Q, Lee K-R, Zhu B-K, Qian J-W, Gao C-J. *Journal of Membrane Science.* 2013;429:181–9.
788. Saleem S, Rangou S, Abetz C, Filiz V, Abetz V. *Polymers.* 2019;12(1):41.
789. DePaz S, Sengupta A, Chiao Y-H, Wickramasinghe SR. *Polymers.* 2021;13(14):2366.
790. Yang Z, Zhang M, Zhao Z, Lan W, Zhang X, Fan M. *International Journal of Hydrogen Energy.* 2022;47(37):16580–96.
791. Hong M, Wang B, Xu X, Bin P, Zhang J, Zhang Q. *Journal of Membrane Science.* 2022;662:121002.

792. Govind B, Singhal P, Tyagi A, Rattan S. *Chemical Data Collections*. 2023;44:101002.
793. Wu Q, Wang B, Zhang L. *Analytical Methods*. 2025;17(12):2577–85.
794. Zhao B, Ke D, Zhang Z, Lv R, Jin S, Su T et al. *ACS Applied Polymer Materials*. 2024;6(10):5608–17.
795. Căprărescu S, Tihan GT, Zgârian RG, Grumezescu AM, Lazau C, Bandas C et al. *Polymers*. 2025;17(4):446.
796. Rathod NH, Upadhyay P, Pal S, Kulshrestha V. *Separation and Purification Technology*. 2023;308:122895.
797. Strack G, McDonald R, Salter WB, Simpson K, Volkov D, Owens J. *Journal of Materials Science*. 2017;52:12902–12.
798. Cheng Y, Zhang H, Yuan L, Shao H, Lin X. *Journal of Membrane Science*. 2025:123813.
799. Mulder M. *Basic Principles of Membrane Technology*: Springer, Dordrecht, 2012.
800. Wandera D, Wickramasinghe SR, Husson SM. *Journal of Membrane Science*. 2010;357(1-2):6–35.
801. Qiu X, Yu H, Karunakaran M, Pradeep N, Nunes SP, Peinemann K-V. *ACS Nano*. 2013;7(1):768–76.
802. Vriezekolk EJ, Nijmeijer K, de Vos WM. *Journal of Membrane Science*. 2016;504:230–9.
803. Clodt JI, Filiz V, Rangou S, Buhr K, Abetz C, Höche D et al. *Advanced Functional Materials*. 2013;23(6):731–8.
804. Nunes SP, Car A. *Industrial & Engineering Chemistry Research*. 2013;52(3):993–1003.
805. Nunes SP, Sougrat R, Hooghan B, Anjum DH, Behzad AR, Zhao L et al. *Macromolecules*. 2010;43(19):8079–85.
806. Dorin RM, Marques Db S, Sai H, Vainio U, Phillip WA, Peinemann K-V et al. *ACS Macro Letters*. 2012;1(5):614–7.
807. Nunes SP, Karunakaran M, Pradeep N, Behzad AR, Hooghan B, Sougrat R et al. *Langmuir*. 2011;27(16):10184–90.
808. Clodt JI, Rangou S, Schröder A, Buhr K, Hahn J, Jung A et al. *Macromolecular Rapid Communications*. 2013;34(2):190–4.
809. Rodriguez AT, Li X, Wang J, Steen WA, Fan H. *Advanced Functional Materials*. 2007;17(15):2710–6.
810. Penabad-Peña L, Herrera-Morales J, Betancourt M, Nicolau E. *ACS Omega*. 2019;4(27):22456–63.
811. Herrera-Morales J, Turley TA, Betancourt-Ponce M, Nicolau E. *Materials*. 2019;12(2):230.
812. Bauer A, Hauschild S, Stolzenburg M, Förster S, Mayer C. *Chemical Physics Letters*. 2006;419(4-6):430–3.
813. Adhikari B, Majumdar S. *Progress in Polymer Science*. 2004;29(7):699–766.
814. Sartori LR, Santos Wd JR, Kubota LT, Segatelli MG, Tarley CRT. *Materials Science and Engineering*: C. 2011;31(2):114–9.
815. Wang J, Rayson GD, Lu Z, Wu H. *Analytical Chemistry*. 1990;62(18):1924–7.
816. Torrence NJ, Moore CM, Minteer SD. *Langmuir*. 2002;18(16):6254–8.
817. Pinaeva U, Lairez D, Oral O, Faber A, Clochard M, Wade T et al. *Journal of Hazardous Materials*. 2019;376:37–47.
818. Abruña HD, Denisevich P, Umana M, Meyer TJ, Murray RW. *Journal of the American Chemical Society*. 1981;103(1):1–5.
819. Denisevich P, Abruna H, Leidner C, Meyer T, Murray RW. *Inorganic Chemistry*. 1982;21(6):2153–61.
820. Ling X, Pritzker M, Burns C, Byerley J. *Macromolecules*. 1998;31(26):9134–40.
821. Tashiro K, Matsushima K, Kobayashi M. *Journal of Physical Chemistry*. 1990;94(7):3197–204.
822. McCarley RL, Thomas RE, Irene EA, Murray RW. *Journal of the Electrochemical Society*. 1990;137(5):1485.

823. Kishioka S. *Electrochemistry*. 2024;92(8):087004-.
824. Qian J, Tan R, Feng M, Shen W, Lv D, Song W. *Chemosensors*. 2024;12(11):230.
825. Hirata M, Inoue S, Yosomiya R. *Die Angewandte Makromolekulare Chemie: Applied Macromolecular Chemistry and Physics*. 1985;131(1):125–34.
826. Sakai Y, Sadaoka Y, Matsuguchi M. *Journal of the Electrochemical Society*. 1989;136(1):171.
827. Li Y, Deng C, Yang M. *Synthetic Metals*. 2012;162(1-2):205–11.
828. Li Y, Hong L, Yang M. *Talanta*. 2008;75(2):412–7.
829. Yang M, She Y, Li Y. *Journal of Materials Science Letters*. 2002;21(19):1477–9.
830. Li Y, Hong L, Chen Y, Wang H, Lu X, Yang M. *Sensors and Actuators B: Chemical*. 2007;123(1):554–9.
831. Jlassi K, Mallick S, Ali AB, Mutahir H, Salauddin SA, Ahmad Z et al. *Applied Physics A*. 2023;129(10):691.
832. Wu S, Zhu Y, Li F, Shen J. *Journal of Applied Polymer Science*. 1999;74(8):1992–6.
833. Krishnamoorthy S, Pugin R, Brugger J, Heinzelmann H, Hinderling C. *Advanced Functional Materials*. 2006;16(11):1469–75.
834. Li Y, Zhao H, Jiao M, Yang M. *Sensors and Actuators B: Chemical*. 2018;257:1118–27.
835. Zhang M, Si X, Jiang D, Yang X, Lu Y, Shen Y. *Colloid and Polymer Science*. 2017;295:2383–93.
836. Rahim S, Khalid S, Bhanger MI, Shah MR, Malik MI. *Sensors and Actuators B: Chemical*. 2018;259:878–87.
837. Ekomo VM, Branger C, Bikanga R, Florea A-M, Istamboulie G, Calas-Blanchard C et al. *Biosensors and Bioelectronics*. 2018;112:156–61.
838. Lorenzo E, Abruña HD. *Journal of Electroanalytical Chemistry*. 1992;328(1-2):111–25.
839. Ghadimi H, Tehrani RM, Basirun WJ, Ab Aziz NJ, Mohamed N, Ab Ghani S. *Journal of the Taiwan Institute of Chemical Engineers*. 2016;65:101–9.
840. Ghadimi H, Tehrani RM, Ali ASM, Mohamed N, Ab Ghani S. *Analytica Chimica Acta*. 2013;765:70–6.
841. MA Tehrani R, Ghadimi H. *Analytical and Bioanalytical Chemistry Research*. 2016;3(1):111–21.
842. e Silva CCC, Breitkreitz MC, Santhiago M, Corrêa CC, Kubota LT. *Electrochimica Acta*. 2012;71:150–8.
843. Thangamuthu R, Kumar SS, Pillai KC. *Sensors and Actuators B: Chemical*. 2007;120(2):745–53.
844. Thangamuthu R, Senthilkumar S, Chandrasekara Pillai K. *Journal of Solid State Electrochemistry*. 2007;11:126–33.
845. Skrovankova S, Mlcek J, Sochor J, Baron M, Kynicky J, Jurikova T. *International Journal of Electrochemical Science*. 2015;10(3):2421–31.
846. Cai X, Liu C, Liu H, Guo Z, Liu J. *Electroanalysis*. 2007;19(16):1689–94.
847. Malepe L, Ndinteh DT, Ndungu P, Mamo MA. *Heliyon*. 2025;11(1):e40993.
848. Oh J, Kim J-K, Gao J, Jung S, Kim W, Park G et al. *ACS Nano*. 2024;18(19):12146–57.
849. Guo W, Umar A, Algadi H, Albargi H, Ibrahim AA, Cui K et al. *Microchemical Journal*. 2021;169:106548.
850. Barroso J, Saa L, Grinyte R, Pavlov V. *Biosensors and Bioelectronics*. 2016;77:323–9.
851. Rahim S, Rauf A, Rauf S, Shah MR, Malik MI. *RSC Advances*. 2018;8(62):35776–86.
852. Inoue KY, Ino K, Shiku H, Kasai S, Yasukawa T, Mizutani F, Matsue T. *Biosensors and Bioelectronics*. 2010;25(7):1723–8.
853. Jiajia L, Azmi AS, Ali F. *Sains Malaysiana*. 2017;46:1097–102.
854. Sumitha M, Xavier T. *Hybrid Advances*. 2023;2:100023.
855. Chung Y-H, Oh JK. *Biosensors*. 2024;14(11):542.
856. Chen C, Xie Q, Yang D, Xiao H, Fu Y, Tan Y, Yao S. *RSC Advances*. 2013;3(14):4473–91.
857. Wang J. *Chemical Reviews*. 2008;108(2):814–25.

858. Csoeregi E, Schmidtke DW, Heller A. *Analytical Chemistry*. 1995;67(7):1240–4.
859. Sun J, Sun Y, Wang Z, Sun C, Wang Y, Zhang X, Shen J. *Macromolecular Chemistry and Physics*. 2001;202(1):111–6.
860. Mugweru A, Clark BL, Pishko MV. *Journal of Diabetes Science and Technology*. 2007;1(3):366–71.
861. Ahmad F, Chow FC, Ho YW, Chin YS, Christenson A, Bainbridge M et al. *Electroanalysis: An International Journal Devoted to Fundamental and Practical Aspects of Electroanalysis*. 2008;20(9):1008–15.
862. Jung S-K, Wilson GS. *Analytical Chemistry*. 1996;68(4):591–6.
863. Degani Y, Heller A. *Journal of the American Chemical Society*. 1989;111(6):2357–8.
864. Aoki A, Rajagopalan R, Heller A. *The Journal of Physical Chemistry*. 1995;99(14):5102–10.
865. Gregg BA, Heller A. *Analytical Chemistry*. 1990;62(3):258–63.
866. Tsai T-W, Heckert G, Neves Ls F, Tan Y, Kao D-Y, Harrison RG et al. *Analytical Chemistry*. 2009;81(19):7917–25.
867. Poyard S, Martelet C, Jaffrezic-Renault N, Cosnier S, Labbe P. *Sensors and Actuators B: Chemical*. 1999;58(1-3):380–3.
868. Li J, Qiu JD, Xu JJ, Chen HY, Xia XH. *Advanced Functional Materials*. 2007;17(9):1574–80.
869. Nallal M, Perumal S, Park KH, Atchudan R, Ahamed M, Lee W. *Journal of Industrial and Engineering Chemistry*. 2025:In Press.
870. Mugweru A, Rusling JF. *Electrochemistry Communications*. 2001;3(8):406–9.
871. Porfireva A, Hianik T, Evtugyn G. *Macro, Micro, and Nano-Biosensors: Potential Applications and Possible Limitations*. 2021:193–212.
872. Liu A, Anzai J-i. *Materials Science and Engineering: C*. 2004;24(4):503–5.
873. Dennany L, Forster RJ, Rusling JF. *Journal of the American Chemical Society*. 2003;125(17):5213–8.
874. Mugweru A, Rusling JF. *Analytical Chemistry*. 2002;74(16):4044–9.
875. Song B, Shen M, Jiang D, Malla S, Mosa IM, Choudhary D, Rusling JF. *Analyst*. 2016;141(20):5722–9.
876. So M, Hvastkovs EG, Schenkman JB, Rusling JF. *Biosensors and Bioelectronics*. 2007;23(4):492–8.
877. Dennany L, Forster RJ, White B, Smyth M, Rusling JF. *Journal of the American Chemical Society*. 2004;126(28):8835–41.
878. Roundhill DM, Fackler Jr JP. *Optoelectronic Properties of Inorganic Compounds*: Springer, New York, 2010.
879. Carrara S, Stringer B, Shokouhi A, Ramkissoon P, Agugiaro J, Wilson DJ et al. *ACS Applied Materials & Interfaces*. 2018;10(43):37251–7.
880. Saremi M, Amini A, Heydari H. *Microchimica Acta*. 2019;186:1–11.
881. Wu T, Zou G, Hu J, Liu S. *Chemistry of Materials*. 2009;21(16):3788–98.
882. Paek K, Yang H, Lee J, Park J, Kim BJ. *ACS Nano*. 2014;8(3):2848–56.
883. Paek K, Chung S, Cho C-H, Kim BJ. *Chemical Communications*. 2011;47(37):10272–4.
884. Maksimov YM, Podlovchenko B, Azarchenko T. *Electrochimica Acta*. 1998; 43(9):1053–9.
885. Jarjes ZA, Samian MR, Ab Ghani S. *Arabian Journal of Chemistry*. 2015;8(5):726–31.
886. Shaidarova L, Gedmina A, Budnikov G. *Journal of Analytical Chemistry*. 2003;58:171–5.
887. Lu M-C, Whang C-W. *Analytica Chimica Acta*. 2004;522(1):25–33.
888. Yang Z, Zhang M, Cao W. *Acta Physico-Chimica Sinica*. 2007;23(1):1–4.
889. Ponnurangam S, Yun CM, Chernyshova IV. *ChemElectroChem*. 2016;3(1):74–82.
890. Zhao F, Zhang J, Abe T, Wöhrle D, Kaneko M. *Journal of Molecular Catalysis A: Chemical*. 1999;145(1-2):245–56.
891. Kramer W, McCrory C. *Chemical Science*. 2016;7(4):2506–15.

892. Shaidarova L, Gedmina A, Chelnokova I, Budnikov G. *Journal of Analytical Chemistry.* 2003;58:886–91.
893. Seta-Wiaderek E, Szaniawska E, Wadas A, Rutkowska IA, Jurczakowski R, Lewera A et al. *Solar RRL.* 2021;5(6):2000705.
894. Kawabata N, Higuchi I, Yoshida J-I. *Bulletin of the Chemical Society of Japan.* 1981;54(11):3253–8.
895. Kawabata N, Ohira K. *Environmental Science & Technology.* 1979;13(11):1396–402.
896. Kawabata N, Tsuchida Y, Nakamori Y, Kitamura M. *Reactive and Functional Polymers.* 2006;66(12):1641–8.
897. Kawabata N, Yoshida J-I, Tanigawa Y. *Industrial & Engineering Chemistry Product Research and Development.* 1981;20(2):386–90.
898. Lee HJ, Xie Y, Koo YM, Wang NHL. *Biotechnology Progress.* 2004;20(1):179–92.
899. Delgado JA, Águeda VI, Uguina MÁ, García Á, Matarredona J, Moral R. *Separation and Purification Technology.* 2018;200:307–17.
900. Pingdong W, Pigui Z. *Studies in Surface Science and Catalysis.* 1993;80:729–35.
901. Dai C-M, Zhou X-F, Zhang Y-L, Liu S-G, Zhang J. *Journal of Hazardous Materials.* 2011;198:175–81.
902. Wu D-J, Xie Y, Ma Z, Wang N-H. *Industrial & Engineering Chemistry Research.* 1998;37(10):4023–35.
903. Xie Y, Farrenburg CA, Chin CY, Mun S, Wang NHL. *AIChE Journal.* 2003;49(11):2850–63.
904. Xie Y, Chin CY, Phelps DSC, Lee C-H, Lee KB, Mun S, Wang N-HL. *Industrial & Engineering Chemistry Research.* 2005;44(26):9904–20.
905. Hritzko BJ, Ortiz-Vega MJ, Wang N-HL. *Industrial & Engineering Chemistry Research.* 1999;38(7):2754–64.
906. Chanda M, McGarvey C, Rempel G. *Reactive Polymers.* 1989;10(1):79–87.
907. Chanda M, Rempel GL. *Industrial & Engineering Chemistry Research.* 1994; 33(3):623–30.
908. Chanda M, Rempel GL. *Industrial & Engineering Chemistry Research.* 1993; 32(4):726–32.
909. Chanda M, Rempel G. *Reactive Polymers.* 1995;24(3):203–12.
910. Mansri A, Ramdani N. *Research on Chemical Intermediates.* 2015;41:1765–76.
911. Prete MC, Tarley CRT. *Chemical Engineering Journal.* 2019;367:102–14.
912. Lasagabáster-Latorre A, Cela-Pérez M, Fernández-Fernández S, López-Vilariño J, González-Rodríguez M, Abad M, Barral-Losada L. *Materials Chemistry and Physics.* 2013;141(1):461–76.
913. Li Q, Pan F, Li W, Li D, Xu H, Xia D, Li A. *Polymers.* 2018;10(10):1136.
914. Maaz M, Elzein T, Dragoe D, Bejjani A, Jarroux N, Poulard C et al. *Journal of Materials Science.* 2019;54(2):1184–96.
915. Kawabata N, Oda M. *Biotechnology and Bioengineering.* 1989;34(8):1045–9.
916. Huang X, Lin J, Yuan D, Hu R. *Journal of Chromatography A.* 2009;1216(16):3508–11.
917. Yu Q-W, Wang X, Ma Q, Yuan B-F, He H-B, Feng Y-Q. *Analytical Methods.* 2011;4(6):1538–45.
918. Mansri A, Memou C, Benabadji K. *Research on Chemical Intermediates.* 2015;41:1727–40.
919. Chanda M, O'Driscoll K, Rempel G. *Reactive Polymers, Ion Exchangers, Sorbents.* 1983;1(4):281–93.
920. Arslan M, Yılmaz MN, Günay K, Ünlü N. *Polymer Bulletin.* 2023;80(3):3135–52.
921. Grochowicz M, Szajnecki Ł, Rogulska M. *Polymers.* 2022;14(10):2080.
922. Hu W, Wang L, Zhu K, Wu F, Qiao H, Li Y. *Journal of Water Process Engineering.* 2022;49:103052.

923. Valentini G, Caon NB, Faita FL, Parize AL. *Journal of the Brazilian Chemical Society.* 2023;34:167–81.
924. Lee S-H, Kang M, Lim H, Moon SY, Kim MJ, Jang SG et al. *Applied Surface Science.* 2021;555:149722.
925. Shin H, Yeverovich E, Kim KS. *Journal of Materials Research.* 2022;37(24):4483–95.
926. Alotaibi FA. *Polymer International.* 2024;73(9):695–704.
927. Alharbi HY, Aljohani MS, Monier M. *Polymer International.* 2024;73(1):50–60.
928. Tayeb H, Kamel Ismet B, Brahim B, Assia Z, Ahmed B. *Physical Chemistry Research.* 2023;11(2):327–39.
929. Coşkun R. *Polymer Bulletin.* 2011;67:125–40.
930. Medjahed K, Tennouga L, Mansri A, Chetouani A, Hammouti B, Desbrieres J. *Research on Chemical Intermediates.* 2013;39:3199–208.
931. Elkholy SS, Khalil KD, Elsabee MZ. *Journal of Applied Polymer Science.* 2006;99(6):3308–17.
932. Sridhar SP, John J, Holmqvist P, Olsson U, Chandran S, Joseph B. *Langmuir.* 2021;37(13):3996–4006.
933. Arslan M, Günay K. *Journal of Polymer Science & Applications.* 2018;2(1):1000110.
934. Martínez I, Santillán R, Fuentes Camargo I, Rodríguez JL, Andraca Adame JA, Gutiérrez HM. *Polymers.* 2022;14(21):4666.
935. Ezzat AO, Tawfeek AM, Rajabathar JR, Al-Lohedan HA. *Molecules.* 2022;27(2):441.
936. Elgamal AM, Abd El-Ghany NA, Saad GR. *Journal of Applied Polymer Science.* 2022;139(47):e53179.
937. Belkaid S, Mahroug H, Mohammed El Amine Z, Lahcene T. *Physical Chemistry Research.* 2024;12(1):61–72.
938. Atif M, He K, Irfan M, Wang Y, Hu F, Chen L, Wang Y. *Progress in Organic Coatings.* 2021;157:106307.
939. Tymetska S, Lalik S, Rysz J, Bernasik A, Brzychczy-Włoch M, Gosiewski T et al. *Applied Surface Science.* 2024;670:160639.
940. Atif M, Balasini A. *Materials Advances.* 2024;5(4):1420–39.
941. Sadok I, Grochowicz M, Krzyszczak-Turczyn A. *Scientific Reports.* 2024;14(1):18622.
942. Clara-Rahola J, Moscoso A, Ruiz-Muelle AB, Laurenti M, Formanek P, Lopez-Romero JM et al. *Journal of Colloid and Interface Science.* 2018;514:704–14.
943. Hou X, Liu B, Deng X, Zhang B, Yan J. *Journal of Biomedical Materials Research Part A.* 2007;83(2):280–9.
944. He K, Hu F, Chen C, Atif M, Wang Y. *Chinese Journal of Chemical Physics.* 2023;36(3):339–48.
945. Tapdiqov SZ. *Macromolecular Research.* 2021;29:120–8.
946. West C, Lemasson E, Nagai K, Shibata T, Franco P, Bertin S et al. *Chromatographia.* 2019;82:143–52.
947. Nagai K, Shibata T, Shinkura S, Ohnishi A. *Journal of Chromatography A.* 2018;1572:119–27.
948. Moein MM, Said R, Bassyouni F, Abdel-Rehim M. *Journal of Analytical Methods in Chemistry.* 2014;2014.
949. Escamilla-Lara KA, Heredia AC, Peña-Alvarez A, Ibarra IS, Barrado E, Rodriguez JA. *Molecules.* 2020;25(12):2924.
950. Fan Y, Zhang M, Feng Y-Q. *Journal of Chromatography A.* 2005;1099(1-2):84–91.
951. Wen Y, Zhou B-S, Xu Y, Jin S-W, Feng Y-Q. *Journal of Chromatography A.* 2006;1133(1-2):21–8.
952. Pandya PA, Shah PA, Shrivastav PS. *Biomedical Chromatography.* 2021;35(4):e5035.
953. Fan C, Quan K, Chen J, Qiu H. *Journal of Chromatography A.* 2022;1661:462690.
954. Tsuchida E, Karino Y, Nishide H, Kurimura Y. *Die Makromolekulare Chemie: Macromolecular Chemistry and Physics.* 1974;175(1):161–9.

955. Kurimura Y, Tsuchida E, Kaneko M. *Journal of Polymer Science Part A-1: Polymer Chemistry.* 1971;9(12):3511–9.
956. Kurimura Y, Yamada K, Kaneko M, Tsuchida E. *Journal of Polymer Science Part A-1: Polymer Chemistry.* 1971;9(12):3521–7.
957. Moffat A. *Journal of Catalysis.* 1970;19(3):322–9.
958. Biedermann HG, Obwandner J, Wichmann K. *Zeitschrift für Naturforschung B.* 1972;27(11):1332–5.
959. Rivas B, Maturana H, Pereira E. *Die Angewandte Makromolekulare Chemie: Applied Macromolecular Chemistry and Physics.* 1994;220(1):61–74.
960. Chanda M, Rempel G. *Reactive Polymers.* 1992;16(2):149–58.
961. Chanda M, Rempel G. *Reactive Polymers.* 1992;18(2):141–54.
962. Chanda M, Rempel G. *Reactive Polymers.* 1992;17(2):159–74.
963. Fang J, Gu Z, Gang D, Liu C, Ilton ES, Deng B. *Environmental Science & Technology.* 2007;41(13):4748–53.
964. Gang D, Banerji SK, Clevenger TE. *Practice Periodical of Hazardous, Toxic, and Radioactive Waste Management.* 2000;4(3):105–10.
965. Jumadilov T, Kondaurov R, Abilov ZA, Grazulevicius J, Akimov A. *Polymer Bulletin.* 2017;74:4701–13.
966. Lebedeva O, Sipkina E, Pozhidaev YN. *Protection of Metals and Physical Chemistry of Surfaces.* 2018;54:603–8.
967. de Oliveira TF, Ribeiro ES, Segatelli MG, Tarley CRT. *Chemical Engineering Journal.* 2013;221:275–82.
968. Shah R, Devi S. *Analyst.* 1996;121(6):807–11.
969. Soleimani M, Afshar MG. *Journal of Analytical Chemistry.* 2015;70:5–12.
970. Ehrlich H, Buslaeva T, Maryutina T. *Russian Journal of Inorganic Chemistry.* 2017;62:1797–818.
971. Tang W-Q, Mao L-H, Zhou Z-F, Wang C-F, Chen Q-L, Chen S. *Colloid and Polymer Science.* 2014;292:2529–37.
972. Kononova O, Kholmogorov A, Mikhlina E. *Hydrometallurgy.* 1998;48(1):65–72.
973. Yanovska E, Vretik L, Nikolaeva O, Polonska Y, Sternik D, Kichkiruk OY. *Nanoscale Research Letters.* 2017;12:1–6.
974. Behbahani M, Bide Y, Bagheri S, Salarian M, Omidi F, Nabid MR. *Microchimica Acta.* 2016;183:111–21.
975. Shah R, Devi S. *Analytica Chimica Acta.* 1997;341(2-3):217–24.
976. Nishide H, Deguchi J, Tsuchida E. *Chemistry Letters.* 1976;5(2):169–74.
977. Chen Y, Zhao W, Zhang J. *RSC Advances.* 2017;7(8):4226–36.
978. Maksimova YA, Dubenskiy AS, Davankov VA, Pavlova LA, Shigapov IV, Seregina IF, Bolshov MA. *Monatshefte für Chemie-Chemical Monthly.* 2020;151:1291–303.
979. Pavlova L, Kononova E, Maksimova Y, Lyubimov S, Davankov V. *Polymer Science, Series B.* 2022;64(4):470–8.
980. Pavlova L, Lyubimov S, Kononova E, Davankov V, Maksimova YA. *Polymer Science, Series B.* 2022;64(6):835–40.
981. Fredric Marsh S. *Solvent Extraction and Ion Exchange.* 1989;7(5):889–908.
982. Yavuz E, Senkal BF, Bicak N. *Reactive and Functional Polymers.* 2005;65(1-2):121–5.
983. Pribyl JG, Taylor-Pashow KM, Shehee TC, Benicewicz BC. *ACS Omega.* 2018;3(7):8181–9.
984. de Almeida FG, do Prado Ferreira M, Segatelli MG, Beal A, Spinosa WA, Cajamarca FAS, Tarley CRT. *Reactive and Functional Polymers.* 2021;164:104934.
985. Chen G, Wang Y, Weng H, Wu Z, He K, Zhang P et al. *ACS Applied Materials & Interfaces.* 2019;11(27):24560–70.
986. Tarley CRT, Diniz KM, Suquila FAC, Segatelli MG. *RSC Advances.* 2017;7(31):19296–304.

987. Cajamarca Suquila FA, Bertoldo LA, Lins E, Tarley CRT. *Separations*. 2023;10(2):122.
988. Tovstenko-Zabelin M, Vretik L, Yanovska E, Kondratenko O, Sternik D, Nikolaeva O, Savchenko I. *Molecular Crystals and Liquid Crystals*. 2023;752(1):8–22.
989. Cao L, Hu Y, Zhang L, Ma C, Wang X, Wang J. *The Journal of Supercritical Fluids*. 2011;58(2):233–8.
990. Liebenberg KE, Ambushe AA, Zinyemba O. *Minerals*. 2021;11(8):884.
991. García-Vargas M, Munguía-Quintero MF, Alcaraz-Cienfuegos J, Rosas-Aburto A, de los Ángeles Valdivia-López M, Hernández-Luna MG, Vivaldo-Lima E. *International Journal of Biological Macromolecules*. 2025;284:137986.
992. Sun XD, Song J, Duan Z, Feng D, Tian Z, Gao D. *Langmuir*. 2023;39(49):18011–21.
993. Zheng R, Mao G, Fu Z, Zhang X, Yu S, Yue Y, Li J. *Journal of Radioanalytical and Nuclear Chemistry*. 2023;332(10):3903–10.
994. Moridi H, Behroozikhah H, Talebi M, Mousavi SE, Abbasizadeh S. *Environmental Science and Pollution Research*. 2025;32(11):6628–57.
995. Haryono A, Binder WH. *Small*. 2006;2(5):600–11.
996. Dement'eva O, Skryleva E, Zaitseva A, Rudoy V. *Colloid Journal*. 2009;71:754–63.
997. Derikov Y, Abetz C, Shandryuk G, Talroze R, Ezhov A, Abetz V et al. *Polymer Science, Series C*. 2018;60:78–85.
998. Dong H, Fey E, Gandelman A, Jones WE. *Chemistry of Materials*. 2006;18(8):2008–11.
999. Kunz MS, Shull KR, Kellock AJ. *Journal of Colloid and Interface Science*. 1993;156(1):240–9.
1000. Shull KR, Kellock AJ. *Journal of Polymer Science Part B: Polymer Physics*. 1995;33(9):1417–22.
1001. Rahim S, Ali SA, Ahmed F, Imran M, Shah MR, Malik MI. *Journal of Nanoparticle Research*. 2017;19:1–11.
1002. Spatz JP, Sheiko S, Möller M. *Macromolecules*. 1996;29(9):3220–6.
1003. Ramírez O, Leal M, Briones X, Urzúa M, Bonardd S, Saldías C, Leiva A. *Polymers*. 2022;14(18):3782.
1004. Kim YC, Hoang S, Winey KI, Composto RJ. *ACS Applied Materials & Interfaces*. 2024;16(44):61083–95.
1005. Ju J, Hayward RC. *ACS Applied Materials & Interfaces*. 2023;15(42):49633–41.
1006. Zhu H, Masson J-F, Bazuin CG. *ACS Applied Nano Materials*. 2024;7(5):5637–51.
1007. Bulut Y, Sochor B, Reck KA, Schummer B, Meinhardt A, Drewes J et al. *Langmuir*. 2024;40(43):22591–601.
1008. Pooja D, Panyaram S, Kulhari H, Reddy B, Rachamalla SS, Sistla R. *International Journal of Biological Macromolecules*. 2015;80:48–56.
1009. Uson R. *A Review of: The Chemistry of Gold*. In: Clark RJH (ed.). Oxford, New York; 1978. p. 503–504
1010. Johnston BD, Kreyling WG, Pfeiffer C, Schäffler M, Sarioglu H, Ristig S et al. *Advanced Functional Materials*. 2017;27(42):1701956.
1011. Rossner C. *Macromolecular Rapid Communications*. 2025:2400851.
1012. Shan J, Tenhu H. *Chemical Communications*. 2007(44):4580–98.
1013. Lu C-H, Huang C-F, Kuo S-W, Chang F-C. *Macromolecules*. 2009;42(4):1067–78.
1014. Moessmer S, Spatz JP, Moeller M, Aberle T, Schmidt J, Burchard W. *Macromolecules*. 2000;33(13):4791–8.
1015. Liu X, Yan N, Jin J, Du Y, Jiang W. *Nano Letters*. 2023;23(17):8022–8.
1016. Derikov YI, Yablanovich A, Litmanovich E, Amarantov S, Chernikova E, Kudryavtsev YV. *Polymer Science, Series B*. 2021;63(6):882–93.
1017. Moutsios I, Manesi G-M, Tselekidou D, Maryasevskaya AV, Rosenthal M, Gioti M et al. *European Polymer Journal*. 2024;203:112668.
1018. Youk JH, Park M-K, Locklin J, Advincula R, Yang J, Mays J. *Langmuir*. 2002;18(7):2455–8.

1019. Liu Y, Cui X, Lee Y-I, Liu H-G. *Langmuir.* 2022;38(15):4589–98.
1020. Popescu M-T, Tsitsilianis C. *Molecules.* 2022;27(12):3764.
1021. Dennany L, Hogan CF, Keyes TE, Forster RJ. *Analytical Chemistry.* 2006;78(5):1412–7.
1022. Bum P, Holmes A, Kraft A, Bradley D, Brown A, Friend R, Gymer R. *Nature.* 1992;356:47–9.
1023. Tessler N, Medvedev V, Kazes M, Kan S, Banin U. *Science.* 2002;295(5559):1506–8.
1024. Cassidy JF, Vos JG. *Journal of the Electrochemical Society.* 1988;135(4):863.
1025. Haas O, Vos JG. *Journal of Electroanalytical Chemistry and Interfacial Electrochemistry.* 1980;113(1):139–49.
1026. Nishiyama S, Tajima M, Nakajima Y, Hanajiri T, Yoshida Y. *Japanese Journal of Applied Physics.* 2008;47(1S):432.
1027. Zeb G, Tri PN, Le XT. *Nanomaterials-Based Coatings.* 2019:119–38.
1028. Zeb G, Tri PN, Palacin S, Le XT. *RSC Advances.* 2016;6(84):80825–9.
1029. Sharma S, Sudhakara P, Omran AAB, Singh J, Ilyas R. *Polymers.* 2021;13(17):2898.
1030. Scharber MC, Sariciftci NS. *Advanced Materials Technologies.* 2021;6(4):2000857.
1031. Belanger D, Wrighton MS. *Analytical Chemistry.* 1987;59(10):1426–32.
1032. Cruz MV, Cyr M, Boughanmi H, García-Calvo J, Brusso J, Torres T, Lessard BH. *Polymer Chemistry.* 2024;15(34):3475–9.
1033. Lee KI. *Journal of Applied Polymer Science.* 1994;51(10):1781–9.
1034. Li X, Xu M, Zhang R, Yu Y, Pei Y, Yu Q et al. *ChemRxiv.* 2022. doi: 10.26434/chemrxiv-2022-d8bfz.
1035. Nishiyama S, Tajima M, Yoshida Y. *Journal of Photopolymer Science and Technology.* 2005;18(6):683–9.
1036. Achira H, Nakamura M, Shidzuma M, Watase S. *European Journal of Inorganic Chemistry.* 2022;2022(36):e202200544.
1037. Zhang M, Rene-Boisneuf L, Hu Y, Moozeh K, Hassan Y, Scholes G, Winnik MA. *Journal of Materials Chemistry.* 2011;21(26):9692–701.
1038. Mustapha K, Metwalli KM. *European Polymer Journal.* 2021;156:110591.
1039. Ebe H, Chiba T, Pu Y-J, Kido J. Solution-processed organic LEDs and perovskite LEDs. In: Ogawa S (ed.). *Organic Electronics Materials and Devices.* Springer, Tokyo, 2024, pp. 239–82.
1040. Pandey P. *Soft Nanoscience Letters.* 2015;5(1), 3–11.
1041. Cummins C, Morris MA. *Microelectronic Engineering.* 2018;195:74–85.
1042. Xue F, Li H, Jiang S. *RSC Advances.* 2017;7(53):33355–63.
1043. Hulkkonen H, Sah A, Niemi T. *ACS Applied Materials & Interfaces.* 2018;10(49):42941–7.
1044. Bräutigam M, Weyell P, Rudolph T, Dellith J, Krieck S, Schmalz H et al. *Journal of Materials Chemistry A.* 2014;2(17):6158–66.
1045. Yang GG, Choi HJ, Han KH, Kim JH, Lee CW, Jung EI et al. *ACS Applied Materials & Interfaces.* 2022;14(10):12011–37.
1046. Gross JH. *European Journal of Mass Spectrometry.* 2021;27(5):191–204.
1047. Lim H, Lee Y, Han S, Yoo Y, Kim K-J. *Bulletin of the Korean Chemical Society.* 1999;20(7):853–6.
1048. Corten C, Kretschmer K, Kuckling D. *Beilstein Journal of Organic Chemistry.* 2010;6(1):756–65.
1049. Chendo C, Phan TN, Marque SR, Gigmes D, Charles L. *International Journal of Mass Spectrometry.* 2016;405:50–8.
1050. Shin DO, Lee DH, Moon HS, Jeong SJ, Kim JY, Mun JH et al. *Advanced Functional Materials.* 2011;21(2):250–4.
1051. Hou P, Han W, Philippi M, Schäfer H, Steinhart M. *ACS Applied Nano Materials.* 2018;1(3):1413–9.
1052. Esmeraldo Paiva A, Baez Vasquez JF, Selkirk A, Prochukhan N, GL Medeiros Borsagli F, Morris M. *ACS Applied Materials & Interfaces.* 2022;14(30):35265–75.

1053. Qiu M, Du W, Zhou S, Cai P, Luo Y, Wang X et al. *Progress in Polymer Science.* 2023;142:101688.
1054. Kao K-C, Nishi H, Tatsuma T. *Physical Chemistry Chemical Physics.* 2018;20(5):3735–40.
1055. Yin S, Tian T, Weindl CL, Wienhold KS, Ji Q, Cheng Y et al. *ACS Applied Materials & Interfaces.* 2022;14(2):3143–55.
1056. Badila M, Brochon C, Hebraud A, Hadziioannou G. *Polymer.* 2008;49(21):4529–33.
1057. Hong H, Sfez R, Vaganova E, Yitzchaik S, Davidov D. *Thin Solid Films.* 2000;366(1-2):260–4.
1058. Panunzi B, Diana R, Caruso U. *Polymers.* 2019;11(10):1712.
1059. Li M, Feng S, Fang S, Xiao X, Li X, Zhou X, Lin Y. *Electrochimica Acta.* 2007;52(14):4858–63.
1060. Singh PK, Bhattacharya B, Nagarale R, Pandey S, Kim K-W, Rhee H-W. *Synthetic Metals.* 2010;160(9-10):950–4.
1061. Masud, Kim YS, Kim HK. *ACS Applied Energy Materials.* 2024;7(12):5226–34.
1062. Zappia S, Di Mauro AE, Mastria R, Rizzo A, Curri ML, Striccoli M, Destri S. *European Polymer Journal.* 2016;78:352–63.
1063. Holmes C. *ECS Transactions.* 2007;6(5):1.
1064. Kim Y, Song S, Liaw D-J, Huang Y-C, Ko Y-G, Ihm K et al. *ACS Omega.* 2018;3(10):13036–44.
1065. Yan Z-L, Liang F-C, Yeh C-Y, Kurniawan D, Benas J-S, Chen W-C et al. *Chemical Engineering Journal.* 2022;444:136518.
1066. Barzic AI. *Polymers for Advanced Technologies.* 2022;33(4):1116–25.
1067. Mao Y, Chen S, Wang G, Zhang H, Zhang Y, Chen X, Wang J. *Physica Status Solidi (a).* 2022;219(16):2100795.
1068. Jiang Q, Ren Y, Cui Z, Li Z, Hu L, Guo R et al. *ACS Applied Nano Materials.* 2023;6(10):8655–67.
1069. Huo N, Rivkin J, Jia R, Zhao Y, Tenhaeff WE. *Advanced Optical Materials.* 2024;12(10):2302201.
1070. Rodrigues DF, Martins J, Sauvage F, Abreu CM, Coelho JF, Serra AC et al. *Surfaces and Interfaces.* 2024;44:103627.
1071. Rodrigues DF, Abreu CM, Sauvage F, Coelho JF, Serra AC, Ivanou D, Mendes A. *ACS Applied Polymer Materials.* 2024;6(15):8939–49.
1072. Borbone F, Caruso U, De Maria A, Fusco M, Panunzi B, Roviello A, *Macromolecular Symposia.* 2004;218(1);313–322.
1073. Elim HI, Ji W, Meng GC, Ouyang J, Goh SH. *Journal of Nonlinear Optical Physics & Materials.* 2003;12(02):175–86.
1074. Caruso U, Centore R, Panunzi B, Roviello A, Tuzi A. *European Journal of Inorganic Chemistry.* 2005;2005(13):2747–2753.
1075. Bera J, Betal A, Sharma A, Shankar U, Rath AK, Sahu S. *ACS Applied Nano Materials.* 2022;5(6):8502–10.
1076. Torun I, Huang C, Kiremitler NB, Kalay M, Shim M, Onses MS. *Small.* 2024;20(45):2405429.
1077. Vaganova E, Yitzchaik S. *Acta Polymerica.* 1998;49(10-11):636–41.
1078. Hong H, Sfez R, Yitzchaik S, Davidov D. *Synthetic Metals.* 1999;102(1-3):1217.
1079. Hagfeldt A, Boschloo G, Sun L, Kloo L, Pettersson H. *Chemical Reviews.* 2010;110(11):6595–663.
1080. Rodrigues DF, Santos F, Abreu CM, Coelho JF, Serra AC, Ivanou D, Mendes A. *ACS Sustainable Chemistry & Engineering.* 2021;9(17):5981–90.
1081. Murai S, Mikoshiba S, Hayase S. Solar *Energy Materials and Solar Cells.* 2007;91(18):1707–12.

1082. Sakaguchi S, Ueki H, Kato T, Kado T, Shiratuchi R, Takashima W et al. *Journal of Photochemistry and Photobiology A: Chemistry*. 2004;164(1-3):117–22.
1083. Wu J, Hao S, Lan Z, Lin J, Huang M, Huang Y et al. *Journal of the American Chemical Society*. 2008;130(35):11568–9.
1084. Chaudhary B, Kulkarni A, Jena AK, Ikegami M, Udagawa Y, Kunugita H et al. *ChemSusChem*. 2017;10(11):2473–9.
1085. Yavari M, Mazloum-Ardakani M, Gholipour S, Tavakoli MM, Taghavinia N, Hagfeldt A, Tress W. *ACS Omega*. 2018;3(5):5038–43.
1086. Kim N, *Macromolecular Symposia*. 1996;101(1):235–242.
1087. Cuetara-Guadarrama F, Vonlanthen M, Sorroza-Martinez K, Gonzalez-Mendez I, Rivera E. *Dyes and Pigments*. 2021;194:109551.
1088. Messier J, Prasad P, Ulrich D. *Nonlinear Optical Effects in Organic Polymers*: Springer, Dordrecht, 2012.
1089. Prasad PN, Williams DJ. *Introduction to Nonlinear Optical Effects in Molecules and* Polymers: Wiley, New York, 1991.
1090. Caruso U, De Maria A, Panunzi B, Roviello A. *Journal of Polymer Science Part A: Polymer Chemistry*. 2002;40(17):2987–93.
1091. Jin Z, Sun X, Xu G, Goh SH, Ji W. *Chemical Physics Letters*. 2000;318(6):505–10.
1092. Jaafar AH, Gee A, Kemp N. *Journal of Physics D: Applied Physics*. 2023;56(50):503002.
1093. Leong WL, Lee PS, Lohani A, Lam YM, Chen T, Zhang S et al. *Advanced Materials*. 2008;20(12):2325–31.
1094. Huang C-M, Liu Y-S, Chen C-C, Wei K-H, Sheu J-T. *Applied Physics Letters*. 2008;93(20).
1095. Kang NG, Cho B, Kang BG, Song S, Lee T, Lee JS. *Advanced Materials*. 2012;24(3):385–90.
1096. Betal A, Bera J, Sharma A, Rath AK, Sahu S. *ACS Applied Electronic Materials*. 2022;4(3):1109–16.
1097. Pradhan RR, Bera J, Betal A, Dagar P, Sahu S. ACS *Applied Materials & Interfaces*. 2021;13(21):25064–71.
1098. Vaganova E, Berestetsky N, Yitzchaik S, Goldberg A. *Molecular Simulation*. 2008;34(10-15):981–7.
1099. Vaganova E, Rozenberg M, Dubnikova F, Danovich D, Yitzchaik S. *New Journal of Chemistry*. 2015;39(8):5920–2.
1100. Berestetsky N, Vaganova E, Wachtel E, Leitus G, Goldberg A, Yitzchaik S. *The Journal of Physical Chemistry B*. 2008;112(12):3662–7.
1101. Rozenberg M, Vaganova E, Yitzchaik S. *New Journal of Chemistry*. 2000;24(3):109–11.
1102. Vaganova E, Damm C, Israel G, Yitzchaik S. *Journal of Fluorescence*. 2002;12:219–24.
1103. Vaganova E, Dubnikov F, Kesselman E, Lokshin V, Khodorkovsky V. *Macromolecular Symposia*. 2019;385(1):1800157.
1104. Vaganova E, Eliaz D, Shimanovich U, Leitus G, Aqad E, Lokshin V, Khodorkovsky V. *Molecules*. 2021;26(22):6925.
1105. Vaganova E, Reisfeld R, Yitzchaik S. *Optical Materials*. 2003;24(1-2):69–76.
1106. Tan K, Tan B, Kang E, Neoh K. *Journal of Applied Physics*. 1989;66(12):5868–71.
1107. Kang E, Ong Y. *Solid State Communications*. 1986;57(8):587–90.
1108. Aydogdu Y, Erol I, Yakuphanoğlu F, Aydogdu A, Ahmedzade M. *Synthetic Metals*. 2003;139(2):327–34.
1109. Nugay N, Küçükyavuz Z, Küçükyavuz S. *Polymer*. 1993;34(22):4649–54.
1110. Lupinski J, Kopple K, Hertz J. *Journal of Polymer Science Part C: Polymer Symposia*. 1967;16(3):1561–78.
1111. Han J, Tiernan E, Lee T, Chiu A, McGuiggan P, Adams N et al. *Advanced Materials*. 2022;34(27):2201062.

1112. Audenaert M, Gusman G, Mehbod M, Deltour R, Noirhomme B, Vander Donckt E. *Solid State Communications*. 1979;30(12):791–3.
1113. Kronick P. *The Journal of Physical Chemistry*. 1965;69(9):3178–9.
1114. Möller M, Lenz RW. *Die Makromolekulare Chemie: Macromolecular Chemistry and Physics*. 1989;190(5):1153–68.
1115. Perrier G, Dao L. *Journal of Applied Polymer Science*. 1988;36(6):1235–43.
1116. Chetri P, Dass NN, Sarma NS. *Polymer Bulletin*. 2008;61:11–7.
1117. Aronson S, Wilensky S. *Journal of Polymer Science Part A: Polymer Chemistry*. 1988;26(4):1259–62.
1118. Mainthia S, Kronick P, Labes M. *The Journal of Chemical Physics*. 1964;41(7):2206–7.
1119. Streinz C, Steckenrider J, Wagner J, Moran P. *Journal of the Electrochemical Society*. 1989;136(10):2811.
1120. Phillips L, Kelly R, Wagner J, Moran P. *Journal of the Electrochemical Society*. 1986;133(1):1.
1121. Rao RV, Ashokan P, Shridhar M. *Materials Chemistry and Physics*. 2000;64(3):222–8.
1122. Khanna VK. *Implantable Medical Electronics: Prosthetics, Drug Delivery, and Health Monitoring*: Springer, Cham, 2016, pp. 167–83.
1123. Zhai P, Liu K, Wang Z, Shi L, Yuan S. *Journal of Power Sources*. 2021;499:229973.
1124. Zhang X, Jia C, Zhang J, Zhang L, Liu X. *Advanced Science*. 2024;11(2):2305201.
1125. Yu H, Wang Z, Zheng R, Yan L, Zhang L, Shu J. *Angewandte Chemie*. 2023;135(46):e202308397.
1126. Shin I, Lee K, Kim E, Kim TH. *Bulletin of the Korean Chemical Society*. 2018;39(9):1058–65.
1127. Shin I, Nam J, Lee K, Kim E, Kim T-H. *Polymer Chemistry*. 2018;9(42):5190–9.
1128. Banerjee A, Ziv B, Shilina Y, Luski S, Aurbach D, Halalay IC. *ACS Energy Letters*. 2017;2(10):2388–93.
1129. Naoi K, Iwamizu Y, Mori M, Naruoka Y. *Journal of the Electrochemical Society*. 1997;144(4):1185.
1130. Huo N, Tenhaeff WE. *Macromolecules*. 2023;56(5):2113–22.
1131. Rawashdeh-Omary MA, Lopez-de-Luzuriaga JM, Rashdan MD, Elbjeirami O, Monge M, Rodriguez-Castillo M, Laguna A. *Journal of the American Chemical Society*. 2009;131(11):3824–5.
1132. Hobbollahi E, List M, Monkowius U. *Monatshefte für Chemie-Chemical Monthly*. 2019;150:877–83.
1133. Lv XC, Lu L, Zuo XX, Achalkumar AS, Zhao AJ, Bermeshev MV et al. *Chemistry–A European Journal*. 2023;29(20):e202204060.
1134. Basu R. *Applied Energy*. 2024;375:124127.
1135. Liu S, Huo B, Guo C-Y. *Materials*. 2024;17(14):3524.
1136. Rahman MM, Karim MR, Alharbi HF, Aldokhayel B, Uzzaman T, Zahir H. *Chemistry–An Asian Journal*. 2021;16(8):902–21.
1137. Noro A, Higuchi K, Sageshima Y, Matsushita Y. *Macromolecules*. 2012;45(19):8013–20.
1138. Wang M, Kumar S, Lee A, Felorzabihi N, Shen L, Zhao F et al. *Journal of the American Chemical Society*. 2008;130(29):9481–91.
1139. Zhang M, Wang M, He S, Qian J, Saffari A, Lee A et al. *Macromolecules*. 2010;43(11):5066–74.
1140. Umoren S. *The Open Corrosion Journal*. 2009;2(1):175–88.
1141. Chetouani A, Azzaoui K, Aouniti A. *Journal of Applied Science and Environmental Studies*. 2022;5(4)200-218.
1142. Goni LK, Mazumder MAJ, Ali SA, Chauhan DS. Water-soluble polymeric corrosion inhibitors. In: Mazumder MJ, Quraishi M, Al-Ahmed A (eds.). *Polymeric Corrosion Inhibitors for Greening the Chemical and Petrochemical Industry*: Wiley, New York, 2022, pp. 97–123.

1143. Katić J, Nikolić J, Juračić T, Klačić T, Namjesnik D, Begović T. *Journal of Materials Chemistry C*. 2024;12(42):17241–53.
1144. Annand R, Hurd R, Hackerman N. *Journal of the Electrochemical Society*. 1965;112(2):138.
1145. Abed Y, Arrar Z, Hammouti B, Taleb M, Kertit S, Mansri A. *Anti-Corrosion Methods and Materials*. 2001;48(5):304–8.
1146. Mansri A, Bouras B, Hammouti B, Warad I, Chetouani A. *Research on Chemical Intermediates*. 2013;39:1753–70.
1147. Larabi L, Harek Y, Traisnel M, Mansri A. *Journal of Applied Electrochemistry*. 2004;34:833–9.
1148. Mostafa AE- KB, Abdel-Wahaab S, Mabrouk E-SM. *Surface and Coatings Technology*. 1986;27(4):317–24.
1149. Khalifa O, Hamid IA, Mokhtar S, Kassab A. *Asian Journal of Chemistry*. 1993;5(3):749.
1150. Chetouani A, Medjahed K, Benabadji K, Hammouti B, Kertit S, Mansri A. *Progress in Organic Coatings*. 2003;46(4):312–6.
1151. Belkaid S, Tebbji K, Mansri A, Chetouani A, Hammouti B. *Research on Chemical Intermediates*. 2012;38:2309–25.
1152. El Ayyoubi S, Chetouani A, Warthan A, Mansri A, Al-Deyab S. *International Journal of Electrochemical Science*. 2012;7(2):1639–55.
1153. Chetouani A, Medjahed K, Sid-Lakhdar K, Hammouti B, Benkaddour M, Mansri A. *Corrosion Science*. 2004;46(10):2421–30.
1154. Das C, Kastania E, Witt J, Ozcan O. *Materials and Corrosion*. 2022;73(3):427–35.
1155. Kumari S, Panigrahi A, Singh S, Pradhan SK. *Journal of Coatings Technology and Research*. 2018;15:583–92.
1156. Mosavian SY, Hamidi Z, Sabbaghi N, Shahabi M, Noroozifar M, Karimi Zarchi MA, Raissi H. *Polymer Bulletin*. 2023;80(7):7569–98.
1157. Ilim I, Jefferson A, Simanjuntak W, Jeannin M, Syah YM, Bundjali B, Buchari B. *Indonesian Journal of Chemistry*. 2016;16(2):198–207.
1158. Ilim I, Bahri S, Simanjuntak W, Syah Y, Bundjali B, Buchari B. *Journal of Materials and Environmental Sciences*. 2017;8(7):2381–2390.
1159. Ilim I, Fitriani R, Prabowo T, Bahri S, Marjunus R, Simanjuntak W. *International Journal of Corrosion and Scale Inhibition*. 2021;10(1):284–301.
1160. Ilim I, Bahri S, Marjunus R, Simanjuntak W. *Journal of Physics: Conference Series*. 2021;1751:012094.
1161. Abed Y, Hammouti B, Touhami F, Aouniti A, Kertit S, Mansri A, Elkacemi K. *Bulletin of Electrochemistry*. 2001;17(3):105–10.
1162. Abdallah M, Soliman KA, Al-Gorair AS, Felaly R, Al-Fahemi JH, Al-Juaid SS et al. *Journal of Dispersion Science and Technology*. 2024:1–17.
1163. Duivenvoorde F, Van Nostrum C, Van Der Linde R. *Progress in Organic Coatings*. 1999;36(4):225–30.
1164. Friberg S. *Journal of Dispersion Science and Technology*. 1985;6(4):497-.
1165. Creutz S, Jérôme R. e-*Polymers*. 2001;1(1):014.
1166. van Den Haak HJW, Schöps M, Adler HJP, Fietzek C, Montagut J. *Macromolecular Symposia*. 1998;126(1):253–65.
1167. Yuan J XJ. Materials. 2000;17(2):330.
1168. Creutz S, Jérôme R. *Langmuir*. 1999;15(21):7145–56.
1169. Duivenvoorde FL, van Nostrum CF, Laven J, van der Linde R. *Journal of Coatings Technology*. 2000;72:145–52.
1170. Duivenvoorde FL, Laven J, van der Linde R. *Progress in Organic Coatings*. 2002;45(2-3):127–37.

1171. Duivenvoorde FL. Pigment dispersing in powder coatings: Synthesis and use of block copolymer dispersing agents. Ph. D. Thesis, Chemical Engineering and Chemistry, Technische Universiteit Eindhoven, 2000.
1172. Duivenvoorde FL, Van Es JS, Van Nostrum CF, Van Der Linde R. *Macromolecular Chemistry and Physics*. 2000;201(6):656–61.
1173. Duivenvoorde F, Jansen K, Laven J, Van der Linde R. *Journal of Coatings Technology*. 2002;74:49–57.
1174. Vallejos S, Trigo-López M, Arnaiz A, Miguel Á, Muñoz A, Mendía A, García JM. *Polymers*. 2022;14(22):4954.
1175. Satchanska G, Davidova S, Petrov PD. *Polymers*. 2024;16(8):1159.
1176. Arslan M, Yiğitoğlu M, Şanlı O, Ünal HI. *Polymer Bulletin*. 2003;51:237–44.
1177. Ünal HI, Şanli O. *Journal of Applied Polymer Science*. 1996;62(8):1161–5.
1178. Yiğitoğlu M, Arslan M. *Polymer Bulletin*. 2007;58:785–98.
1179. Chakraborty S, Kar S, Dasgupta S, Mukhopadhyay R, Chauhan NP, Ameta SC, Bandyopadhyay S. *Polymer Engineering & Science*. 2011;51(8):1675–81.
1180. Chakraborty S, Sajith P, Agrawal S, Bandyopadhyay S, Mukhopadhyay R, Deuri A. *Journal of Elastomers & Plastics*. 2006;38(3):249–60.
1181. Ciullo PA, Hewitt N. *The Rubber Formulary*. William Andrew Inc, Norwich, 1999.
1182. Rinne W, Rose J. *Industrial & Engineering Chemistry*. 1948;40(8):1437–40.
1183. Mao Y, Wen S, Chen Y, Zhang F, Panine P, Chan TW et al. *Scientific Reports*. 2013;3(1):2508.
1184. Tang Z, Wu X, Guo B, Zhang L, Jia D. *Journal of Materials Chemistry*. 2012;22(15):7492–501.
1185. Hu R, Zhao S, Chen F, Shangguan Y, Zheng Q. *Journal of Polymer Science*. 2022;60(15):E1–E12.
1186. Vyas MK, Schneider K, Nandan B, Stamm M. *Soft Matter*. 2008;4(5):1024–32.
1187. Rathinamoorthy R. *Fibers and Polymers*. 2019;20:1218–25.
1188. Shih JS, Yap ET, Hornby JC, Srinivas B, Product and Process for Making Quaternized, Water Soluble Vinylpyridine Carboxylate Polymers. US Patent 6271386B1; 2001.
1189. Fredj A, Johnston JP, Thoen CA, Dye Transfer Inhibiting Compositions Comprising Polymeric Dispersing Agents. US Patent 5470507A, 1995.
1190. Shih JS, Srinivas B, Hornby JC, Water Soluble Dye Complexing Polymers. US Patent 5929175A, 1999.
1191. Oakes J, Dixon S. *Coloration Technology*. 2003;119(3):140–9.
1192. Oakes J, Gratton PL, Paul PK. *Coloration Technology*. 2003;119(3):150–7.
1193. Beneito-Cambra M, Herrero-Martínez JM, Ramis-Ramos G. *Electrophoresis*. 2008;29(15):3245–52.
1194. Binks BP, Murakami R, Armes SP, Fujii S, Schmid A. *Langmuir*. 2007;23(17):8691–4.
1195. Fujii S, Nakamura Y. *Langmuir*. 2017;33(30):7365–79.
1196. Hunter TN, Pugh RJ, Franks GV, Jameson GJ. *Advances in Colloid and Interface Science*. 2008;137(2):57–81.
1197. Ito M, Takano K, Hanochi H, Asaumi Y, Yusa S-i, Nakamura Y, Fujii S. *Frontiers in Chemistry*. 2018;6:269.
1198. Dupin D, Howse JR, Armes SP, Randall DP. *Journal of Materials Chemistry*. 2008;18(5):545–52.
1199. Dupin D, Armes SP, Connan C, Reeve P, Baxter SM. *Langmuir*. 2007;23(13):6903–10.
1200. Fujii S, Armes SP, Binks BP, Murakami R. *Langmuir*. 2006;22(16):6818–25.
1201. Fujii S, Kameyama S, Armes SP, Dupin D, Suzaki M, Nakamura Y. *Soft Matter*. 2010;6(3):635–40.
1202. Berk H, Balci B, Ertan S, Kaya M, Cihaner A. *Materials Today Communications*. 2019;19:336–41.

1203. Lin H-K, Lai Y-S, Liu Y-L. *ACS Sustainable Chemistry & Engineering*. 2019;7(4):4515–22.
1204. Westerman CR, Jenkins CL. *Macromolecules*. 2018;51(18):7233–8.
1205. Diaz C, Valenzuela ML, Laguna-Bercero MÁ. *International Journal of Molecular Sciences*. 2022;23(3):1093.
1206. Rao CNR, Müller A, Cheetham AK. *The Chemistry of Nanomaterials: Synthesis, Properties and Applications*: Wiley, Germany; 2006.
1207. Diaz C, Barrera G, Segovia M, Valenzuela ML, Osiak M, O'Dwyer C. *Journal of Nanomaterials*. 2015:105157.
1208. Diaz C, Valenzuela M, Zepeda L, Herrera P, Valenzuela C. *Journal of the Chilean Chemical Society*. 2021;66(4):5380–6.
1209. Han Y-F, Chen F, Zhong Z, Ramesh K, Chen L, Widjaja E. *The Journal of Physical Chemistry B*. 2006;110(48):24450–6.
1210. Laurent S, Forge D, Port M, Roch A, Robic C, Elst L, Muller R. *Chemical Reviews*. 2008;108:2064.
1211. Teja AS, Koh P-Y. *Progress in Crystal Growth and Characterization of Materials*. 2009;55(1-2):22–45.
1212. Patra AK, Kundu SK, Bhaumik A, Kim D. *Nanoscale*. 2016;8(1):365–77.
1213. Zheng Y, Cheng Y, Wang Y, Bao F, Zhou L, Wei X et al. *The Journal of Physical Chemistry B*. 2006;110(7):3093–7.
1214. Hua J, Gengsheng J. *Materials Letters*. 2009;63(30):2725–7.
1215. Liu Y, Yu C, Dai W, Gao X, Qian H, Hu Y, Hu X. *Journal of Alloys and Compounds*. 2013;551:440–3.
1216. Diaz C, Valenzuela ML, Cifuentes-Vaca O, Segovia M, Laguna-Bercero MA. *Nanomaterials*. 2020;10(12):2470.
1217. Diaz C, Valenzuela ML, Baez R, Segovia M. *Journal of the Chilean Chemical Society*. 2015;60(4):2716–20.
1218. Díaz C, Valenzuela A, Soto K, Laguna-Bercero M. *Journal of the Chilean Chemical Society*. 2019;64(3):4502–6.
1219. Díaz C, Valenzuela ML, Cifuentes-Vaca O, Segovia M, Laguna-Bercero M. *Materials Chemistry and Physics*. 2020;252:123276.
1220. Díaz C, Platoni S, Molina A, Valenzuela M, Geaney H, O'Dwyer C. *Journal of Nanoscience and Nanotechnology*. 2014;14(9):6748–53.
1221. Malpani D, Srivastava RK, Nandan B. *Polymer*. 2020;199:122594.
1222. Malpani D, Majumder A, Samanta P, Srivastava RK, Nandan B. *ACS Omega*. 2018;3(11):15666–78.

2 Functionalized Polymers

2.1 THE HISTORICAL DEVELOPMENT OF FUNCTIONALIZED POLYMERS

The IUPAC definition of functional polymers includes those polymers that possess particular chemical groups or those that have specific chemical, physical, pharmacological, biological, or other uses, based on their distinct chemical or functional groups [1]. Indeed, functional polymers are macromolecules that are functionalized and incorporate the green benefits of small molecules possessing the same functional groups. The use of polymer-functionalized materials in organic transformations dates back to 1946 [2], although key contributions were reported somewhat later by Merrifield, Lestinger, Frechet, Sherrington, Hodge, Hunt, and among others [3–9]. To my knowledge, the first report on the use of functional polymers in the synthesis of organic compounds was made by Merrifield in 1963, when he used the chloromethylated polystyrene as a solid support in multistep synthesis of polypeptides (Scheme 2.1) [3]. In 1984, Merrifield received the Nobel Prize in chemistry for his groundbreaking achievement. In the Merrifield's synthesis of peptides, first, the covalent attachment of an *N*-protected amino acid to the solid support, followed by deprotection of amine group, yields polymer (I) which further reacts with another *N*-protected amino acid molecule that its carboxylic acid group is activated by *N,N'*-dicyclohexylcarbodiimide (DCC). The successive addition of further *N*-protected amino acids to the resulted polymer (II) and deprotection in an alternating manner produce the desired peptide which is then removed from

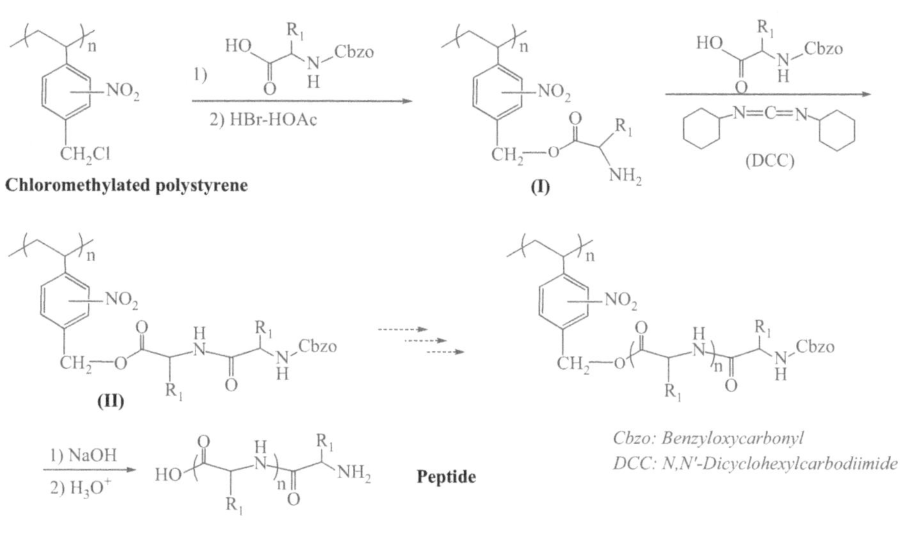

SCHEME 2.1 Merrifield's synthesis of peptides [3].

DOI: 10.1201/9781003656104-2

the support by cleavage of the ester linkage. Finally, the peptide was isolated from the polymer by filtration. In the Merrifield's method of peptide synthesis, his solid phase technique was used for the preparation of oligodeoxyribonucleotides [6], oligosaccharides [7,10], oligonucleotide [11], poly(amidoamines) [12], and natural peptides or their analogues [13].

Since their discovery, the development of functionalized polymeric materials has attracted tremendous research interest in organic synthesis and related chemical fields. Today, there are several books and reviews in the literature on different types of functionalized polymers [14–38]. Interest in the field is being enhanced by the possibility of creating systems that combine the advantageous properties of the functional groups and those of their polymeric nature. Functionalized polymers have been used as reagents; catalysts; acid acceptor; polymer-bound nucleophiles; protecting groups; redox reagents; polymeric acylating, alkylating, and halogenating reagents; drug delivery carriers; microencapsulated materials; etc.

2.2 DIFFERENT TYPES OF MACROMOLECULES FOR PREPARATION OF FUNCTIONALIZED POLYMERS

As mentioned above, a functionalized polymer in the present context is a macromolecule with chemically bound active groups. The polymer can be a linear type, which is soluble in appropriate solvents, or alternatively a cross-linked type with porous structure networks, which is swellable in a suitable solvent and insoluble in any media. The fundamental properties of a polymeric support to be functionalized and used in a variety of useful transformations can be summarized as follows:

1. Typically, the support should be completely insoluble in common solvents, as this facilitates the recovery and handling of the polymer support. This is commonly accomplished through the utilization of cross-linked polymers.
2. It should be able to be functionalized to achieve a high loading capacity of reactive sites with a uniform distribution throughout the polymer.
3. The chemical inertness and stability of the polymer backbone toward the reagents.
4. It must exhibit strong mechanical stability throughout the immobilization process and during handling.

The functionalized linear macromolecules have been utilized as polymer-supported reagents and catalysts, especially when the separation of the polymer after its application is not necessary. For example, microcapsules are constructed from linear polymers, because the poor solvation of cross-linked polymers in microencapsulation can be practically problematic. The main benefits that arise from the use of functionalized linear soluble polymers are as follows:

1. Reactions can occur uniformly within a single phase, thereby reducing the diffusion constraints of substrates to and from the active sites.

2. The rates of reaction and conversions remain unaffected by the bulkiness of the backbone of the polymer chain.
3. The flexible side chains can provide improved and equal accessibility to the functional groups involved.

However, the use of linear polymers as support has been limited, and their application may be accompanied by some disadvantages such as difficulty in its separation and recovery from low-molecular weight contaminants.

Various strategies for separation and recovery of soluble polymer-bound catalysts have been developed [39]. Solvent precipitation is a common strategy for recovery of soluble polymers. However, this method suffers from the drawback of green chemistry such as the use of large amounts of expensive or toxic solvents. Moreover, since the polymer support is separated via precipitation from an appropriate solvent, some low-molecular weight species may be precipitated with the polymer support which sometimes makes their separation difficult. Although separation can also be accomplished by centrifugation, ultrafiltration, or dialysis, however, these methods may not be quantitative.

The two approaches, separation of polymer-bound catalysts with thermomorphic solvent mixtures and latent biphasic separation, have also attracted the attention of organic chemists for separation of soluble polymer catalysts [40]. The former method can be accomplished by polar/or nonpolar polymers or by dendrimers as thermomorphic biphasic catalyst supports. For example, when a polar-phase soluble polymer-bound catalyst is used, first a biphasic system is produced, in which the polymer catalyst is retained within the polar phase (Figure 2.1). Then, the initially immiscible mixture is heated to become miscible. Upon cooling the mixture, two distinct phases emerge: a polar phase that contains the polymer catalyst and a nonpolar phase that holds the reaction product. After phase separation, the polymer catalyst is recovered and can be utilized in a future reaction [39]. Poly(ethylene oxide), some dendrimers, and poly(*N*-isopropylacrylamide) (PNIPAM) are among supports used in thermomorphic systems [40–44]. Hydrogenations of alkenes with a copolymer constituted of *N*-isopropylacrylamide and *N*-acryloxysuccinimide bounded to Rh(I) catalyst (as a polar polymer catalyst) in biphasic mixtures of heptane and ethanol/water (90% ethanol) is a typical example of reactions under thermomorphic conditions [42].

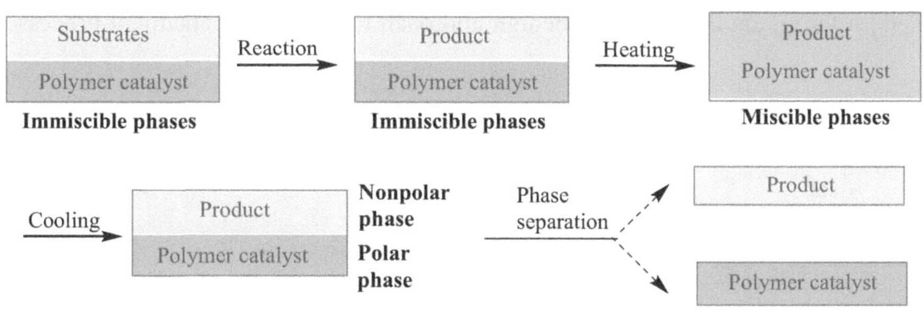

FIGURE 2.1 Separation of a thermomorphic soluble polymeric catalyst [39].

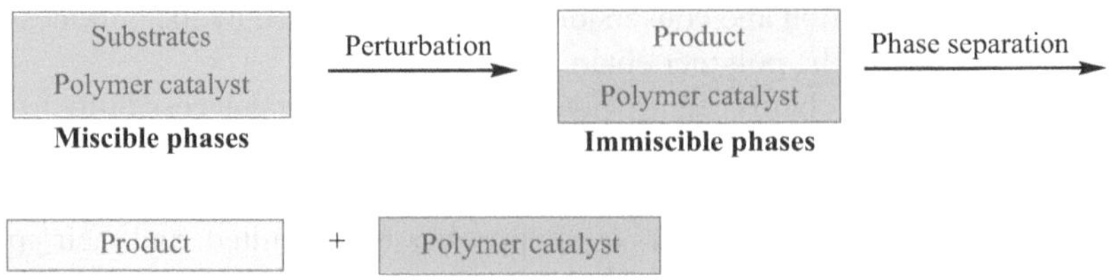

FIGURE 2.2 Illustration of latent biphasic process [45].

On the other hand, in another alternative method, namely latent biphasic separation, the reaction solvents are at the threshold of immiscibility. After completion of the reaction, a small amount of another solvent or other perturbing agents was added into the reaction media to yield a biphasic mixture (Figure 2.2). Following the separation of two immiscible phases, the polymer catalyst is obtained, which can be reused [45]. Precise control of solvents miscibility is a key factor in this method. For example, while the binary solution of ethanol/hexane (1:1 in volume) is miscible in all proportions, but it exhibits phase separation upon adding a small amount of water. Also, a mixture of heptane, water, and ethanol in a volume ratio of 10:0.5:9.5 is fully miscible, whereas a mixture in a volume ratio of 10:1:9 exhibits biphasic behavior. A typical example for latent biphasic process is a nucleophilic Michael reaction in the presence of poly(4-*tert*-butylstyrene) bearing triarylphosphine catalyst in the mixed solvent heptane/ethanol (1:1 in volume) [46]. After completion of the reaction, a small amount (<10 vol.%) of water was added to induce phase separation. The polar phase exhibited a significant concentration of product, whereas the nonpolar phase (heptane) was rich in polymer catalyst.

Another approach for separation of soluble polymer-bound catalysts is the use of polymers with inverse temperature-dependent solubility. The polymer catalyst is soluble at lower temperatures and becomes insoluble upon heating above these temperatures (Figure 2.3). Most systems with the above property have been observed in aqueous solutions, typically within the temperature range of 0°C–100°C, although it is also known to present in other solvents as well. The temperature at which this phase separation occurs is referred to as the lower critical solution temperature (LCST). PNIPAM derivatives as supports for catalysts are well known for their inverse temperature-dependent solubility [47–50].

Cross-linked polymers as support, when used for replacing linear soluble polymers, offer an environmentally benign approach toward green chemical processes.

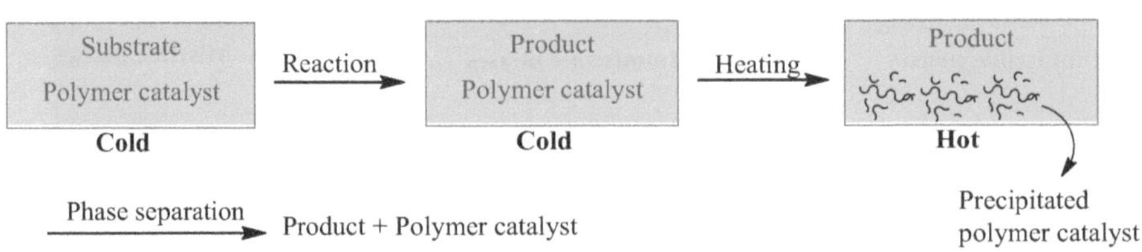

FIGURE 2.3 Separation of a soluble polymer catalyst which exhibit inverse temperature-dependent solubility [39].

The distinct advantages in choosing cross-linked polymers as support for the preparation of functionalized polymers are as follows:

1. They are insoluble in the reaction mixture, enabling ease of product separation by a simple filtration. This prevents the need for a chromatographic separation and laborious work-up procedures. For example, when they are used as spherical bead forms, they do not coalesce in solvents and low-molecular weight contaminants and can be isolated from the reaction media on a filter and washed with various solvents. Meanwhile, easy separation of cross-linked polymer supports allows rapid isolation of unstable products from the reaction media without exposing them to water or other solvents during work-up procedures.
2. The limited ability of cross-linked polymers to absorb solvents and their tendency to swell makes them promising support materials for various catalysts and reagents. In fact, polymer beads characterized with a low degree of cross-linking or even those highly cross-linked polymers with well-defined porosity of structure can be swelled in conventional organic solvents, thereby exposing their functional groups to soluble substrates.

Although cross-linked polymers have attracted widespread interest in the field of functional polymers, they have a number of disadvantages yet. These may include the following:

1. The physical and chemical properties of cross-linked polymers are influenced by the preparation method and the extent of cross-linking, resulting in notable differences in their performance.
2. The relatively limited accessibility of active sites in cross-linked functional polymers to reagents significantly affects their efficiency.
3. The methods used to characterize the synthesized cross-linked functional polymers, as well as the analytical techniques for identifying structural changes occurring at each stage of their synthesis, are frequently difficult and can be challenging to perform.
4. Accurately managing the loading of the resin poses significant challenges, which make it difficult for researchers to carefully design the functional polymers that are required for a special application.

Cross-linked polymers display considerable differences in properties and behavior depending on their preparation method and the degree of cross-linking. Three main types of cross-linked polymers as support can be identified for the preparation of functionalized polymers [15,18]:

1. **Microporous or gel-type resins**: Typically, these samples are characterized by their clarity, transparency, and glass-like appearance, and they are produced through the suspension polymerization of the relevant monomer by <10% cross-linking levels, without the addition of any solvent. Generally, microporous polymers can swell in suitable solvents and are utilized for their ability to withstand sudden impacts and their high loading capacity

during functionalization processes. However, the mechanical stability of these resins diminishes significantly as the amount of cross-linking agent decreases; resins with less than 1% cross-linking demonstrate low mechanical stability and tend to break down into linear fragments during handling. In the context of chemical transformations, a cross-linking degree of 2% in the resin was found to be the best, allowing adequate penetration of reagents and catalysts into porous support while maintaining sufficient mechanical characteristics [18].

2. **Macroporous resins**: Typically, these materials are amorphous and opaque, produced through the suspension polymerization technique with a significant quantity of cross-linking agent (approximately 20%). The polymerization process involves dissolving the monomer in an inert solvent. The solvent may serve good for both the comonomers and the resultant polymer (non-cross-linked), or it may function as a good solvent solely for the monomer while acting as a precipitant for the polymer. Utilizing poor solvents results in the formation of large pores; however, a combination of solvents and non-solvents yields the desired pore sizes. Macroporous resins exhibit greater mechanical stability compared to microporous resins but like microporous resins, they can swell adequately in different solvents [18].
3. **Macroreticular resins**: These resins exhibit a truly macroporous structure. When the solvent used in suspension polymerization serves as a good solvent for the monomer, yet acts as a precipitant for the polymer (non-cross-linked), the term macroreticular is typically utilized. The extent of polymerization and cross-linking can be controlled by adjusting the feed ratio of the non-solvent. In comparison with the microporous and macroporous resins, macroreticular resins possess a greater surface area with a permanent internal porous structure, and are not susceptible to swelling in typical organic solvents due to their high rigidity and degree of cross-linking. They have poor loading capacity and are prone to brittleness. Easy separation from the reaction medium by simple filtration and enhanced mechanical stability are the obvious advantages of the macroreticular resins [18].

Since the advent of functional polymers, various physical forms of polymers have been utilized as support for different species. Among these, cross-linked polymer beads are among the most commonly used polymers which can be synthesized through suspension polymerization methods. Generally, an organic phase comprising a vinyl monomer, a free radical initiator, and a cross-linking agent is gradually added to an aqueous phase and the suspension is heated. After a certain time, polymer particles or beads are formed.

The "loading" or the "capacity", referring to the level of functional groups, is one of the most critical factors in functionalized polymers which plays a major role in enhancing rates of chemical processes and improving product yields. It is usually expressed as mmol/g and possess a wide range of values, based on the desired application of the functionalized polymer. Gel-type resins, especially those with minimal cross-linking (<2%), can achieve a loading level of 100%. In contrast, macroporous materials exhibit variable loading levels since all active parts of the polymer are inaccessible for chemical reagents, because they are enclosed in a network structure.

The maximum effective loading for macroporous resins is approximately 40%. Generally, an optimal loading of around 0.1–0.7 mmol/g is suitable for polymeric catalysts, about 0.5–2 mmol/g for solid-phase synthesis, and ~>2 mmol/g for polymer-supported reagents and scavengers.

2.3 REPRESENTATIVE EXAMPLES OF POLYMER SUPPORT FOR THE FUNCTIONALIZATION PROCESSES

One of the earliest and most prevalent cross-linked polymers for the functionalization processes is chloromethylated polystyrene, known as the Merrifield resin. It has been prepared by direct chloromethylation of polystyrene using chloromethyl methyl ether in the presence of zinc chloride, stannic chloride, or boron trifluoride etherate (Scheme 2.2) [3,51–53]. However, due to the carcinogenic nature of chloromethyl ether, other alternative methods have been used for the preparation of the chloromethylated polystyrene These methods include utilizing less volatile chloromethyl ethers with unknown toxicity, employing convenient halogenating agents, and the copolymerization of (chloromethyl)styrenes (70/30 meta/para mixture) [15,54]. Also, the polymerization of poly(*p*-methylstyrene) and subsequent chlorination of methyl group using sodium hypochlorite under phase transfer conditions has been reported for the preparation of chloromethylated polystyrene (Scheme 2.3) [55]. Chloromethylated polystyrene is a key intermediate for the preparation of ion-exchange resins (Scheme 2.4) and many polymer-supported reagents and catalysts [28,30].

Poly(vinylpyridine) is another polymer support frequently used for the preparation of functionalized polymers [15,30,36,56,57]. In the following chapters, I will explain this polymer and its corresponding derivatives in detail.

It is difficult to imaging an aspect of functional polymers not touched by polyacrylamide [58,59]. Its hydrophilic characteristics allow it to dissolve in both aqueous media and various organic solvents. The cross-linked polyacrylamide can absorb

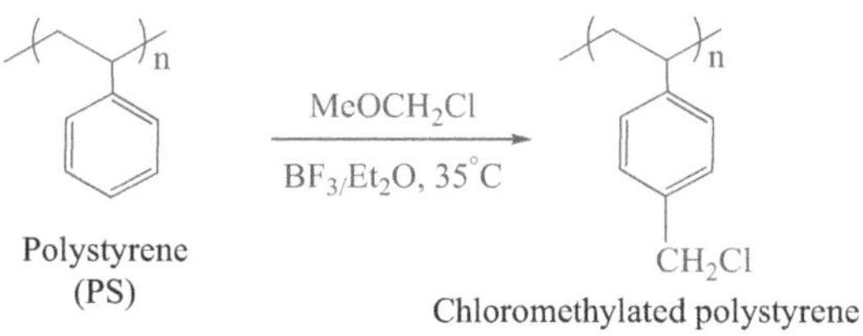

SCHEME 2.2 The chloromethylation of polystyrene [52,53].

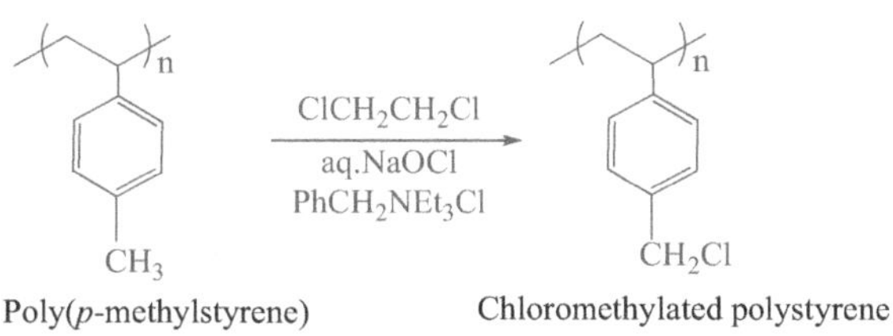

SCHEME 2.3 The chlorination of poly(*p*-methylstyrene) [55].

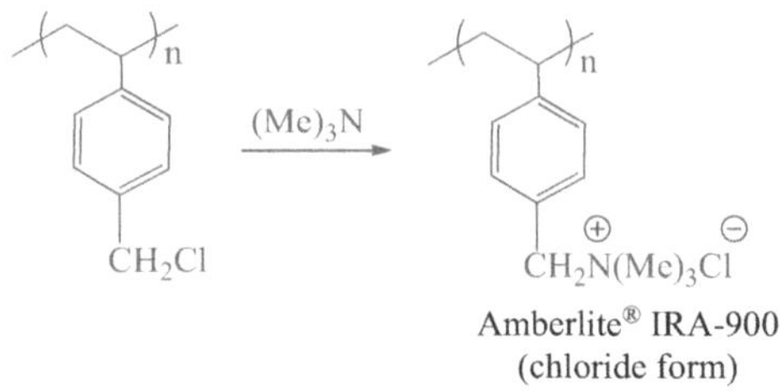

SCHEME 2.4 The preparation of Amberlite® IRA-900 (chloride form) [30].

water molecules and swell without dissolving. The amide group moieties of the polymer facilitate straightforward chemical modifications, which can change the polymer properties and broaden its range of applications. Polyacrylamide and its derivatives have been used as polymeric reagents and catalysts, ion-exchange resins, drug delivery systems, and many laboratory as well as industrial materials [30,60,61]. For example, a copolymer of *N*-(acryloyloxy)succinimide (NASI) and PNIPAM has been used as support for cationic Rh(I) catalyst which is used in substitution and hydrogenation reactions (Scheme 2.5) [49]. The inverse temperature-dependent solubility of PNIPAM is a prominent characteristic of this polymer. In essence, precipitation of polymer from solution occurs upon heating above its LCST [62].

The preparation of biologically significant compounds based on polyacrylamide is another application of this useful polymer in the field of functional polymers. The synthesis of cross-linked polymeric β-cyclodextrin (β-CD) as a water-swellable and biocompatible carrier has been reported (Scheme 2.6) [63]. It possesses significant potential for utilization in bilirubin adsorption to eliminate excess bilirubin from the bloodstream through hemoperfusion. Table 2.1 shows a representative selection of the most commonly utilized polymer supports and examples along with their applications.

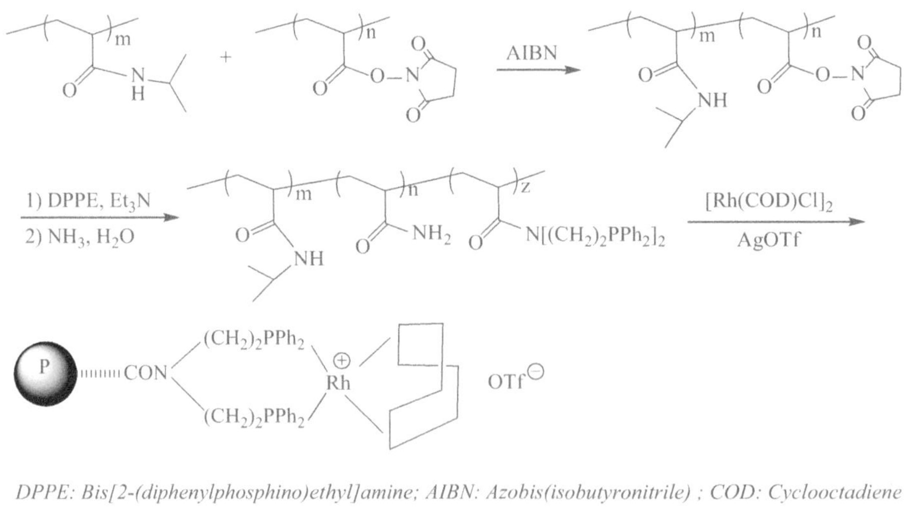

SCHEME 2.5 Modification of PNIPAM to attach Rh(I) catalyst [49].

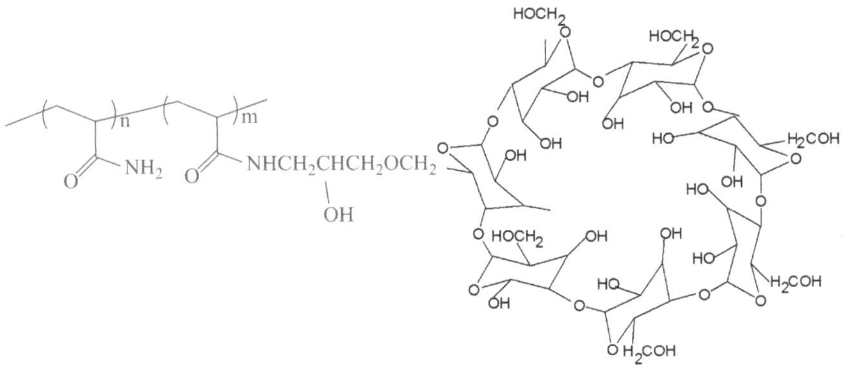

SCHEME 2.6 Polyacrylamide-supported β-CD [63].

TABLE 2.1
Different Types of Polymeric Supports with Representative Examples of their Applications

Polymer Support	Representative Examples
Polystyrene	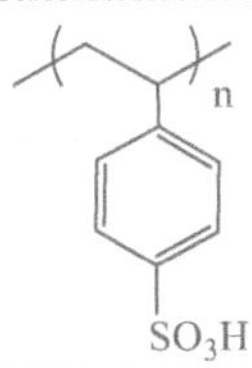 Amberlyst-15 As catalyst in synthesis of β-enaminones and β-enaminoesters [64,65]
	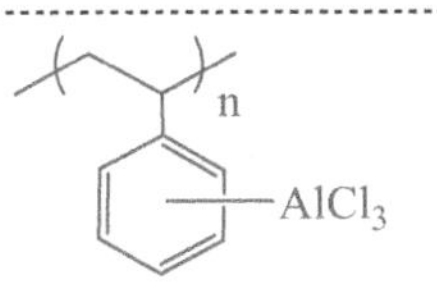 Polystyrene-$AlCl_3$ (PS-$AlCl_3$) As catalyst in Friedel–Crafts sulfonylation of aromatic compounds, thioacetalization of carbonyl compounds, and preparation of diaryl sulfoxide from arenes [66–68]
	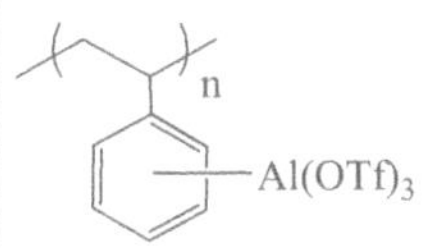 Polystyrene-$Al(OTf)_3$ (PS-$Al(OTf)_3$) As catalyst in thioacetalization of carbonyl compounds [69], Friedel–Crafts acylation of aromatic compounds [70], and sulfonylation of arenes [71]

(Continued)

TABLE 2.1 (*Continued*)
Different Types of Polymeric Supports with Representative Examples of their Applications

Polymer Support	Representative Examples
	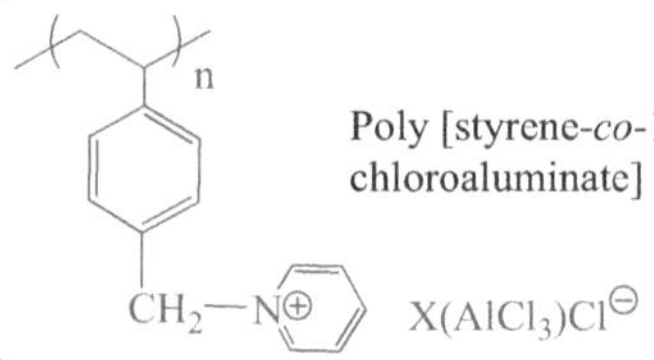 Poly [styrene-*co*-1-((4-vinylphenyl)methyl)pyridinium) chloroaluminate] (PS-PyCl-$XAlCl_3$) As catalyst in selective synthesis of benzimidazoles [72]
	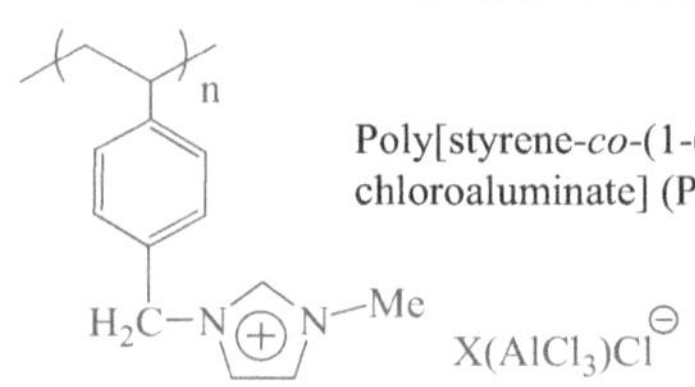 Poly[styrene-*co*-(1-((4-vinylphenyl)methyl)-3-methylimidazolium chloroaluminate] (PS-[MeIm]Cl-$XAlCl_3$) As catalyst for Knoevenagel condensation [73]
Poly(vinylpyridine)	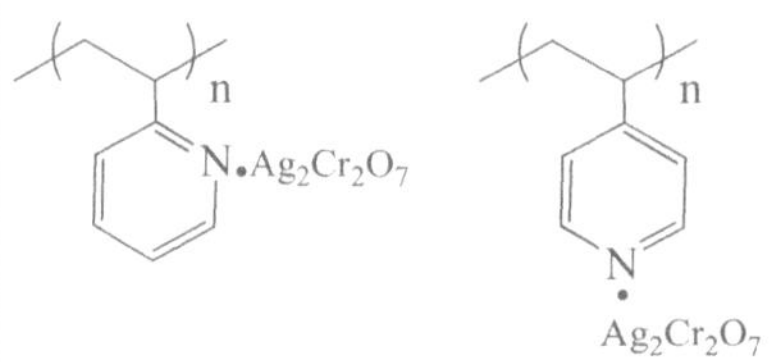 Poly(2-vinylpyridine)-silver dichromates Poly(4-vinylpyridine)-silver dichromates For oxidation of alcohols, oximes, thiols, and aromatic hydrocarbons [74]
	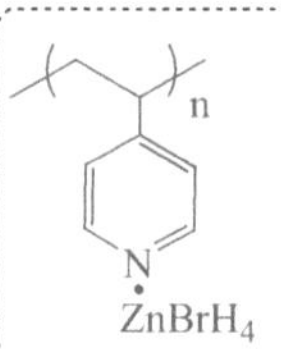 Poly(4-vinylpyridine)-zirconium borohydride For reduction of carbonyl compounds [75]
Poly(*N*-vinyl-2-pyrrolidone)	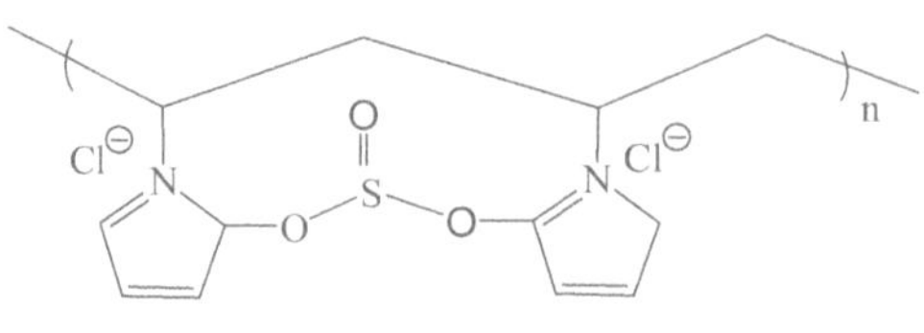 Polyvinylpyrrolidone-thionyl chloride For dehydration of aldoximes to nitriles [76]

(*Continued*)

TABLE 2.1 (*Continued*)
Different Types of Polymeric Supports with Representative Examples of their Applications

Polymer Support	Representative Examples
Cellulose	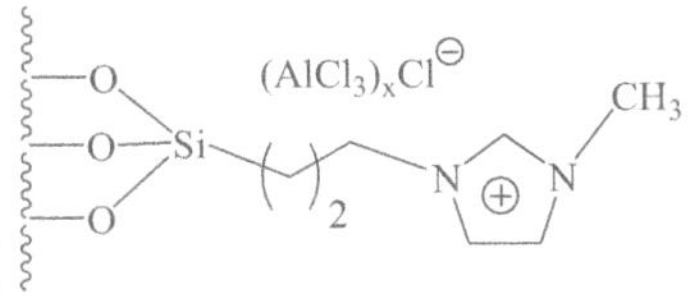 Cellulose aluminum oxide composite-supported imidazolium chloroaluminate ionic liquid (Cell/Al_2O_3-[MeIm]Cl-$XAlCl_3$) As catalyst in synthesis of 1,8-dioxo-octahydroxanthenes [77]
Polyacrylonitrile	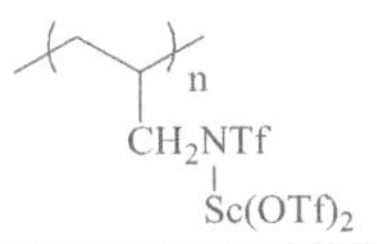 (Polyallyl)scandium trifylamide ditriflate (PA-Sc-TAD) As catalyst in quinoline synthesis [78]
Nafion-H	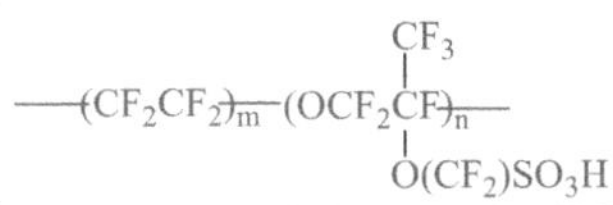 Nafion-H As catalyst in tetrahydropyranylation [79] and thioacetalization [80]
Polyacrylamide	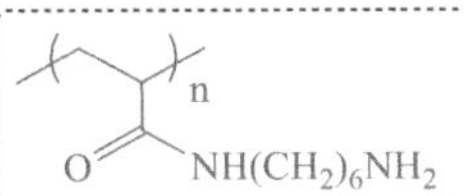 Poly[*N*-(6-aminohexyl)-acrylamide) As catalyst for Knoevenagel reaction [81]
Poly(thiophene)	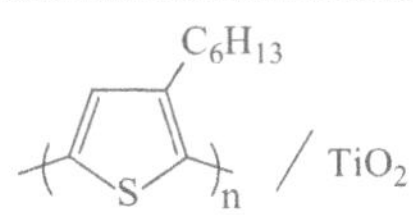 Poly(3-hexylthiophene)/TiO_2 As photocatalyst [82]
Polysulfone	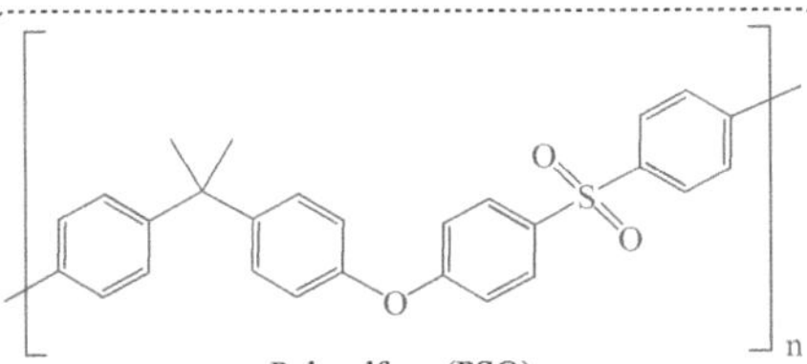 Polysulfone (PSO) As H_2TiO_3 support for Li^+ adsorption [83]

(*Continued*)

TABLE 2.1 (*Continued*)
Different Types of Polymeric Supports with Representative Examples of their Applications

Polymer Support	Representative Examples
Chitosan	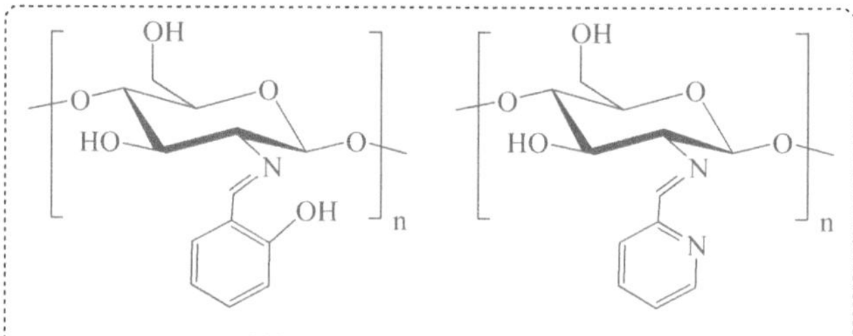 Chitosan Schiff base complexes A support for Pd catalysts in C–C coupling reaction [84]
Hypercross-linked polymer with triphenylphosphine skeleton	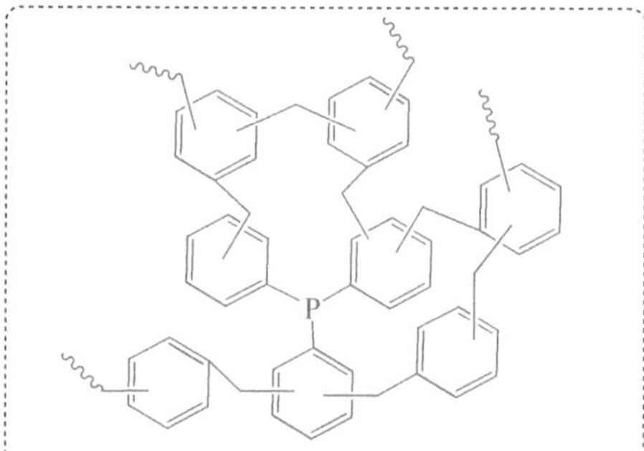 Triphenylphosphine-hyper-crosslinked polymer A support for ruthenium as a catalyst for hydrogen evolution reaction and ammonia borane hydrolysis [85]
Fluoropolymers	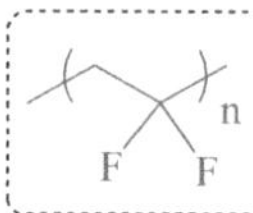 Polyvinylidene fluoride (PVDC) As Fe_3O_4-PVDC membranes for filtration and benzyl alcohol oxidation [86]
Polyhydroxybutyrate (PHB)	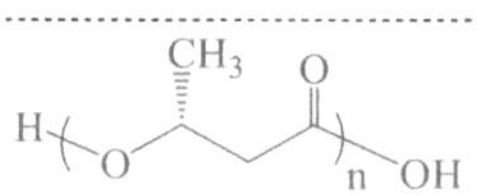 Polyhydroxybutyrate (PHB) The photocatalytic activity and biodegradation of PHB films containing titanium dioxide (TiO_2) [87]

(*Continued*)

TABLE 2.1 (*Continued*)
Different Types of Polymeric Supports with Representative Examples of their Applications

Polymer Support	Representative Examples
Poly(dimethylsiloxane)	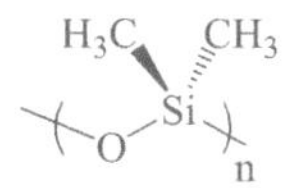 Poly(dimethylsiloxane) (PDMS) The photocatalytic activity of poly(dimethylsiloxane) films containing titanium dioxide (TiO_2) [88]
Poly(1-naphthylamine)	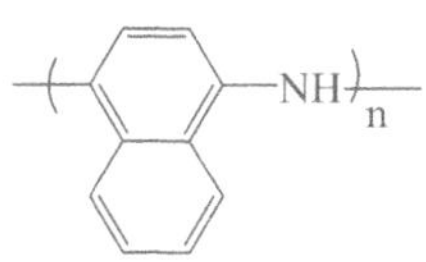 Poly(1-naphthylamine) (PNA) The photocatalytic activity of poly(1-naphthylamine)/SiO_2 containing TiO_2 [89]
Poly(methyl methacrylate)	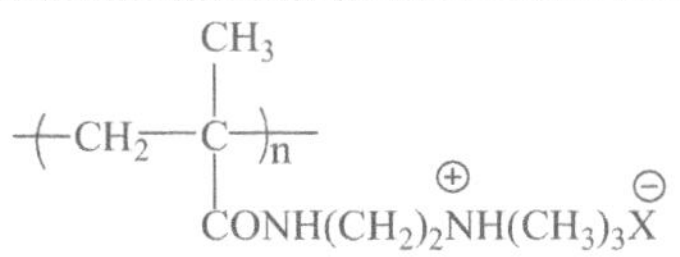 Poly(methyl methacrylate-*co*-divinylbenzene)-dichromate and chlorochromate For oxidation of alcohols [90]
Alginic acid	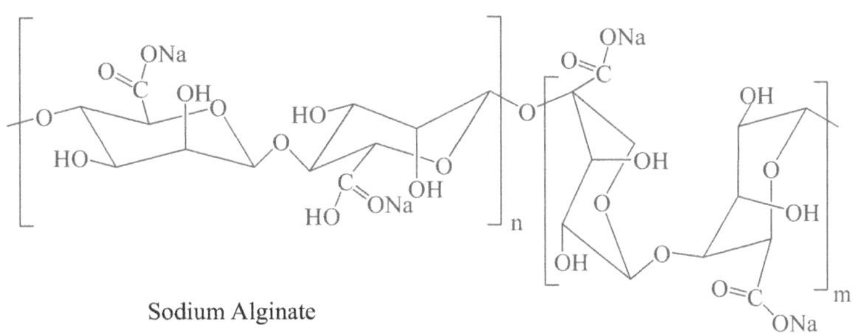 Sodium Alginate The zero-valent iron and palladium-doped zero-valent iron immobilized in the alginate bead for degradation of trichloroethylene [91]

(*Continued*)

TABLE 2.1 (*Continued*)
Different Types of Polymeric Supports with Representative Examples of their Applications

Polymer Support	Representative Examples
Polyvinyl alcohol	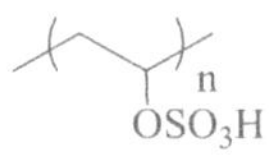 Sulfonated polyvinyl alcohol (PVA) As catalyst in synthesis of α-aminonitriles [92]
Polyethylene	 / TiO_2 NPs Polyethylene (PE) containing TiO_2 nanoparticles For photocatalytic degradation of organic dyes [93]
Poly(ethylene glycol)	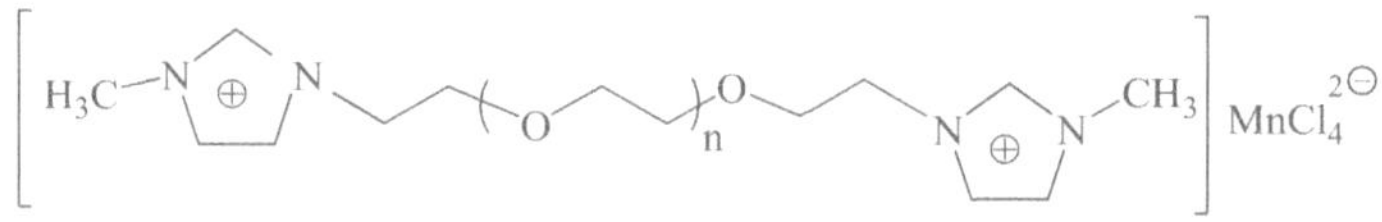 Poly(ethylene glycol) dicationic ionic liquid-based $MnCl_4^{2-}$ As a phase transfer catalyst for nucleophilic substitution reactions of benzyl halides [94]
Polypropylene	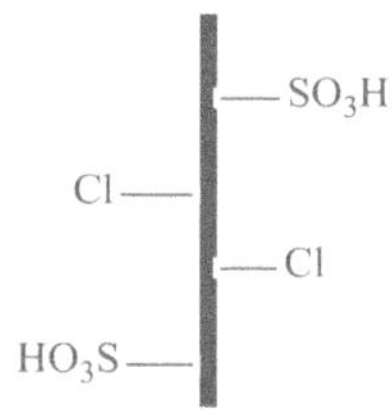 Polypropylene fiber-SO_3H As catalyst for Biginelli reactions [95]
Polyurethane	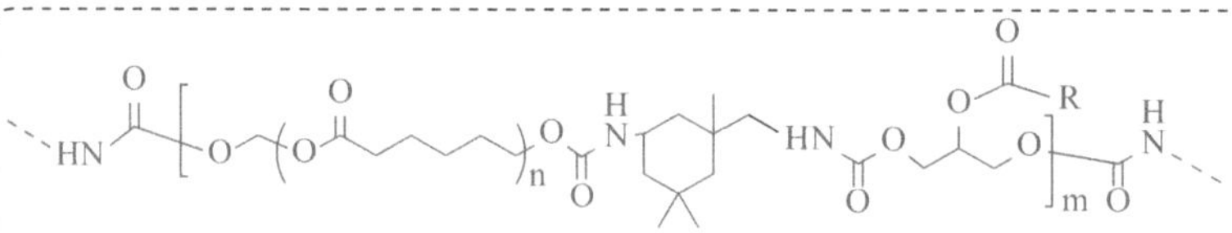 Hyperbranched polyurethane-supported palladium-silver-carbon dots As catalyst for oxidative *ipso*-hydroxylation of aryl boronic acids [96]

(*Continued*)

TABLE 2.1 (*Continued*)
Different Types of Polymeric Supports with Representative Examples of their Applications

Polymer Support	Representative Examples
Polyvinyl chloride	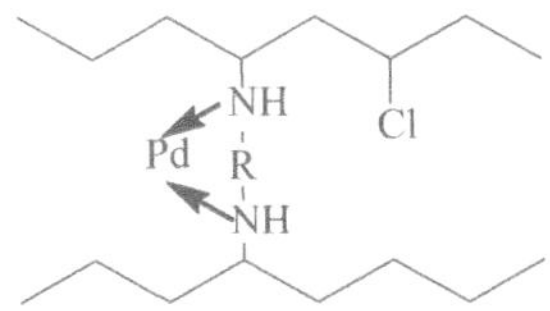 Polyvinyl chloride–polyethylene–polyamine supported palladium complexes As catalyst for Heck reaction [97]

REFERENCES

1. McNaught AD, Wilkinson A. *Compendium of Chemical Terminology*: Blackwell Science, Oxford, UK; 1997.
2. Sussman S. *Industrial and Engineering Chemistry*. 1946;38(12):1228–30.
3. Merrifield RB. *Journal of the American Chemical Society*. 1963;85(14):2149–54.
4. Merrifield R. *Biochemistry*. 1964;3(9):1385–90.
5. Feinberg R, Merrifield R. *Tetrahedron*. 1974;30(17):3209–12.
6. Letsinger RL, Mahadevan V. *Journal of the American Chemical Society*. 1966;88(22):5319–24.
7. Schuerch C, Frechet JM. *Journal of the American Chemical Society*. 1971;93(2):492–6.
8. MacKenzie W, Sherrington D. *Polymer*. 1981;22(4):431–3.
9. Hodge P, Hunt B, Shakhshier I. *Polymer*. 1985;26(11):1701–7.
10. Seeberger PH, Danishefsky SJ. *Accounts of Chemical Research*. 1998;31(10):685.
11. Gravert DJ, Janda KD. *Chemical Reviews*. 1997;97(2):489–510.
12. Hartmann L, Krause E, Antonietti M, Börner HG. *Biomacromolecules*. 2006;7(4):1239–44.
13. Gordon K, Balasubramanian S. *Journal of Chemical Technology & Biotechnology: International Research in Process, Environmental & Clean Technology*. 1999;74(9):835–51.
14. Labadie JW. *Current Opinion in Chemical Biology*. 1998;2(3):346–52.
15. Hodge P, Sherrington DC. *Polymer-Supported Reactions in Organic Synthesis*: John Wiley, England; 1980.
16. Takemoto K, Ottenbrite RM, Kamachi M. *Functional Monomers and Polymers*: Marcel Dekker, New York; 1997.
17. Sherrington DC, Hodge P. *Synthesis and Separation Using Functional Polymers*: John Wiley, England; 1988.
18. Akelah A, Moet A. *Functionalized Polymers and their Applications*: Thomson, New Delhi, St Edmundsbury, Suffolk; 1990.
19. Smith K. *Solid Supports and Catalysts in Organic Synthesis*: Ellis Horwood, Chichester, UK; 1992.
20. Oğuz O, Zanjani JSM, Soytaş SH, Menceloğlu YZ. 3-specific interactions and self-organization in polymer/functionalized nanoparticle systems. In: K. Pielichowski & T. M. Majka (eds.). *Polymer Composites with Functionalized* Nanoparticles: Elsevier, Amsterdam; 2019, pp. 85–117.

21. Girija AR. 12-medical applications of polymer/functionalized nanoparticle systems. In: K. Pielichowski & T. M. Majka (eds.). *Polymer Composites with Functionalized Nanoparticles*. Elsevier, Amsterdam; 2019, pp. 381–404.
22. Wei C, Akinwolemiwa B, Yu L, Hu D, Chen GZ. Polymer composites with functionalized carbon nanotube and graphene. In: K. Pielichowski & T. M. Majka (Eds.), *Polymer Composites with Functionalized Nanoparticles*: Elsevier, Amsterdam; 2019, pp. 211–48.
23. Gun'ko VM. Polymer composites with functionalized silica. In: K. Pielichowski & T. M. Majka (Eds.), *Polymer Composites with Functionalized Nanoparticles*: Elsevier, Amsterdam; 2019, pp. 119–48.
24. Kong I. *Design and Applications of Nanostructured Polymer Blends and Nanocomposite Systems*: William Andrew, Boston; 2016, pp. 125–54.
25. Akelah A, Sherrington DC. *Chemical Reviews*. 1981;81(6):557–87.
26. Fréchet JM. *Tetrahedron*. 1981;37(4):663–83.
27. Früchtel JS, Jung G. *Angewandte Chemie International Edition in English*. 1996;35(1):17–42.
28. Hermkens PH, Ottenheijm HC, Rees D. *Tetrahedron*. 1996;52(13):4527–54.
29. Shuttleworth SJ, Allin SM, Wilson RD, Nasturica D. *Synthesis*. 2000;2000(8):1035–74.
30. Ley SV, Baxendale IR, Bream RN, Jackson PS, Leach AG, Longbottom DA et al. *Journal of the Chemical Society, Perkin Transactions 1*. 2000(23):3815–4195.
31. Hetemi D, Pinson. *Journal of Chemical Society Reviews*. 2017;46(19):5701–13.
32. Benaglia M, Puglisi A, Cozzi F. *Chemical Reviews*. 2003;103(9):3401–30.
33. Das A, Theato P. *Chemical Reviews*. 2016;116(3):1434–95.
34. Voit BI, Lederer A. *Chemical Reviews*. 2009;109(11):5924–73.
35. Zhao X, Lv L, Pan B, Zhang W, Zhang S, Zhang Q. *Chemical Engineering Journal*. 2011;170(2–3):381–94.
36. Tamami B, Parvanak Boroujeni K. *Iranian Polymer Journal*. 2009;18(3):191–206.
37. Leadbeater NE, Marco M. *Chemical Reviews*. 2002;102(10):3217–74.
38. De Vos D, Vankelecom IF, Jacobs PA. *Chiral Catalyst Immobilization and Recycling*. Wiley-VCH, New York; 2008.
39. Bergbreiter DE. *Chemical Reviews*. 2002;102(10):3345–84.
40. Bianga J, Künnemann KU, Gaide T, Vorholt AJ, Seidensticker T, Dreimann JM, Vogt D. *Chemistry–A European Journal*. 2019;25(50):11586–608.
41. Bergbreiter DE, Osburn PL, Wilson A, Sink EM. *Journal of the American Chemical Society*. 2000;122(38):9058–64.
42. Bergbreiter DE, Liu Y-S, Osburn PL. *Journal of the American Chemical Society*. 1998;120(17):4250–1.
43. Bergbreiter DE, Osburn PL, Frels JD. *Journal of the American Chemical Society*. 2001;123(44):11105–6.
44. Albert S. *Chemical Communications*. 2002(15):1570–1.
45. Bergbreiter DE, Osburn PL, Smith T, Li C, Frels JD. *Journal of the American Chemical Society*. 2003;125(20):6254–60.
46. Bergbreiter DE, Li C. *Organic Letters*. 2003;5(14):2445–7.
47. Lanzalaco S, Mingot J, Torras J, Alemán C, Armelin E. *Advanced Engineering Materials*. 2023;25(4):2201303.
48. Costa MC, Silva SM, Antunes FE. *Journal of Molecular Liquids*. 2015;210:113–8.
49. Bergbreiter DE, Case BL, Liu Y-S, Caraway JW. *Macromolecules*. 1998;31(18):6053–62.
50. Bergbreiter DE, Liu Y-S. *Tetrahedron Letters*. 1997;38(45):7843–6.
51. Sparrow JT. *Tetrahedron Letters*. 1975;16:4637–8.
52. Warshawsky A, Deshe A. *Journal of Polymer Science: Polymer Chemistry Edition*. 1985;23(6):1839–41.
53. Olah GA, Da B, Sh Y, Ja O. *Synthesis*. 1974:560–1.

54. Balakrishnan T, Ford WT. *Journal of Applied Polymer Science*. 1982;27(1):133–8.
55. Mohanraj S, Ford WT. *Macromolecules*. 1986;19(10):2470–2.
56. Fréchet JM, de Meftahi MV. *British Polymer Journal*. 1984;16(4):193–8.
57. Li Y, Wang Y, He L, Meng L, Lu H, Li X. *Journal of Hazardous Materials*. 2020;383:121144.
58. Xiong B, Loss RD, Shields D, Pawlik T, Hochreiter R, Zydney AL, Kumar M. *NPJ Clean Water*. 2018;1(1):17.
59. Zhang Y, Ren Y, Zhang Y. *Chemical Reviews*. 2009;109(9):4375–97.
60. Galaev IY. *Russian Chemical Reviews*. 1995;64(5):471.
61. Caulfield MJ, Qiao GG, Solomon DH. *Chemical Reviews*. 2002;102(9):3067–84.
62. Bergbreiter DE, Mijalis AJ, Fu H. *Journal of Chemical Education*. 2012;89(5):675–7.
63. Wang H, Ma J, Zhang Y, He B. *Reactive and Functional Polymers*. 1997;32(1):1–7.
64. Venkat Narsaiah A, R Reddy A, VS Reddy B, S Yadav. *The Open Catalysis Journal*. 2011;4(1).
65. Pal R, Sarkar T, Khasnobis S. *Arkivoc*. 2012;2012(1):570–609.
66. Parvanak Boroujeni K, Tamami B. *Catalysis Communications*. 2007;8(8):1191–6.
67. Tamami B, Parvanak Boroujeni K. *Iranian Polymer Journal*. 2003;12(6):507–13.
68. Parvanak Boroujeni K. *Phosphorus, Sulfur, and Silicon*. 2010;185(10):2085–91.
69. Parvanak Boroujeni K, Massah AR. *Reactive and Functional Polymers*. 2006;66(10):1126–31.
70. Parvanak Boroujeni K. *Bulletin of the Korean Chemical Society*. 2010;31(11):3156–8.
71. Parvanak Boroujeni K. *Bulletin of the Korean Chemical Society*. 2010;31(7):1887–90.
72. Parvanak Boroujeni K, Zhianinasab A, Jafarinasab M. *Journal of the Serbian Chemical Society*. 2013;78(2):155–64.
73. Parvanak Boroujeni K, Jafarinasab M. *Chinese Chemical Letters*. 2012;23(9):1067–70.
74. Tamami B, Hatam M, Mohadjer D. *Polymer Bulletin*. 1989;21:531–3.
75. Tamami B, Goudarzian N. *Journal of the Chemical Society, Chemical Communications*. 1994(9):1079–80.
76. Tamami B, Kiasat A. *Journal of Chemical Research, Synopses*. 1999(7):444–5.
77. Parvanak Boroujeni K, Tahani P. *Inorganic and Nano-Metal Chemistry*. 2017;47(8):1150–6.
78. Kobayashi S, Nagayama S. *Journal of the American Chemical Society*. 1996;118(37):8977–8.
79. Olah GA, Husain A, Singh BP. *Synthesis*. 1983;1983(11):892–5.
80. Surya Prakash G, Ramaiah P, Krishnamurti R. *Catalysis Letters*. 1991;9:59–63.
81. Tamami B, Fadavi A. *Catalysis Communications*. 2005;6(11):747–51.
82. Wang D, Zhang J, Luo Q, Li X, Duan Y, An J. *Journal of Hazardous Materials*. 2009;169(1–3):546–50.
83. Lawagon CP, Nisola GM, Cuevas RAI, Kim H, Lee S-P, Chung W-J. *Journal of Industrial and Engineering Chemistry*. 2019;70:124–35.
84. Leonhardt SE, Stolle A, Ondruschka B, Cravotto G, De Leo C, Jandt KD, Keller TF. *Applied Catalysis A: General*. 2010;379(1–2):30–7.
85. Xu C, Wang H, Wang Q, Wang Y, Zhang Y, Fan G. *Applied Surface Science*. 2019;466:193–201.
86. Gumus H. *Chinese Journal of Chemical Engineering*. 2019;27(2):314–21.
87. Yew S-P, Tang H-Y, Sudesh K. *Polymer Degradation and Stability*. 2006;91(8):1800–7.
88. Iketani K, Sun R-D, Toki M, Hirota K, Yamaguchi O. *Journal of Physics and Chemistry of Solids*. 2003;64(3):507–13.
89. Ameen S, Akhtar MS, Kim YS, Shin HS. *Applied Catalysis B: Environmental*. 2011;103(1–2):136–42.
90. Hassanein M, Aly E-SA, Abbas Y, El-Sigeny SM. *European Polymer Journal*. 1992;28(4):411–3.

91. Kim H, Hong H-J, Jung J, Kim S-H, Yang J-W. *Journal of Hazardous Materials.* 2010;176(1–3):1038–43.
92. Maleki A, Rahimi J, Hajizadeh Z, Niksefat M. *Journal of Organometallic Chemistry.* 2019;881:58–65.
93. Naskar S, Pillay SA, Chanda M. *Journal of Photochemistry and Photobiology A: Chemistry.* 1998;113(3):257–64.
94. Goodajdar BM, Akbari F, Davarpanah J. *Applied Organometallic Chemistry.* 2019;33(2):e4647.
95. Shi X-L, Yang H, Tao M, Zhang W. *RSC Advances.* 2013;3(12):3939–45.
96. Bayan R, Karak N. *ChemistrySelect.* 2018;3(40):11210–8.
97. Cui Y, Zhang L. *Journal of Molecular Catalysis A: Chemical.* 2005;237(1–2):120–5.

3 Polymeric Catalysts Based on Poly(vinylpyridine)

3.1 POLYMERIC CATALYST

Green chemistry (environment-friendly chemistry) has been defined as a chemical technology that reduces or eliminates the use or generation of dangerous or hazardous waste materials [1]. Organic chemistry can have key roles for achieving the major goals of "green chemistry". For this purpose, many attempts have been done by organic chemists to replace perilous and toxic materials and methods with the environmentally friendly ones [2]. Among these, heterogenization of homogeneous catalysts by polymers is recognized to play a leading part in the development of green chemistry, and extensive studies have been carried out on synthesis and application of various kinds of polymer supported catalysts [3–6]. The utilization of polymeric catalysts offers several advantages such as enhanced stability, higher selectivity, easier handling, the possibility to use an excess of catalyst to force the reaction completion without causing work-up problems, non-toxicity, non-corrosiveness, and mildness of the reaction conditions. More importantly, the use of polymer catalyst provides the advantages of the easier product isolation and catalyst recycling, resulting in the elimination of environmentally harmful wastes.

Incorporation of the catalytic species on the polymer support is usually done by two general methods. The first method is the functionalization of a polymer, usually cross-linked polymer support. The conventional catalytic species can be attached to it by appropriate reactions. The second method is the polymerization or copolymerization of a functionalized monomer(s) already containing the necessary catalytic species (Scheme 3.1). Of the two methods, the former is much more established, and various kinds of polymeric catalysts have been synthesized by this method.

The first method, the chemical modification approach, has the following advantages:

1. Its operational simplicity.
2. Microporous or macroporous polymer beads with good physical size and form and any appreciable extent of cross-linking are commercially available.
3. After chemical modification, the obtained functionalized polymer mostly has the same physical form as the pristine polymer.

DOI: 10.1201/9781003656104-3

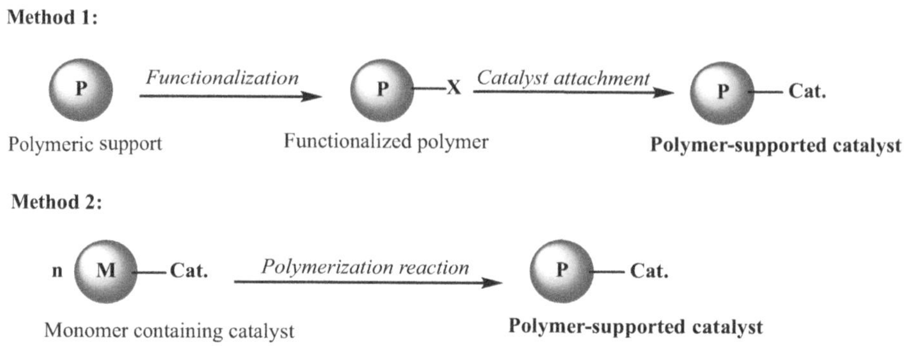

SCHEME 3.1 Preparation of polymeric catalysts.

However, this method suffers from several limitations such as follows:

1. Sometimes the resulted polymer catalyst requires further purification steps which may be laborious.
2. In order to prevent polymer degradation, the modification process should be performed in mild conditions.
3. The obtained polymer catalysts prepared from different modification methods may have dissimilar functional group distribution and exhibit diversity of behavior.
4. Sometimes uniform distribution of the functional groups is not easily accessible because of the poor swelling of the used macroporous beads and the diffusion limitations of mass transport to and from active sites.

The second method can have several advantages, such as follows:

1. Distribution of functional groups in polymer chains is fairly uniform.
2. Characterization of functional groups by analysis of the corresponding pristine monomers.
3. The easier control of the loading of functional groups, e.g., by controlling the feed monomer content or by copolymerization.
4. It is a clean way to obtain polymeric catalyst with high purity because the supports are not contaminated by the functional groups remaining from the chemical modifications.

However, there are a number of disadvantages in using this method, such as follows:

1. Monomers with desired functional groups are not commercially available and their preparation is often difficult.
2. The incompatibility of various functional groups in reactive monomers with polymerization conditions.
3. Monomers of high purity are needed to obtain polymeric catalysts with high molecular weight.

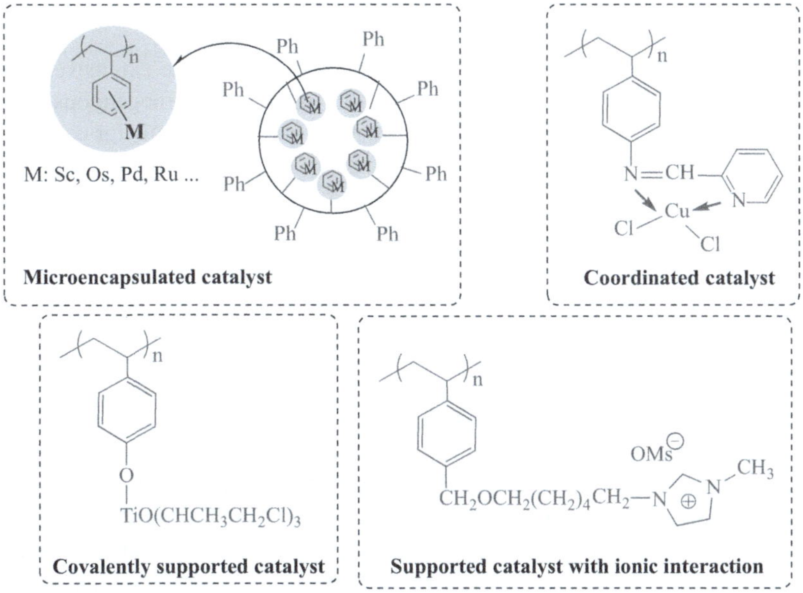

SCHEME 3.2 The anchoring of the catalyst species on polymer supports through entrapment [7], coordinative binding [8], covalent binding [9], and ionic interactions [10].

There are four main interactions between the catalyst species and the polymer backbone, namely entrapment [7], coordinative binding [8], covalent binding [9], and ionic interactions [10] (Scheme 3.2). Of these, the latter two are the most commonly used because no significant leaching of the catalytic species occurred from polymer support during reactions due to it being chemically bonded to the support or because of strong electrostatic interactions.

As mentioned in Chapter 2, some of the main considerations when choosing which polymer to use as support for catalysts include loading capacity and solvent compatibility (good swelling of the polymer by the solvent). Usually, higher loading is preferred, since this reduces the amount of both polymer and solvent needed. Also, high loading level can obviously enhance reaction rates. The loading capacity of polymer-supported catalysts usually varies, mostly ranging from less than 1 to about 5 mmol/g. On the other hand, the swelling of polymer in reaction solvents plays a key role in the catalyst accessibility and activity. Catalytic selectivity could be tailored by controlling the catalyst loading capacity of the polymer support and also by the solvent used to swell the polymer. In general, the selectivity increases with decreasing swelling ratio of the polymer. Indeed, when the reaction rates are governed by molecular diffusion, polymer catalysts can exhibit significant catalytic selectivity depending on the degree of substrate penetration into the support's pore architecture.

3.2 PVPy AS SUPPORT FOR CATALYST SPECIES

Numerous types of organic and inorganic polymers have been extensively used in heterogeneous catalysis as supports for immobilization of homogeneous catalysts. Among these, PVPy is potentially one of the most useful materials for catalytic processes, either directly using the pyridine rings or by immobilization of the active catalytic species on the pyridine nitrogen. The catalytic performance of PVPy is because of its unique physicochemical characteristics, high loading capacity, easy availability, remarkable stability to neutral and basic conditions, nonhygroscopic, facile functionalization, good swelling of the polymer by the solvents, and metal–ion affinity interactions. A somewhat protonated PVPy serves as adsorbent for negatively charged species, consequently leading to an increase in the local concentration of species close to catalytic active sites and enhancement of the activity. Also, PVPy with many pyridine groups can inhibit acidic condition and provide a bufferic environment for many catalysts, reagents, and metal complexes. As mentioned in Chapter 1, in addition to having catalytic applications, PVPy has unique capabilities in different areas of polymer science and applied chemistry. For example, the PVPy salts are a conductive and polyelectrolyte, and it is stable in the presence of powerful oxidants [11]. PVPy has become significant in the field of chemical engineering owing to its stable basic pyridine group and its capacity to create a charge transfer complex with acidic dopants [12]. Also, it has been employed as molecular imprinting [13], PCM films (phase change materials) [14], ion-exchange resins [15], plutonium sorption [16], and membranes [17].

The most utilized polymer support based on PVPy in heterogeneous catalysis is a copolymer of vinylpyridine and DVB (as cross-linking agent) monomers [18]. Copolymerization (emulsion polymerization method) is usually performed using AIBN initiator and different combinations of suspension agents to assist the generation of porous beads [19,20]. To produce insoluble beads (microporous or gel-type resins) with sufficient mechanical strength, generally a minimum level (usually 1%–5%) of DVB is needed. As mentioned previously, the chemo-physical characteristics of the polymer catalyst were shown to be strongly affected by the degree of cross-linking. More highly cross-linked polymer exhibits increased brittleness, and although it can be suitable for use as a support material, the polymer chains are prone to degradative scission in handling. In addition, increasing the amount of cross-linking agent causes a decrease in the swelling of polymer network, causing less penetration of reagents into the networks of polymer chains, and thus decreasing the reaction rates. One of the most significant categories of cross-linked PVPy polymers that has been utilized consistently as a support for different species, including oxidizing and reducing agents, is P_4VPy, which is cross-linked with 2% DVB (Reillex® 402 ion-exchange resin). BET (Brunauer, Emmett, and Teller) specific surface area of P_4VPy with 2, 20, and 25% DVB has found to be 17 [21], 16.7 [22], and 52 m^2g^{-1}[23], respectively.

P_4VPy and P_2VPy are two commercially available isomers of PVPy [24–26]. Study on evaluation of Flory constant *K*, the molecular weight–viscosity relationship, and the light scattering and intrinsic viscosity of P_4VPy have been documented [27–30]. Also, the effect of the molecular weight and the presence of other polymers on the viscosimetric parameters in the P_4VPy mixtures have been investigated [31,32]. A literature survey indicates that reagents and catalysts supported by P_4VPy demonstrate significantly greater effectiveness compared to those derived from P_2VPy. With few exceptions, nearly all polymeric reagents and catalysts utilizing (PVPy)s have been synthesized from P_4VPy [4,18]. The enhanced efficiency of P_4VPy-supported reagents and catalysts with respect to those of P_2VPy is likely attributable to a higher local concentration and/or improved accessibility of the reagent or catalytic species on this polymer support (due to less steric crowding), allowing for better interaction with the substrates [33–36]. In the following discussion in this chapter, I want to introduce several types of polymeric catalysts based on PVPy in four categories as acidic catalysts, basic catalysts, metal-complex catalysts, and polysoaps and phase transfer catalysts (PTCs).

3.2.1 Acidic Catalysts Based on PVPy

Polymeric acidic catalysts are one of the economically and ecologically important species in the field of catalysis due to their environmental compatibility combined with good yield and selectivities, ease of handling, and easy recovery and reusability [37,38]. Among polymer materials used as a support for acidic catalysts, PVPy has gained more attention. Several types of PVPy-supported acidic catalysts have been synthesized and employed as alternatives to common mineral liquid acids (e.g., H_2SO_4, HCl) for catalyzing chemical transformations (Table 3.1). In the following subsections, they are discussed under two subgroups: Brønsted acid catalysts and Lewis acid catalysts. I try to explain some of these catalysts by the related reactions and mechanisms presented in schemes.

3.2.1.1 Brønsted Acid Catalysts Based on PVPy

Among the catalysts shown in Table 3.1, Brønsted acid catalysts are gaining much attention. Representative examples of applications of these types of catalysts are shown in Scheme 3.3. Most of these catalysts show activities comparable to those of the corresponding unsupported counterparts. For example, it has been reported that either partially protonated P_4VPy or non-protonated P_4VPy can act as catalysts for solvolyses of nitrophenyl acetates such as potassium 3-nitro-4-acetoxybenzenesulfonate (NABS) and 2,4-dinitrophenyl acetate (DNPA) (Table 3.1, entry 1, Scheme 3.3 (i, ii)) [39]. The results showed that the catalytic reactivity of P_4VPy-HCl is more than that of non-protonated P_4VPy. Also, when P_4VPy-HCl was used as catalyst, the rate of solvolyses of NABS is more than that of DNPA. As mentioned previously, the faster solvolyses of NABS compared to DNPA can be ascribed to the greater tendency of the protonated PVPy segments with positive charge to attract NABS with negative charge, quite near to basic nitrogen units which are responsible for the solvolysis. However, the neutral DNPA substrate does not follow the same behavior as NABS.

Also, P_4VPy-supported HCl, HBr, HI, $HClO_4$, and TsOH have been prepared and used for the synthesis of drug nifuroxazide, from the reaction of the corresponding aldehydes or O,O-acetals with 4-hydroxybenzhydrazide (Table 3.1, entry 7, Scheme 3.3 (iv)) [50]. It was found that the catalytic activity of polymer-supported

TABLE 3.1
Polymeric Acid Catalysts Based on PVPy

Entry	Polymer Catalysts with Their Applications
1	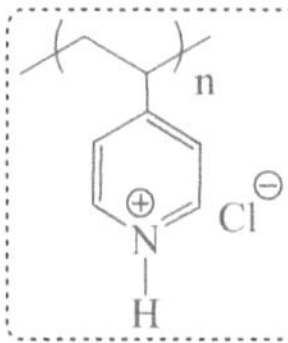 Poly(4-vinylpyridinium)chloride
	Solvolyses of acetates [39]
	Acetalization and esterification of carbonyl compounds [40]
	Tetrahydropyranylation of alcohols [41]
	Aromatic nitration [42]
2	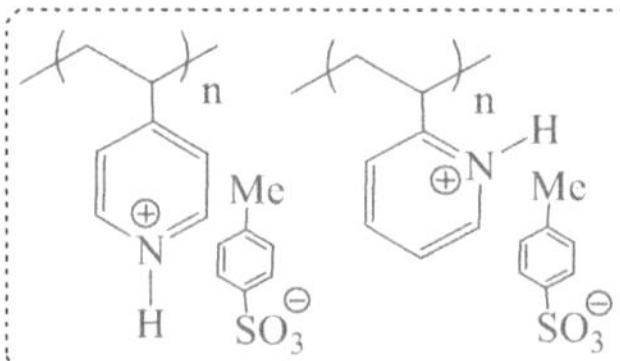 Poly(4-vinylpyridinium) *p*-toluenesulfonate and Poly(2-vinylpyridinium) *p*-toluenesulfonate
	Tetrahydropyranylation of alcohols [43]
	Hydrolysis of tetrahydropyranyl ethers [44]
	Preparation of sulfonamides from amines [45]
3	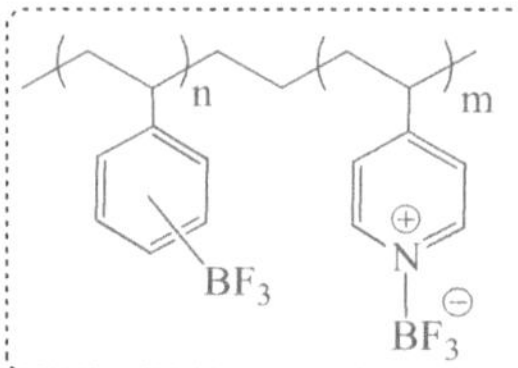 Poly(styrene-*co*-4-vinyl pyridine)-supported boron trifluoride
	Preparation of acetals, esters, and ethers [46]
4	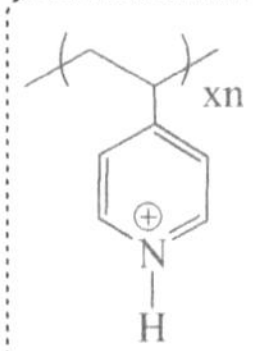 $PMo_{12}O_{40}^{3-}$ (X= 3), $SiMo_{12}O_{40}^{4-}$ (X= 4) or $H_2[CoW_{12}O_{40}]^{4-}$ (X= 4) P4VPy-supported Keggin-type heteropolyacids
	Oxidation of 1,4-diols [47]
5	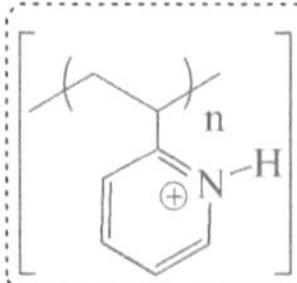 $Ce(NO_3)_4^{2-}$ P_2VPy-supported ceric nitrate
	Ring opening of epoxides [48]

(Continued)

TABLE 3.1 (*Continued*)
Polymeric Acid Catalysts Based on PVPy

Entry	Polymer Catalysts with Their Applications
6	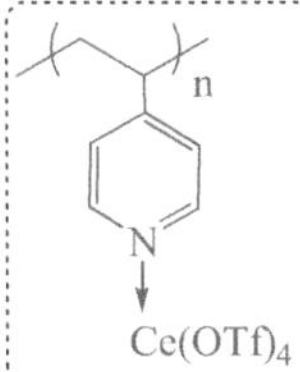 P_4VPy-supported ceric triflate Synthesis of thiiranes [49]
7	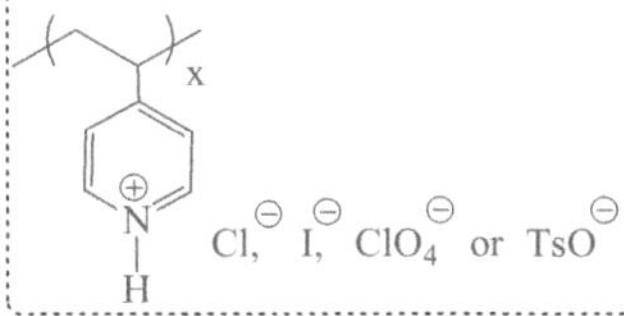 Poly(4-vinylpyridinium) cholride, Poly(4-vinylpyridinium) iodide, Poly(4-vinylpyridinium) prechlorate and Poly(4-vinylpyridinium) *p*-toluenesulfonate Synthesis of nifuroxazide drug [50]
8	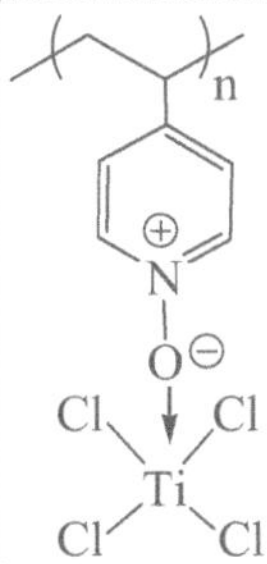 Poly(4-vinylpyridine-*N*-oxide)-supported titanium tetrachloride Esterification reactions [51]
9	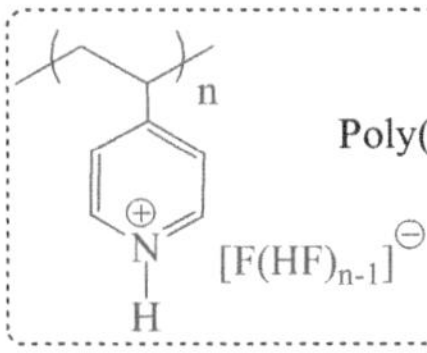 Poly(4-vinyl pyridinium) poly(hydrogen fluoride) Isoparaffin-olefin alkylation [52]
10	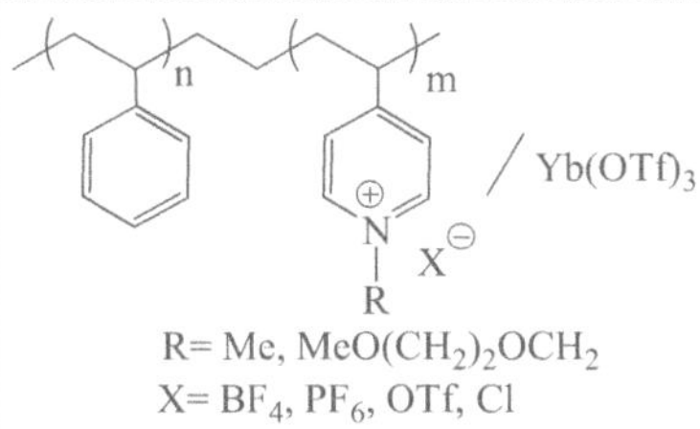 Quaternized poly(styrene-*co*-4-vinyl pyridine)- supported ytterbium triflate *β*-Amino ketone synthesis [53]

(*Continued*)

TABLE 3.1 (*Continued*)
Polymeric Acid Catalysts Based on PVPy

Entry	Polymer Catalysts with Their Applications
11	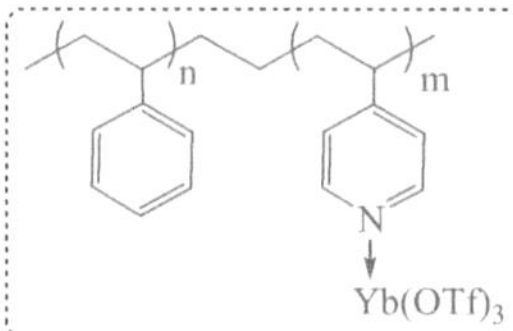 Poly(styrene-*co*-4-vinyl pyridine)-supported ytterbium triflate *β*-Amino ketone and tetrahydroquinoline synthesis [54] *β*-Amino ketone synthesis [55]
12	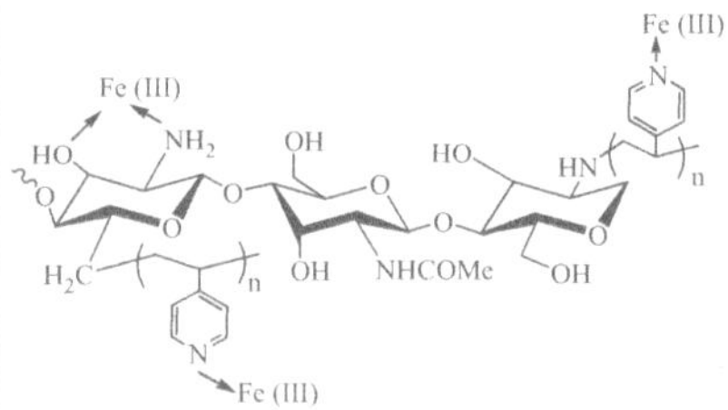 Chitosan-*g*-P_4VPy-Fe (III) Michael additions [56,57]
13	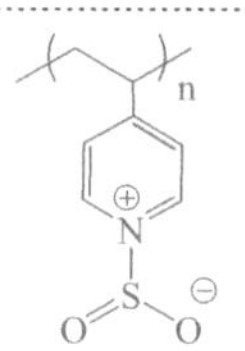 $P_4VPy\text{-}SO_2$ Synthesis of *α*-aminonitriles [58]
14	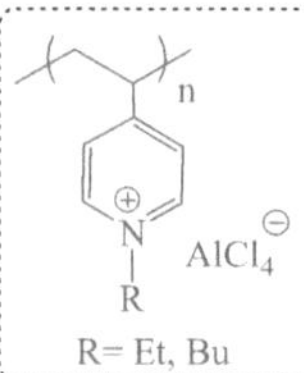 Poly(4-*n*-butyl-4-vinylpyridinium) chloroaluminate (p([BVIm]$AlCl_4$)) Diels–Alder reaction [59]
15	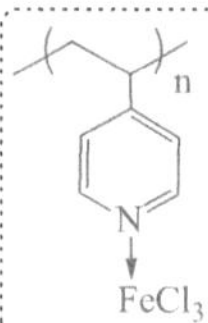 P_4VPy- supported ferric chloride Oxidation of cyclohexane [60] Deprotection of aldoximes and hydrazine [61] Oxidation of toluene [62] Oxidation of benzylic methylenes and alcohols [63]

(*Continued*)

TABLE 3.1 (*Continued*)
Polymeric Acid Catalysts Based on PVPy

Entry	Polymer Catalysts with Their Applications
16	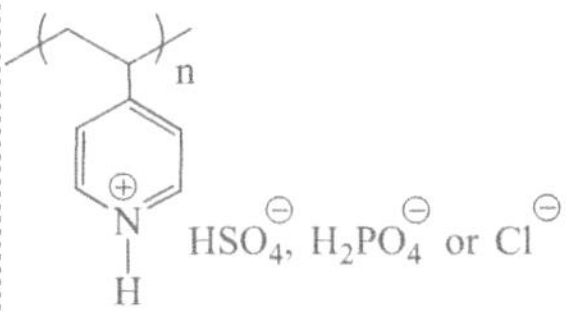 Poly(4-vinyl pyridinium) hydrogen sulphate, Poly(4-vinyl pyridinium) hydrogen phosphate, and Poly(4-vinyl pyridinium) chloride
	Acetylation of amines and alcohols [64]
17	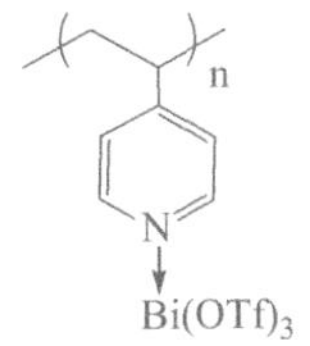 P_4VPy- supported bismuth(III) triflate
	Silylation of alcohols [65]
18	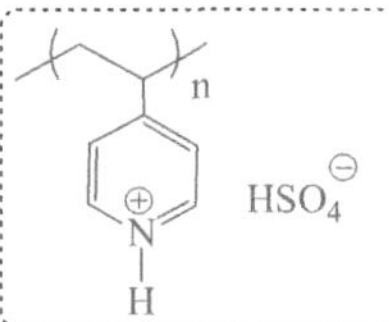 Poly(4-vinylpyridinium) hydrogen sulfate
	Synthesis of coumarin derivatives [66,67]
	Synthesis of 1,1-diacetates [68,69]
	Synthesis of xanthene derivatives [70–72]
	Preparation of azidohydrins from epoxides [73]
	Protection of aldehydes to acylals [74]
	Conversion of epoxides to thiocyanohydrins [75]
	Synthesis of indenonaphthapyrans [76]
	Synthesis of sulfoxides and *gem*-dihydroperoxides [77]
	Synthesis of benzo[*b*]fluoren-9,11-diones [78]
	Aromatic nitration [79]
19	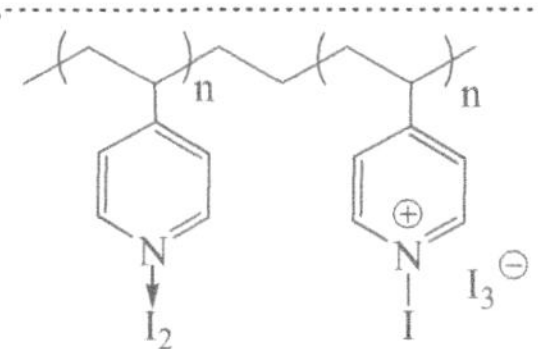 Poly(*N*-iodo-4-vinylpyridinium) triiodide
	Ring opening of epoxides [80]
	Preparation of cyclic carbonates [81]

(*Continued*)

TABLE 3.1 (*Continued*)
Polymeric Acid Catalysts Based on PVPy

Entry	Polymer Catalysts with Their Applications
20	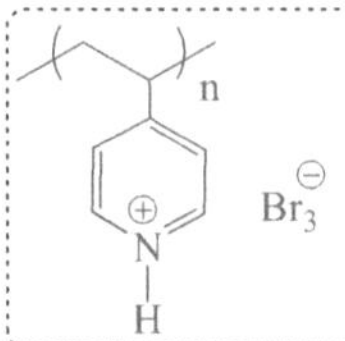 Poly(4-vinylpyridinium) tribromide Preparation of acetate derivatives [82]
21	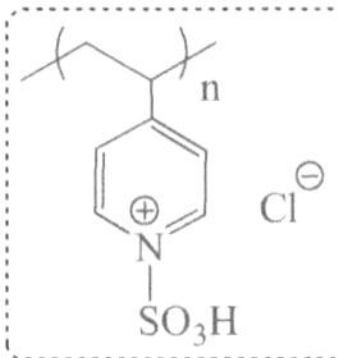 Poly (*N*- sulfonic acid-4-vinylpyridinium) chloride Preparation of 1,1-diacetates [83] Preparation of xanthene derivatives [84] *N*-BOC protection of amines [85]
22	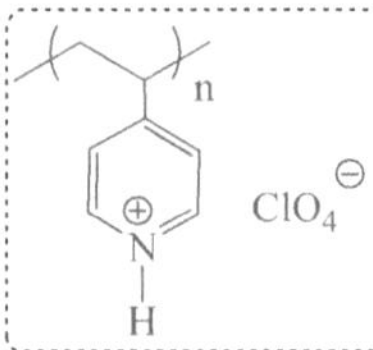 Poly(4-vinylpyridinium) perchlorate Acetylation of amines and alcohols [86] Synthesis of substituted coumarins [87] Preparation of 1,1-diacetates [88] Synthesis of biscoumarins and bis(indolyl)methanes [89]
23	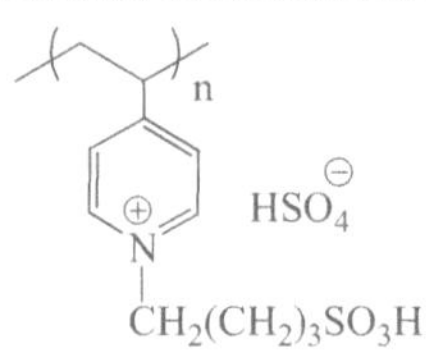 Poly(4-*n*-butane sulfonic acid-4-vinylpyridinium) hydrogen sulfate Preparation of bis(pyrazolyl)methanes [90] 1,8-Dioxo-octahydroxanthene synthesis [91] Synthesis of β-acetamido ketones [92] Synthesis of biscoumarin derivatives [93]

(*Continued*)

TABLE 3.1 (*Continued*)
Polymeric Acid Catalysts Based on PVPy

Entry	Polymer Catalysts with Their Applications
24	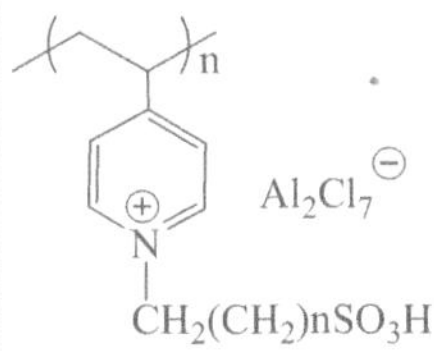 n: 3, Poly(4-*n*-butane sulfonic acid-4-vinylpyridinium) chloroaluminate n: 2, Poly(4-*n*-propane sulfonic acid-4-vinylpyridinium) chloroaluminate *n*: 3, Synthesis of biscoumarins [94] *n* = 2. The alkylation reaction using a real refinery feedstock of paraffins/olefins [95]
25	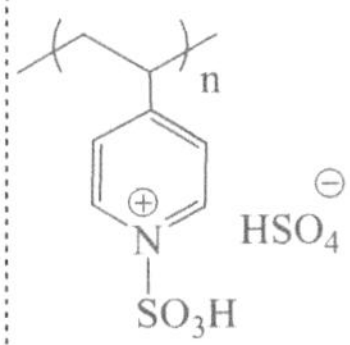 Poly (*N*- sulfonic acid-4-vinylpyridinium) hydrogen sulfate Synthesis of xanthene derivatives [96] Acetylation and 1,1-diacetylation reactions [97]
26	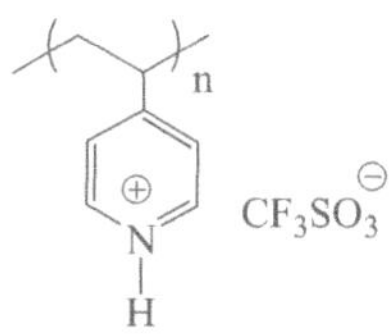 Poly(4-vinylpyridinium) poly(triflic acid) Friedel–Crafts reactions [98]
27	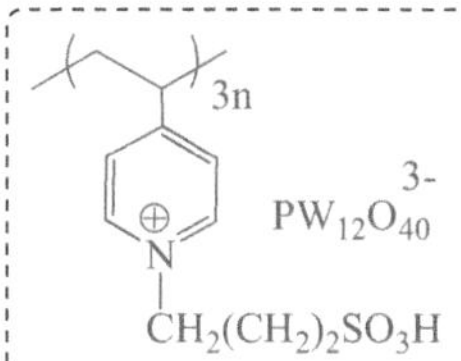 Poly(4-*n*-propane sulfonic acid-4-vinylpyridinium)PW$_{12}$O$_{40}$ $PW_{12}O_{40}^{3-}$: Dodecatungstophosphate anion Synthesis of *β*-amino carbonyl compounds [99] Preparation of 2,3-dihydroquinazolin-4(1*H*)-ones [100]

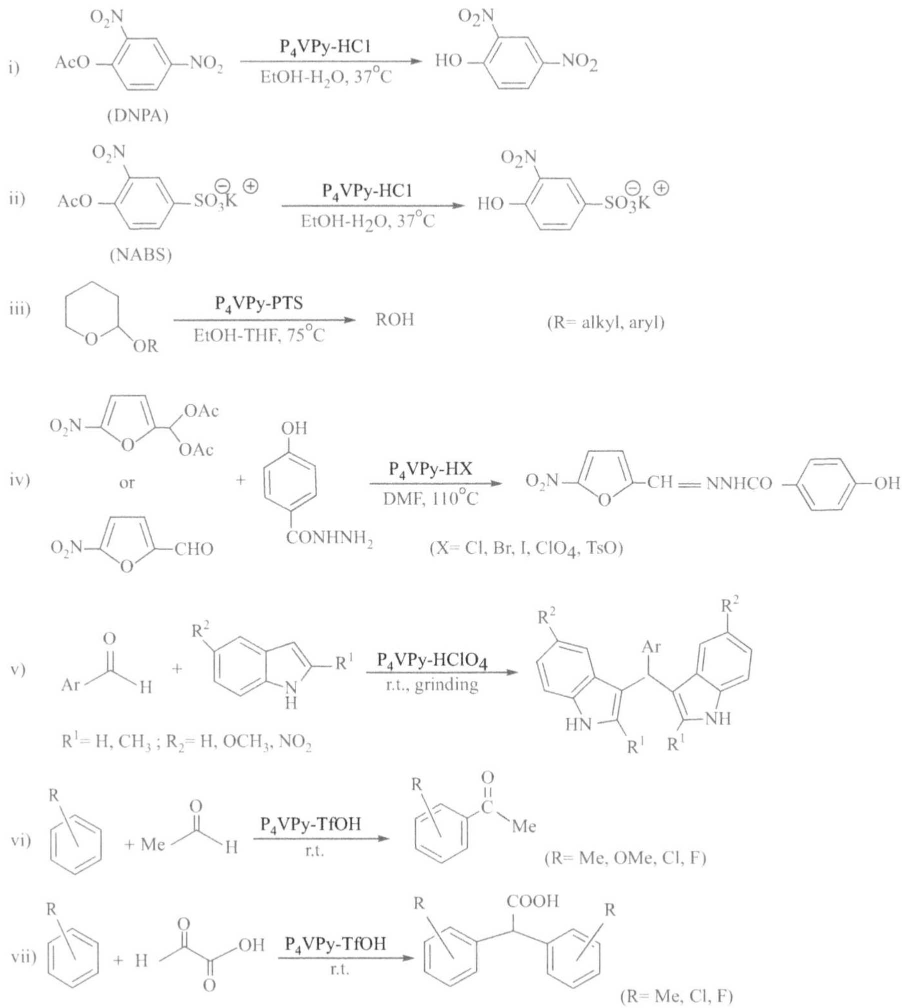

SCHEME 3.3 P_4VPy-supported Brønsted acidic catalysts for the solvolyses of esters (i, ii) [39], hydrolysis of tetrahydropyranyl ethers (iii) [44], the synthesis nifuroxazide (iv) [50], the synthesis of bis(indolyl)methanes (v) [89], and Friedel–Crafts hydroxyalkylation and acylation reactions (vi, vii) [98].

acids is strongly influenced by counteranions. The order of reaction rates based on the pseudo-first-order rate constants obtained from kinetic studies were as follows: iodide > perchlorate > chloride = bromide > tosylate. Also, the results show that the rate of the reaction was affected by the pK_a value of the acidic catalysts as well as the extent of their deactivation by the formed by-product. The catalysts were reusable for several times without appreciable change in their efficiency.

P_4VPy-supported triflic acid (P_4VPy-TfOH) is another Brønsted acid catalyst, which has been investigated and employed for organic transformations (Table 3.1, entry 26, Scheme 3.3 (vi, vii)) [98]. It is well known that triflic acid is a very reactive protic acid with broad applications including catalysis [101–103]. However, it has serious problems of high moisture and air sensitivity, highly corrosive, and the production of waste, and thus, its heterogenization is of vital importance in chemistry [104,105]. In contrast to triflic acid, P_4VPy-TfOH is nonvolatile and less hygroscopic with easy handling without any corrosion problems. The results showed that the polymer to triflic acid loading ratio of 1:10 proved to be favorable for catalytic applications. Also, this solid acid catalyst is regenerable.

Other types of PVPy-supported Brønsted acids are sulfonated materials which have been extensively used as strong solid acid catalysts in organic synthesis. The first such sulfonated catalysts were prepared by the sulfonation of P_4VPy with H_2SO_4 or $ClSO_3H$. Scheme 3.4 shows the application of some of these type catalysts. For example, poly(*N*-sulfonic acid-4-vinylpyridinium) hydrogen sulfate (P_4VPy-[SO_3H][HSO_4]) has been employed as a dual acidic catalyst in the preparation of different types of xanthenes (Table 3.1, entry 25, Scheme 3.4 (v)) [96]. The polymeric catalyst was synthesized by the reaction of P_4VPy and $ClSO_3H$, resulting in the formation of P_4VPy-[SO_3H]Cl which subsequently undergoes anion exchange with H_2SO_4. The catalyst is reusable.

Another kind of sulfonated catalysts based on PVPy were obtained by quaternization of PVPy with propanesultone or butanesultone followed by reaction with H_2SO_4, $AlCl_3$, or $H_3PW_{12}O_{40}$ (Table 3.1, entries 23, 24, and 27, Scheme 3.5). Recently, [P_4VPy-BuSO_3H]Cl-X($AlCl_3$) has been reported as a supported ionic liquid catalyst (Table 3.1, entry 24, Scheme 3.5 (iii)) [94]. It was synthesized from P_4VPy, butanesultone, and excess amounts of $AlCl_3$ (Scheme 3.6) and used in biscoumarin synthesis. Various kinds of this catalyst containing different amounts of $AlCl_3$ were prepared, and the results showed that those with higher amounts of $AlCl_3$ were more active and both Lewis and Brønsted acid units ($AlCl_3$ and $-SO_3H$, respectively) were involved in catalyzing the reactions. Unlike chloroaluminate ionic liquids, which are susceptible to hydrolysis, [P_4VPy-BuSO_3H]Cl-X($AlCl_3$) is resistant to moisture and can be hold for several months. Probably, the hydrophobic property of P_4VPy causes entrapment of $AlCl_3$ until the catalyst is swollen in a suitable solvent and then the Lewis acid is exposed to the reactants. It is worth nothing that a very similar structure of the aforementioned catalyst, termed as poly(4-vinyl pyridine *N*-propyl sulfonic acid) heptachloroaluminate (P_4VPyPS-Al_2Cl_7), was recently reported as catalyst for the alkylation reaction between isoparaffins and olefins [95]. This process is significantly important in the production of high-octane fuels [106].

Structures similar to the above example, except that heteropolyacid has replaced aluminum chloride, have also been reported (Table 1, entry 27) [99,100]. Finally, the results of a very recent report on the investigation of morphology and rheology of composites of PS-*co*-P_2VPy and phosphotungstic acid, one kind of polyoxometalates (POMs), revealed that the incorporated phosphotungstic acid successfully protonates the 2-VPy monomers, leading to phase segregation due to the significant ionic attraction between the protonated 2-VPy monomers and the deprotonated POM molecules [107].

3.2.1.2 Lewis Acid Catalysts based on P_4VPy

Following investigations on PVPy-supported Brønsted acid catalysts, P_4VPy-supported Lewis acids have been developed. Representative examples of

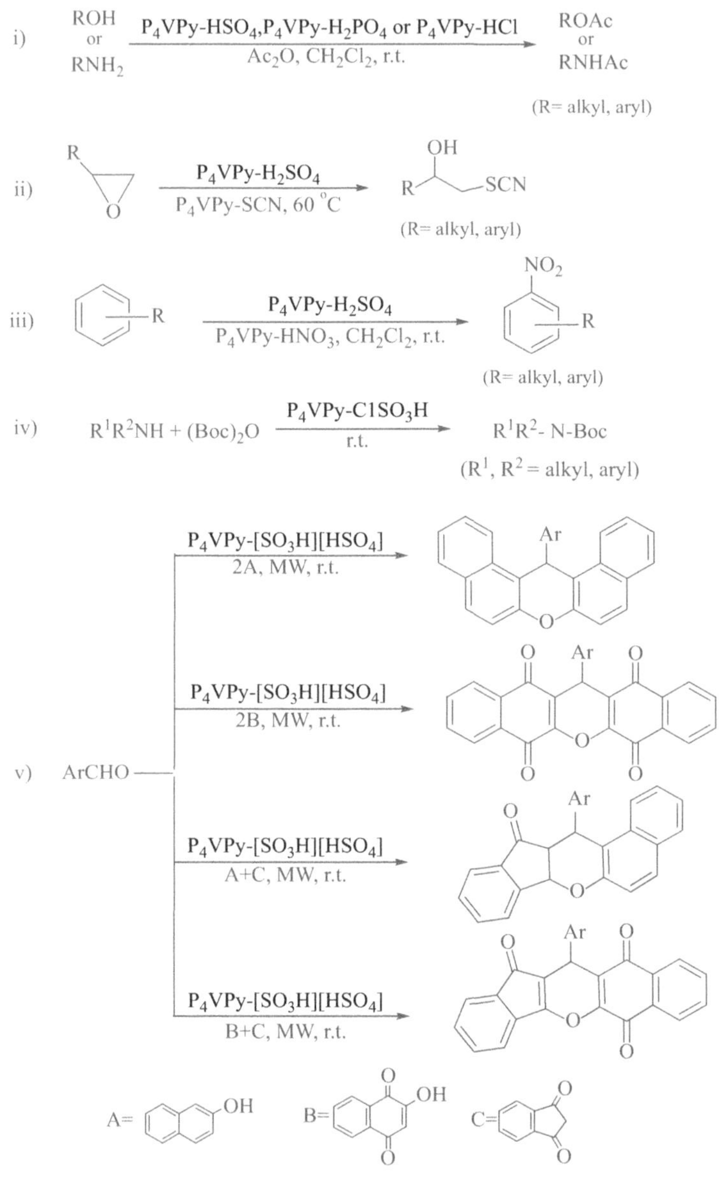

SCHEME 3.4 Sulfonated P_4VPy as a catalyst for acetylation reactions (i) [64], the ring opening epoxides (ii) [75], nitration of aromatic compounds (iii) [79], *N*-BOC protection of amines (iv) [85], and the synthesis of xanthene derivatives (v) [96].

application of these catalysts are shown in Scheme 3.7. For example, P_4VPy and P_2VPy-supported $Ce(OTF)_4$ have been reported as heterogeneous catalysts for the preparation of thiiranes (Table 3.1, entry 6, Scheme 3.7 (iii)) [49]. It is worth noting that in this reaction, the reactivity of P_4VPy and P_2VPy-$Ce(OTF)_4$ is more than

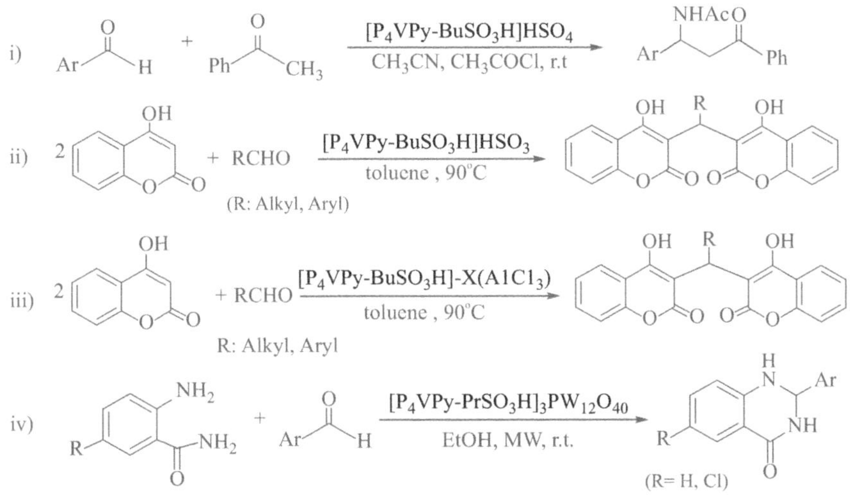

SCHEME 3.5 Quaternized P_4VPy-supported Brønsted acid catalysts for the preparation of *β*-acetamido ketones (i) [92], biscoumarins (ii, iii) [93,94], and 2,3-dihydroquinazolin-4(1*H*)-ones (iv) [100].

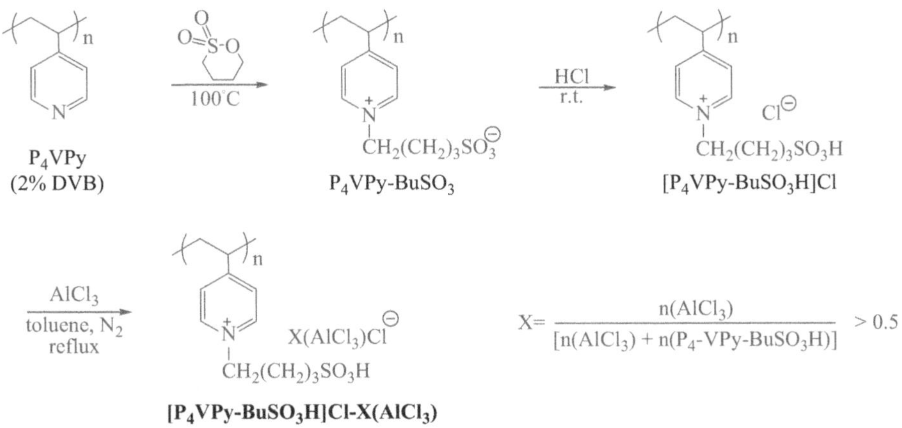

SCHEME 3.6 Synthesis of $[P_4VPy\text{-}BuSO_3H]Cl\text{-}X(AlCl_3)$ [94].

that of the $Ce(OTF)_4$ alone. This can be ascribed to the greater affinity of $Ce(OTF)_4$ toward anionic nucleophiles and complex formation, consequently losing its reactivity. Also, it was found that $P_2VPy\text{-}Ce(OTF)_4$ has lower capacity and less reactivity than $P_4VPy\text{-}Ce(OTF)_4$, due to steric hindrance. One notable achievement of this solid acid catalyst is that it can act as a stereospecific catalyst. For example, R-(+)-styrene oxide and NH_4SCN in the presence of $P_4VPy\text{-}Ce(OTf)_4$ produced S-(−)-styrene sulfide with 87% enantiomeric excesses.

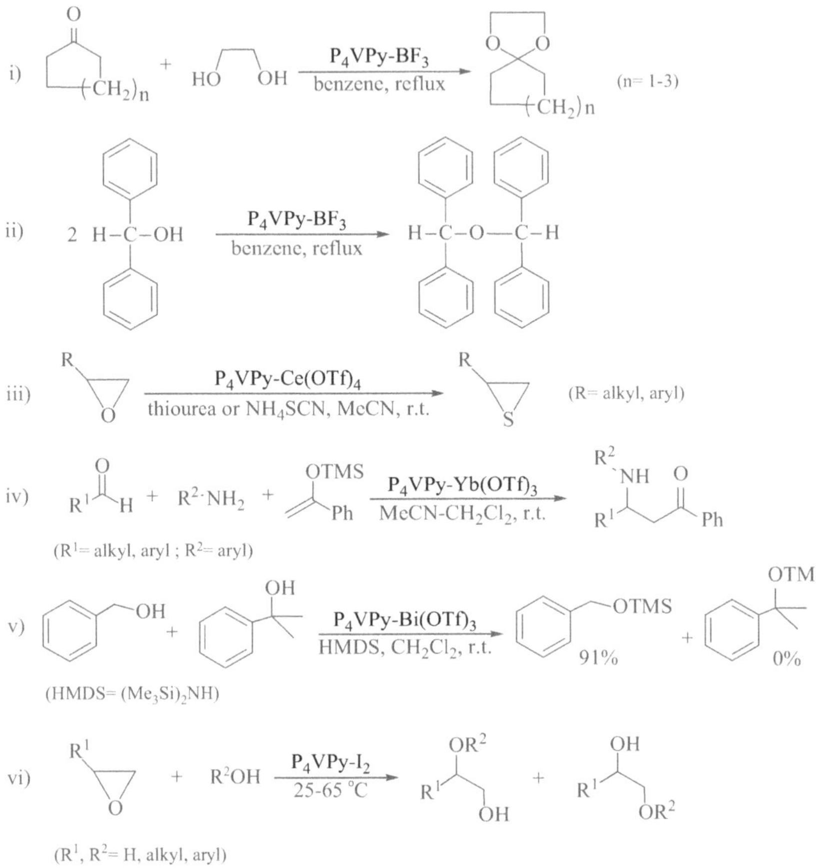

SCHEME 3.7 P_4VPy-supported Lewis acid catalysts for the protection of carbonyl compounds and conversion of diphenylmethanol into ether (i, ii) [46], synthesis of thiiranes (iii) [49], β-amino ketone synthesis (iv) [54], silylation of alcohols and phenols (v) [65], and ring opening of epoxides (vi) [80].

Poly(1-butyl-4-vinylpyridinium) chloroaluminate (abbreviated as P([BVPy]$AlCl_4$)) is another reported heterogeneous Lewis acid catalyst (Table 3.1, entry 14) [59]. Scheme 3.8 shows the catalytic application and the synthetic pathways of P([BVPy]$AlCl_4$) from its starting materials, P_4VPy, *n*-butyl chloride, and $AlCl_3$. The results revealed that in Diels–Alder reactions, P([BVPy]$AlCl_4$) catalyst is as active and stereoselective as its homogeneous counterpart and $AlCl_3$ while having the advantages of heterogeneous catalysts, e.g., easy isolation and recyclability.

Very recently, poly(4-vinylpyridine)-supported iodine (P_4VPy-I_2) has been reported as a polymer catalyst for the preparation of cyclic carbonates through the reaction between oxiranes and CO_2 (Table 3.1, entry 19) [81]. Scheme 3.9 provides a plausible mechanism for this reaction. In the first step, P_4VPy reacts with I_2 to give complex (I) through two consecutive reactions. After that, the nucleophilic attack of I_3^- counteranion to oxirane gave the intermediate (II). On the other hand, the

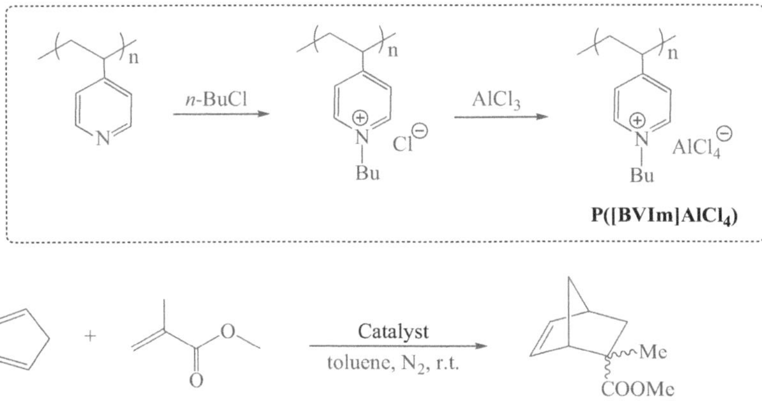

Catalyst	Yield (%)
P([BVIm]AlCl$_4$)	97.7% , 60.6:39.4 (Endo/Exo)
[BVPy]AlCl$_4$ (1-butyl-3-vinylpyridinium chloroaluminate)	98.3% , 60.8:39.2 (Endo/Exo)
AlCl$_3$	97.2% , 60.3:39.8 (Endo/Exo)

SCHEME 3.8 Preparation of P([BVPy]AlCl$_4$) as a heterogeneous catalyst for the Diels–Alder reaction [59].

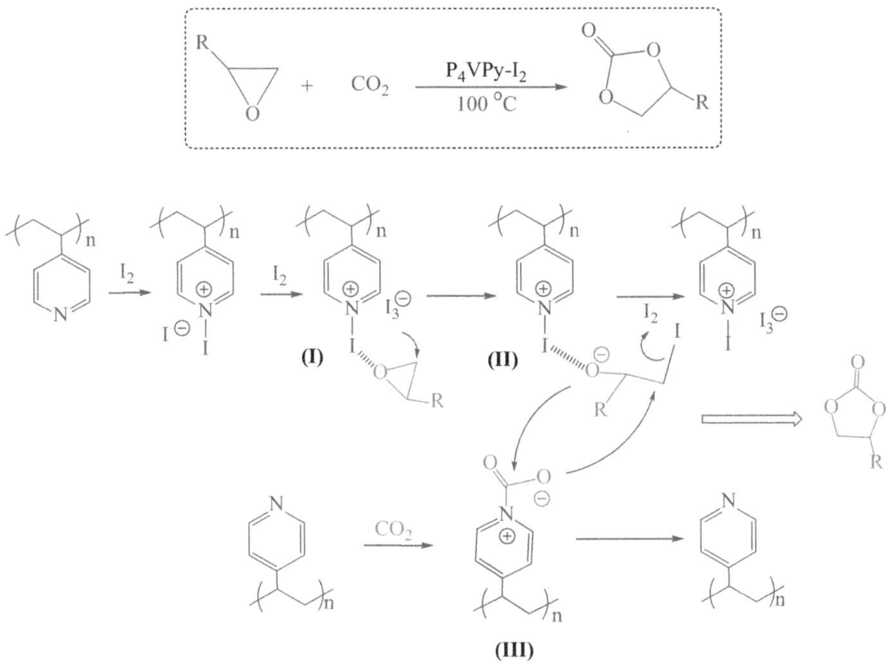

SCHEME 3.9 A plausible mechanism for cyclic carbonate synthesis using P$_4$VPy-I$_2$ catalyst [81].

remaining unreacted pyridine rings in P_4VPy react with CO_2 to produce intermediate (III). Finally, a coupling reaction between intermediates (II) and (III), intermolecular nucleophilic ring closure, resulted in the clean formation of cyclic carbonate product. It is notable that after four successful uses of the catalyst, P_4VPy can be obtained by treating recovered catalyst with $Na_2S_2O_3$.

3.2.2 Basic Catalysts Based on PVPy

Polymeric basic catalysts are one of the economically and ecologically important species in catalysis, and the replacement of the conventional mineral bases with solid bases is gaining importance in organic and industrial chemistry [108]. Among polymer materials used as a base catalyst or as support for basic catalysts, PVPy has gained more attention. Representative polymeric basic catalysts based on PVPy with their applications were summarized in Table 3.2.

TABLE 3.2
Polymeric Basic Catalysts Based on PVPy

Entry	Polymer Catalysts with Their Applications
1	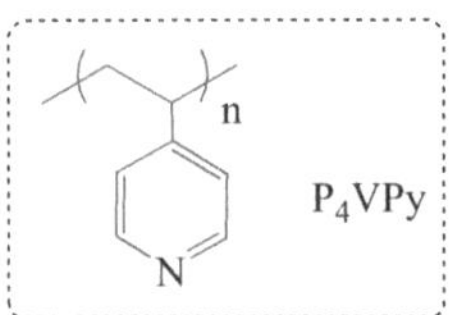
	Synthesis of thioesters [109]
	Synthesis of symmetrical anhydrides [110]
	As reagent/catalyst for the synthesis of acid anhydrides [111]
	aza-Michael reaction of secondary amines and carbamates with α, β-unsaturated esters, cyanides, and ketones [112]
	Synthesis of dimethyl carbonate (DMC) [113]
	Selective methanolysis of methyl and methylene bromides [114]
	Hydrolysis of methyl bromide to methanol and dimethyl ether [115]
	Knoevenagel condensation reaction [116]
	Monoacylation of 2-O-α-D-glucopyranosyl-L-ascorbic acid [117]
	O-TMS protection of alcohols and *N*-BOC protection of amines [118]
	Synthesis of chromene derivatives [119]
	Synthesis of tetrahydrobenzo[b]pyran derivatives [120]
	Isomerization of maleic acid to fumaric acid [121]
	Synthesis of pyrano[2,3-c]pyrazoles and dihydropyranochromenes [122]
	Synthesis of pyrano[2,3-c]pyrazoles [123]
	Synthesis of benzoyl fumarate [124]
	Synthesis of biscoumarin and dihydropyrano[3,2-c]chromene derivatives [125]
	Preparation of cyclic carbonate via cycloaddition of CO_2 to epoxides [126,127]

(Continued)

TABLE 3.2 (*Continued*)
Polymeric Basic Catalysts Based on PVPy

Entry	Polymer Catalysts with Their Applications
2	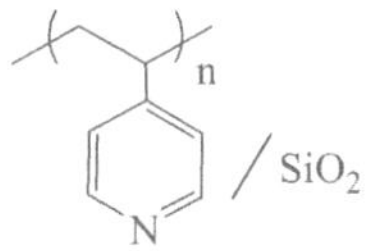 P_4VPy-supported silica gel Synthesis of cyclic carbonates [128]
3	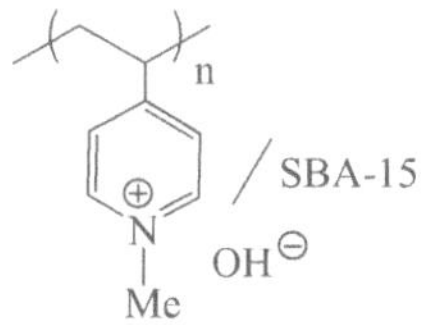 Poly(4-methy-4-vinyl pyridinium) hydroxide-supported SBA-15 Knoevenagel condensation reaction [129]
4	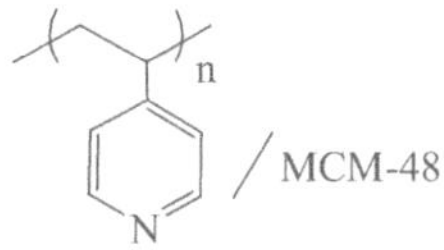 P_4VPy-supported MCM-48 One-pot synthesis of 4H-chromenes [130]
5	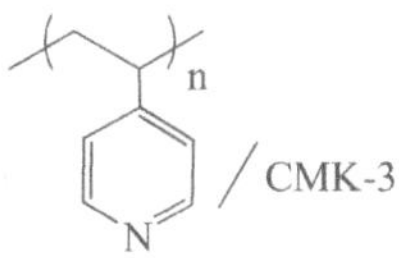 P_4VPy-supported CMK-3 Knoevenagel condensation reaction [131]
6	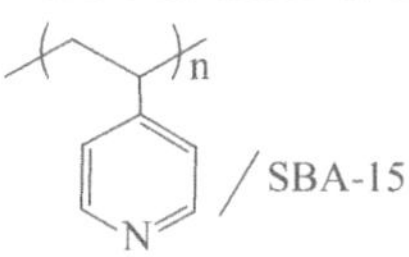 P_4VPy-supported SBA-15 Knoevenagel condensation reaction [132]
7	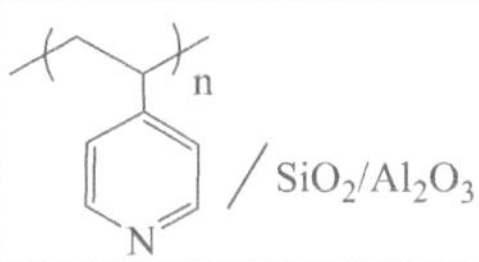 P_4VPy-supported SiO_2/Al_2O_3 Knoevenagel condensation reaction [133]

(*Continued*)

TABLE 3.2 (*Continued*)
Polymeric Basic Catalysts Based on PVPy

Entry	Polymer Catalysts with Their Applications
8	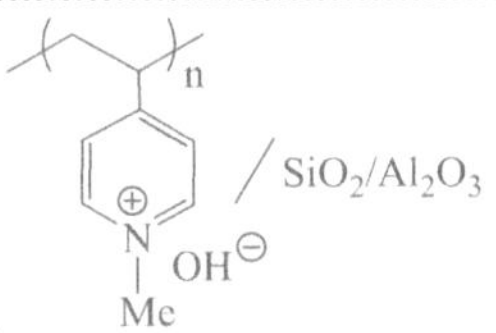 **Poly(4-methyl-4-vinyl pyridinium) hydroxide-supported SiO_2/Al_2O_3** Knoevenagel condensation reaction [134]
9	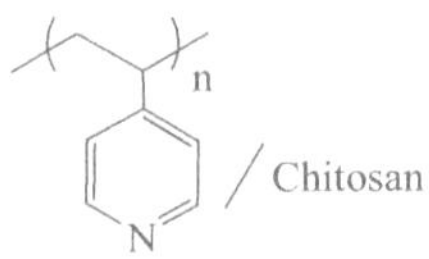 P_4VPy- supported chitosan Synthesis of thiazoles, 1,3-thiazines, and thiazolidine [135] Synthesis of arylazothiazoles and 1,3,4-thiadiazoles [136] Synthesis of coumarin derivatives [137]

As shown in Table 3.2 (entry 1), P_4VPy alone has been employed as an effective base catalyst in organic chemistry. In comparison with other commercially available solid base catalysts, P_4VPy is cheap, commercially available, safe, easy to handle, and environmentally benign. Usually, cross-linked P_4VPy (2% DVB) with a surface area of 17 m^2/g and a particle size of ~60 mesh was employed for the reactions. The basicity of P_4VPy (pK_a of pyridinium ion = 5.23) originates from the nitrogen center of the pyridyl groups bearing a lone pair of electrons that remains nonoverlapping with the aromatic heterocyclic ring [114,138,139]. Scheme 3.10 shows some base-catalyzed reactions using P_4VPy. For example, the synthesis of dimethyl carbonate via transesterification of ethylene carbonate with methanol using P_4VPy as a recyclable base catalyst with high yield has been reported (Table 3.2, entry 1, Scheme 3.10 (ii)) [113]. Also, a comparative analysis utilizing polyvinylpyrrolidone as a catalyst under the same reaction conditions revealed that it produced a lower yield of dimethyl carbonate.

Fumaric acid serves as a significant industrial intermediate compound, which is often synthesized through the isomerization of maleic acid. To my knowledge, thiourea remains the most commonly utilized catalyst in the isomerization of maleic acid for the industrial synthesis of fumaric acid, due to its affordability. However, the produced fumaric acid is of low purity and necessitates additional purification. The catalyst is homogeneous, and hence its separation can be tedious. Recently, P_4VPy (25% DVB) was employed as a substitute for thiourea in the catalytic isomerization of maleic acid (Table 3.2, entry 1, Scheme 3.10 (vi)) [121]. An impressive isomerization conversion rate of 86% was achieved, attributed to the high nucleophilicity of

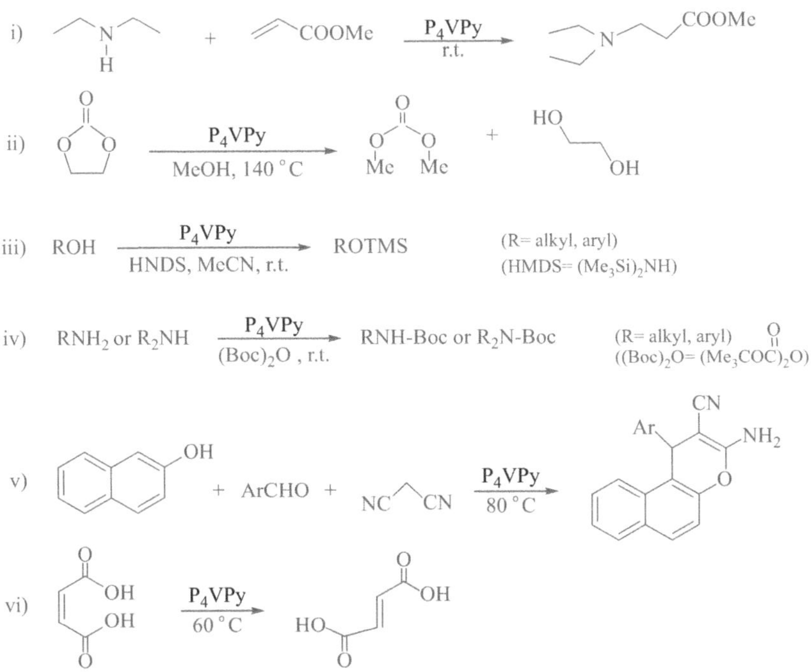

SCHEME 3.10 P_4VPy as a base catalyst for the aza-Michael reaction (i) [112], the synthesis of dimethyl carbonate (ii) [113], chemoselective *O*-TMS protection of alcohols and phenols and *N*-BOC protection of amines (iii, iv) [118], the synthesis of chromene derivatives (v) [119], and isomerization of maleic acid to fumaric acid (vi) [121].

P_4VPy. Additionally, the regenerated polymer exhibited no decline in catalytic activity. It is anticipated that P_4VPy will prove to be a promising candidate for the industrial production of fumaric acid through the isomerization of maleic acid.

Very recently, P_4VPy and P_2VPy were used as catalysts for the cycloaddition of CO_2 with epichlorohydrin to produce chloroprene carbonate (Table 3.2, entry 1, Scheme 3.11) [126]. Three proposed mechanistic pathways (A,B, and C in Scheme 3.11) were introduced for this process as follows: (A) P_4VPy or P_2VPy initially reacts with epichlorohydrin to yield an adduct with zwitterionic structure (I), which is then treated with CO_2 to afford another ionic structure that converts to cyclic carbonate product through an intramolecular cyclization reaction; (B) treatments of P_4VPy with epichlorohydrin produce a quaternized form of P_4VPy (II), which then reacts with second molecule of epichlorohydrin, resulting in the formation of ionic structure (III) which undergoes an intramolecular cyclization reaction to yield cyclic carbonate product; and (C) P_2VPy can directly react with CO_2 molecule to form a carbamate anion (IV) which then reacts with epichlorohydrin to produce a zwitterionic adduct (V) that finally creates cyclic carbonate product through an intramolecular cyclization reaction. It is worth noting that during reactions of P_4VPy, a large amount of bulky dark green solid precipitate was observed, which is related to the formation of quaternized forms of this polymer (mechanism (B), structure (II)). In the case of

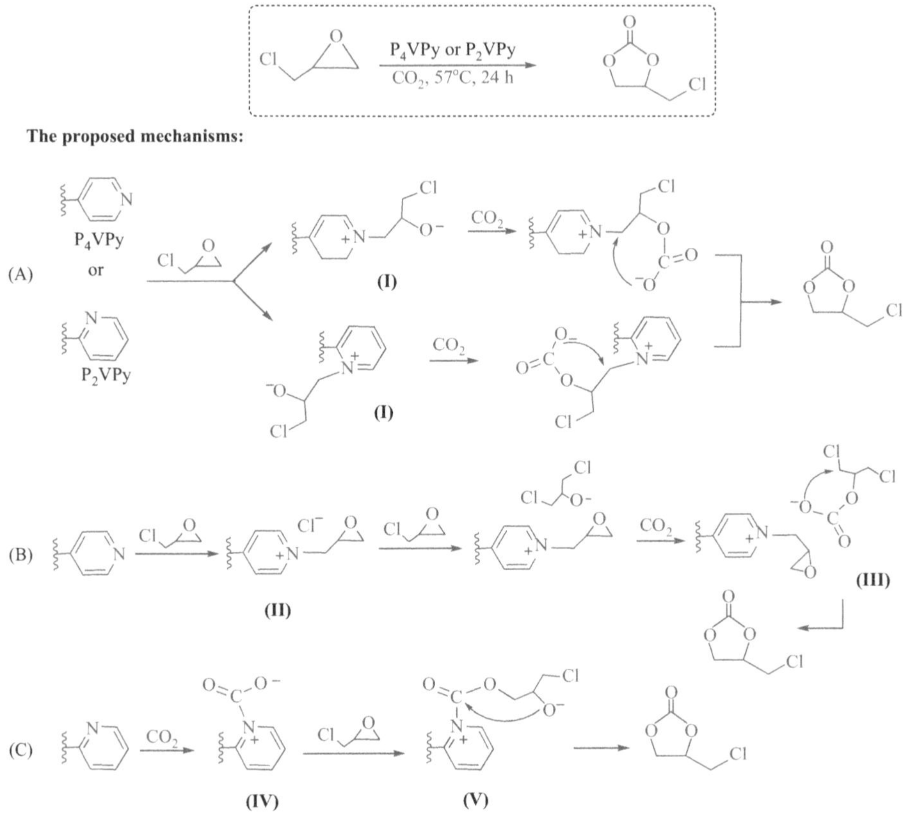

SCHEME 3.11 The preparation of chloroprene carbonate from the cycloaddition of CO_2 with epichlorohydrin in the presence of P_4VPy and P_2VPy [126].

P_2VPy, the reaction proceeded with little or no precipitate formation because it does not undergo the quaternization reaction (mechanism (B)) due to steric crowding at the nitrogen site of its pyridine ring. This study also examined carbonate formation reactions using 2-VPy, 4-VPy, and pyridine (Py) for comparative purposes. The findings indicated that the rate of chloroprene carbonate production follows the order of P_4VPy>P_2VPy>4-VPy>Py>2-VPy, which can be, within experimental error, corresponded to the increasing basicity/nucleophilicity order, i.e., P_4VPy (pK_a= 5.98)>P_2VPy (pK_a= 5.96)>4-VPy (pK_a= 5.5)>Py (pK_a= 5.22)>2-VPy (pK_a= 4.98). Moreover, the results indicated that the catalysts with high steric hindrance (P_2VPy and 2-VPy) display a similar mechanism and epichlorohydrin conversion, just as P_4VPy, 4-VPy, and Py, which are much less sterically hindered.

In addition to the employment of P_4VPy alone as a base catalyst, it has been reported that P_4VPy or its quaternized form with hydroxyl group as counteranion could be immobilized on solid materials (by the sol–gel method) such as SiO_2, MCM-48, CMK-3, SBA-15, and SiO_2/Al_2O_3 and used as solid base catalysts for organic transformations. Typically, Scheme 3.12 shows the applications of P_4VPy/

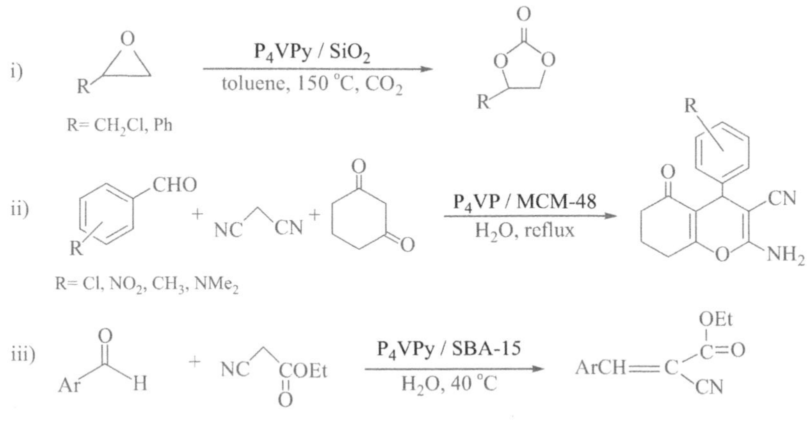

SCHEME 3.12 Immobilized P_4VPy on solid supports as base catalysts for the synthesis of cyclic carbonates (i) [128], synthesis of 4H-chromenes (ii) [130], and Knoevenagel condensation (iii) [132].

SiO_2 (Table 3.2, entry 2) [128], P_4VPy/MCM-48 (Table 3.2, entry 4) [130], and P_4VPy/SBA-15 (Table 3.2, entry 6) [132] as basic catalysts for several organic reactions. Poly(4-methyl-4-vinylpyridinium)hydroxide-supported SiO_2/Al_2O_3 ($P_4MVPH/SiO_2/Al_2O_3$) (Table 3.2, entry 8) is another example; its synthetic pathway and its application for Knoevenagel condensation reaction are shown in Scheme 3.13 [134]. In this report, it has been claimed that the catalyst has higher surface area that it causes more accessibility and good dispersion of the active sites in this composite material compared to P_4VPy alone. Also, the combination of P_4VPy with inorganic solid support can help to achieve easier separation and recovery of the catalyst.

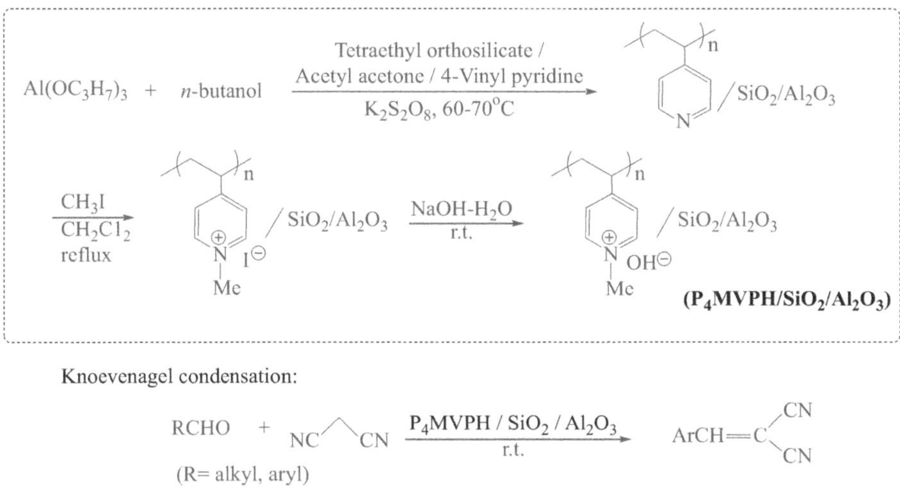

SCHEME 3.13 Synthesis of $P_4MVPH/SiO_2/Al_2O_3$ by sol–gel method and its application for Knoevenagel condensation [134].

3.2.3 Polymer-Bound Metal-Complex Catalysts Based on PVPy

Although numerous homogeneous transition metal complexes demonstrate outstanding catalytic efficiency, their separation from the reaction medium poses significant challenges, complicating their reuse and leading to contamination of the reaction products [140]. Thus, the heterogenization of metal complexes is always a toxicological and environmental challenge, and many attempts have been made by organic chemists for the substitution of homogeneous transition metal complexes with their heterogeneous counterparts. The most practical and widely employed strategy for the heterogenization of a homogeneous metal complex is immobilization into insoluble polymeric supports by covalent bonding or complexation. Among numerous polymer materials used as support for metal-complex catalysts, nitrogen-containing polymers such as PVPy occupy an important place due to the strong affinity of nitrogen atoms to metals and their ability to undergo hydrogen bonding with polar species [141–143]. PVPy is capable of coordinating with a range of transition metals, resulting in the formation of numerous polymer–metal complexes that exhibit significant catalytic activity in organic reactions. In addition to having the catalytic properties, as mentioned in Chapter 1, PVPy-supported metal complexes have also been used in biomolecules separation, as ion conductors and ion-exchange membranes, and for the preparation of composite materials in which colloidal gold was dispersed in the polymer matrix. Polymer-bound metal-complex catalysts based on PVPy are depicted in Table 3.3.

TABLE 3.3
Polymer-Bound Metal-Complex Catalysts Based on PVPy

Entry	Polymer Catalysts with Their Applications
1	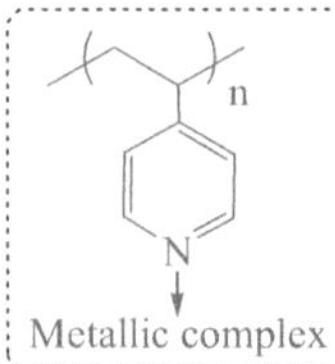 P_4VPy- supported metallic complexes
	copper (II) methoxychloride ($CuOCH_3Cl$)
	Oxidative coupling of diphenylmethanimine [144]
	Ru(2,2′-bipyridine)$_2Cl_2$
	Catalytic reduction of Ce(IV) and oxidation of Fe(II) and also as photocatalysts [145,146]
	Catalytic oxidation of Fe(II) into Fe(III) [147]
	$CuSO_4$
	Catalytic oxidation of ascorbic acid (DH_2) [148]
	Catalytic oxidation of thiosalts by molecular oxygen [149–151]
	Preparation of 3,4-dihydropyrimidin-2(1*H*)ones [152]
	Catalytic phenol oxidation [153]
	Synthesis of 1,4-disubstituted 1*H*-1,2,3-triazoles [154]
	$CuCl_2$

(Continued)

TABLE 3.3 (*Continued*)
Polymer-Bound Metal-Complex Catalysts Based on PVPy

Entry	Polymer Catalysts with Their Applications
	Oxidative cleavage of 4-methyl catechol [155]
	Reduction of nitrobenzene to aniline and azobenzene [156,157]
	Ni(acetylacetonate)$_2$
	Cyclotrimerization of butadiene to cyclododecatriene [158]
	Air oxidation of ethylbenzene (with P_4VPy and P_2VPy) [159]
	OSO_4
	Catalytic dihydroxylation and oxidative cleavage of olefins [160–162]
	$PdCl_2$
	Hydrogenation of unsaturated aldehydes [163]
	Catalytic hydrogenation of 2-ethyl-9,10-anthraquinone [164]
	Catalytic hydrogenation of isophorone [165,166]
	Olefins hydrogenation [141,167]
	Hydrogenation of cyclohexene [141,168]
	Reduction of benzaldehyde and various nitroarenes [141,169]
	H_2PtCl_6
	Catalytic hydrogenation of 2-propen-l-ol and nitrobenzene [170]
	Suzuki cross-coupling reactions [171]
	Mo_6Cl_{12}
	Catalytic photooxidation reaction of alkenes [172]
	RuO_4
	Catalytic oxidation of primary and internal alcohols [173,174]
	[Rh(1,5-cyclooctadiene)(amine)$_2$](PF$_6$) (amine = pyridine, 2-picoline, 3-picoline, 4-picoline, 3,5-lutidine, or 2,6-lutidine)
	As catalysts in water gas shift reaction [175–177]
	Reduction of nitrobenzene [178]
	Hydroformylation, isomerization, and hydroesterification of 1-hexene [179,180]
	$PdCl_2$, K_2PdCl_6 (with P_4VPy and P_2VPy)
	Suzuki-, Heck-, and Stille-type coupling reactions [181]
	[(Cyclopentadienyl)$_2$ZrMe][B(C_6F_5)$_4$]
	As catalyst for ethylene polymerization [182]
	Ru-(2,2′-bipyridine)$_2^{2+}$
	As a catalytic film for square-wave voltammetric detection of chemical DNA damage and DNA hybridization [183,184]
	[Rh(1,5-cyclooctadiene)(amine)$_2$] (PF$_6$) (amine = 4-picoline, 3-picoline, 2-picoline)
	Hydroformylation of 1-hexane [185]
	Hydroxycarbonylation of 1-hexene [186]
	Reduction of some 2-substituted-5-nitrofurans [187]
	Carbonylation of naphtha [188]
	$RuCl_3$

(*Continued*)

TABLE 3.3 (*Continued*)
Polymer-Bound Metal-Complex Catalysts Based on PVPy

Entry	Polymer Catalysts with Their Applications
	Hydrogenation of aromatic hydrocarbons [189]
	Hydrogenation of quinoline [190]
	ZnX_2 (X: Cl, Br, I)
	Coupling reaction of CO_2 and epoxides [191]
	$MeReO_3$ (methyltrioxorhenium, MTO)
	Catalytic epoxidation of olefins [192,193]
	Catalytic oxidation of phenol and anisole [194]
	Epoxidation-methanolysis of glycals [195]
	Oxy-functionalization of C–H bonds using H_2O_2 [196]
	Catalytic epoxidation of monoterpenes [197]
	Na_2PdCl_4
	Catalytic hydrogenation of acetophenone [198]
	One-second allylic arylation [199]
	Suzuki–Miyaura coupling and C–H arylation [200]
	Pd NPs
	Suzuki–Miyaura cross-coupling reactions [201,202]
	Synthesis of alkynyl ketones via acyl Sonogashira reactions [203]
	$CuCl_2$
	Hydrodechlorination of chlorobenzene [204]
	CeO_2
	Synthesis of 3,4-dihydropyrimidin-2(1H)-ones [205]
	$Co(OAc)_2$
	Chemoselective hydrogenation of citral [206]
	$Co_3(NO_3)_2$
	Catalytic oxidation of cyclohexane with air [207]
	$CuCl_2$, $NiCl_2$, $CoCl_2$ (with P_4VPy, P_2VPy, and poly(2,5-methyl-vinylpyridine))
	As enzyme-like complexing polymer sorbents and metal-polymer complex catalysts [208]
	$[Rh(1,5\text{-cyclooctadiene})(4\text{-picoline})_2](PF_6)$
	Hydroesterification and hydroformylation-acetalization of 1-hexene [209]
	$PdCl_2(CH_3CN)_2$
	Heck C–C coupling reaction of aryl halides with alkyl acrylates [210,211]
	$[Cu(NCCH_3)_6][B(C_6F_5)_4]_2$, $[Cu(NCCH_3)_6][B\{C_6H_3(CF_3)_2\}_4]_2$
	Cyclopropanation of olefins [212]
	$Ru(II)(salen)(PPh_3)_2$
	Olefination of aldehydes [213]
	$Mo(CO)_3(CH_3CN)_3$
	Heterogeneous hydrogenation of styrene to ethylbenzene [214]
	$Mo(O_2)_2O{\cdot}2DMF$

(*Continued*)

TABLE 3.3 (*Continued*)
Polymer-Bound Metal-Complex Catalysts Based on PVPy

Entry	Polymer Catalysts with Their Applications
	Catalytic epoxidation of *cis*-cyclooctene [215,216]
	$Co_3(NO_3)_2$, CrO_3
	Cyclohexane oxidation and cyclohexyl hydroperoxide decomposition [217]
	$AlPO_4$–11
	Knoevenagel condensation reaction [218]
	CuI
	Chemoselective synthesis of pentaerythritol diacetals [219]
	Synthesis of coumarin derivatives [220]
	Synthesis of 1,8-dioxooctahydroxanthenes [221]
	Amination of aryl halides [222]
	Synthesis of 1,4-disubstituted-1*H*-1,2,3-triazoles [223,224]
	Synthesis of aryl-14H-dibenzo[*a*-*o*]xanthenes [225]
	Synthesis of 3,4-dihydropyrimidin-2(1*H*)-ones/thiones [226]
	Synthesis of various 4*H*-pyran derivatives [227]
	MnO_x
	A biomimetic and very efficient water oxidizing catalyst [228]
	$HAuCl_4$
	Catalytic reduction of 4-nitrophenol [229,230]
	Aerobic oxidations of alcohols [231]
	$CoCl_2$
	Cyclotrimerization of aryl alkynes [232]
	$MnCl_2$
	Catalytic oxidation of *n*-heptane mainly to the corresponding alcohols and, in lower amounts, aldehydes and ketones [233]
2	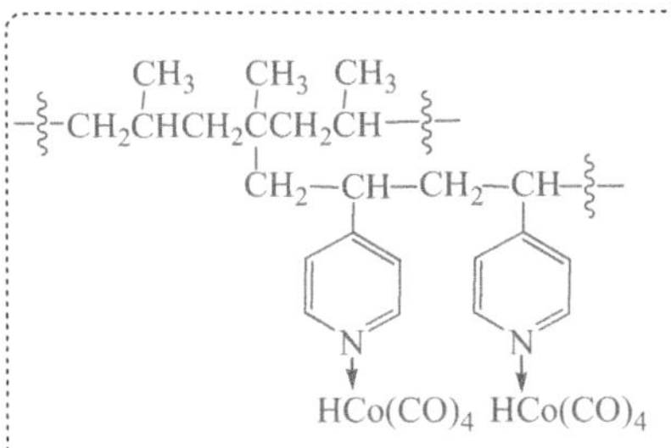 P_4VPy-supported metallic complex of $Co_2(CO)_8$ grafted on polypropylene
	Hydroformylation of 1-hexene [234]

(*Continued*)

TABLE 3.3 (*Continued*)
Polymer-Bound Metal-Complex Catalysts Based on PVPy

Entry	Polymer Catalysts with Their Applications
3	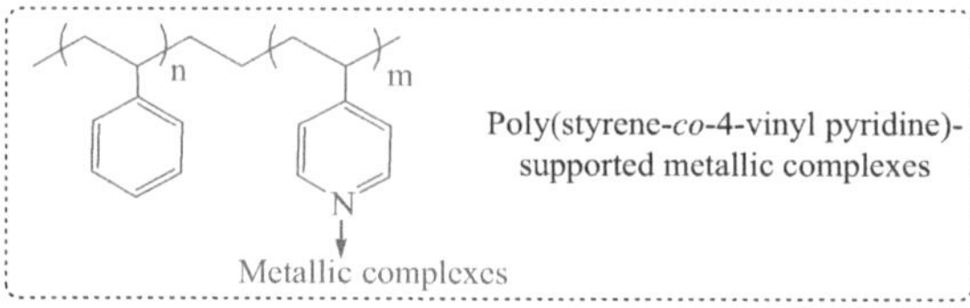 Cu(II) chloride Oxidative coupling of 2,6-disubstituted phenols [235] $HAuCl_4$, $AuCl_3$, Na_2PdCl_4, $Pd(CH_3COO)_2$, $AgNO_3$, $AgClO_4$ Hydrogenation of cycloalkenes [236] Heck reaction [237] CuCl, $Cu(OAc)_2$ 1D nanomaterial growth-CNT synthesis [238] $HAuCl_4$ Catalytic oxidation of alcohols [239]
4	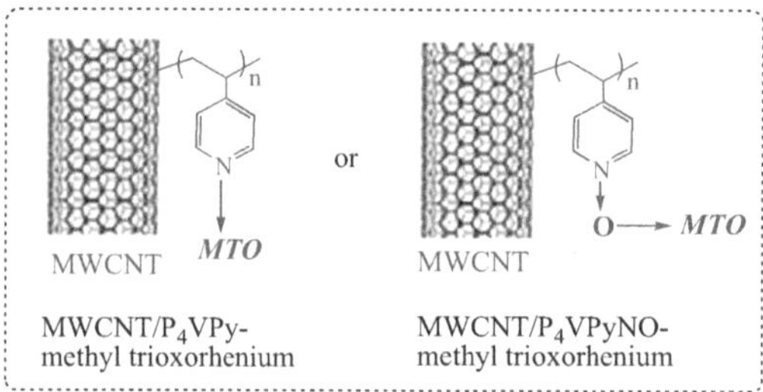 As heterogeneous catalysts for the oxidative desulfurization of model and synthetic diesel fuel [240]
5	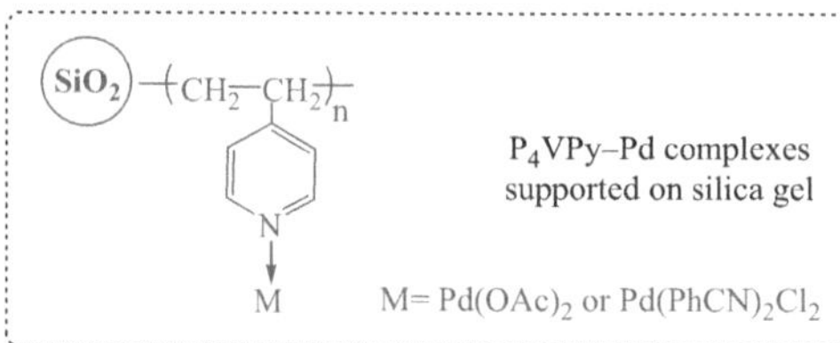 Reduction of nitro and azo groups [241]
6	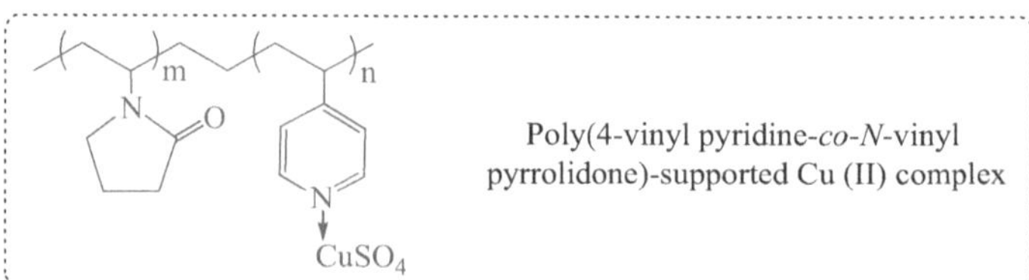 Catalytic oxidation of hydroquinone [242–244]

(*Continued*)

TABLE 3.3 (*Continued*)
Polymer-Bound Metal-Complex Catalysts Based on PVPy

Entry	Polymer Catalysts with Their Applications
7	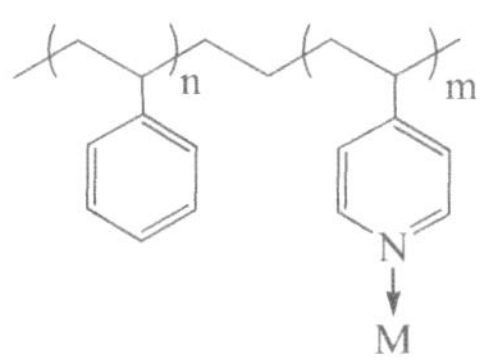 Poly(styrene-*co*-4-vinyl pyridine)-supported bimetallic nanoparticles (M= PdAu, PdPt, or PdZn) Selective hydrogenation of dehydrolinalool [245,246]
8	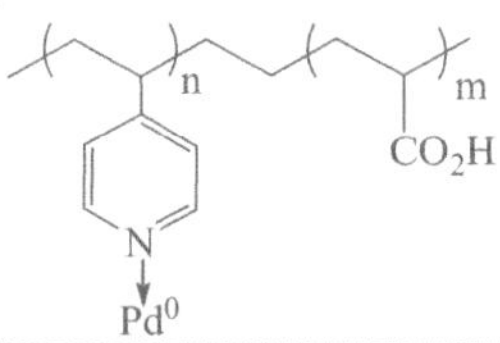 Palladium complex of poly(4-vinylpyridine-*co*-acrylic acid) Catalytic hydrogenation of aromatic nitro compounds [247]
9	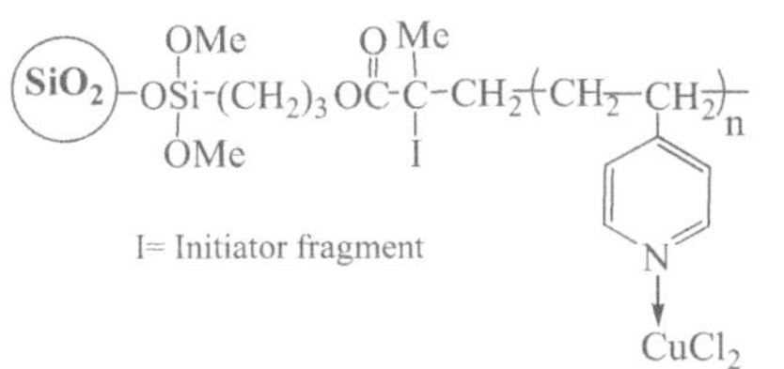 P_4VPy–Cu(II) complex supported on silica gel Oxidation of ethylbenzene with molecular oxygen [248]
10	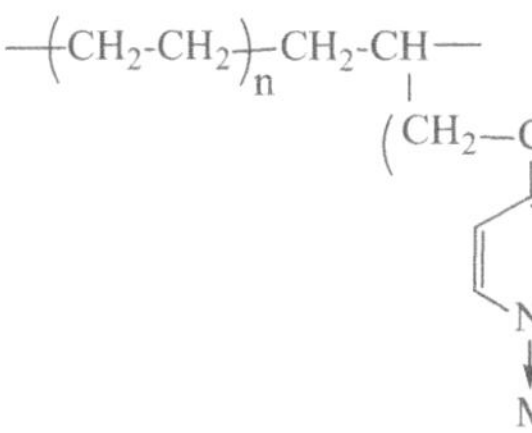 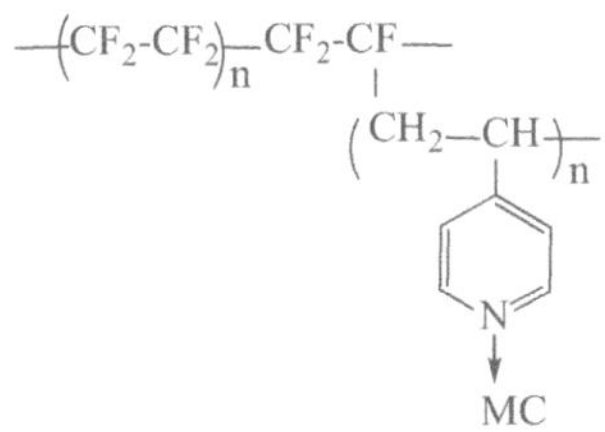 Poly(4-vinylpyridine-*co*-ethylene) and poly(4-vinylpyridine-*co*-tetrafluoroethylene)-supported metal complexes (MC: $FeCl_3$. $6H_2O$, $CoCl_2.6H_2O$, $VOSO_4.5H_2O$, $Na_2MoO_4.2H_2O$, $Na_2WO_4.2H_2O$) Catalytic oxidation of cyclohexene [249]
11	Core–shell PS-*co*-P_4VP-Pd microspheres Suzuki reactions [250,251]

(*Continued*)

TABLE 3.3 (*Continued*)
Polymer-Bound Metal-Complex Catalysts Based on PVPy

Entry — **Polymer Catalysts with Their Applications**

12

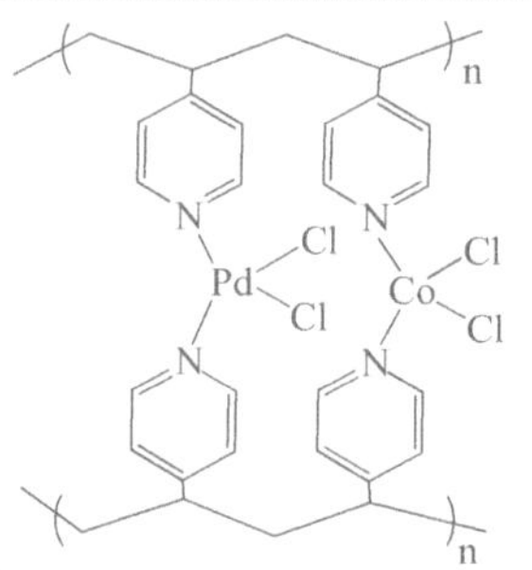

P_4VPy-supported palladium–cobalt bimetal catalyst

Oxidative carbonylation of bisphenol A [252]

13

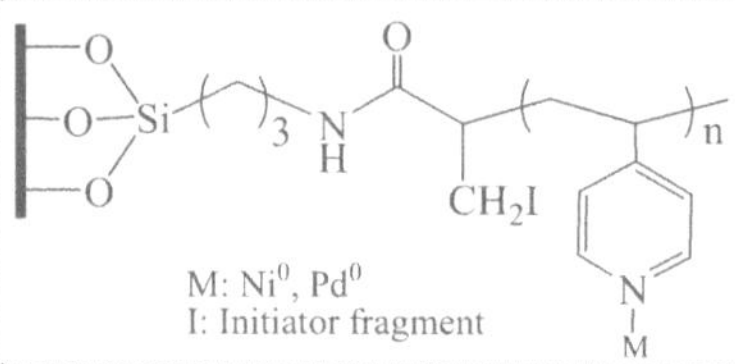

P_4VPy-grafted silica containing palladium nanoparticles

Heck and Suzuki coupling reactions [253]

Sonogashira reaction of phenylacetylene with aryl halides [254]

14

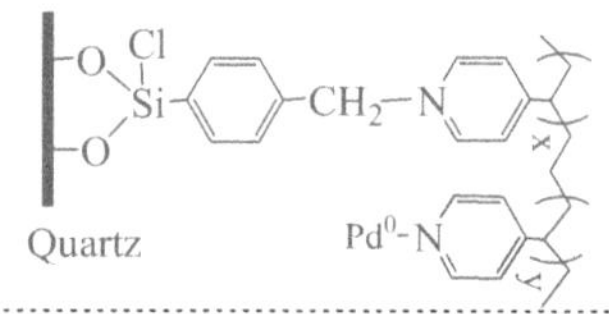

P_4VPy-supported Pd immobilized on quartz

Hydrogenation of aromatic conjugated alkenes [255]

15 Rh(acetylacetonate)$(CO)_2$ catalyst microencapsulated into P_4VPy

Hydroformylation of 1-hexene [256]

16 $[Rh(CO)_3Cl_2]_2$ catalyst microencapsulated into P_4VPy

Hydroformylation of 1-hexene [257]

17

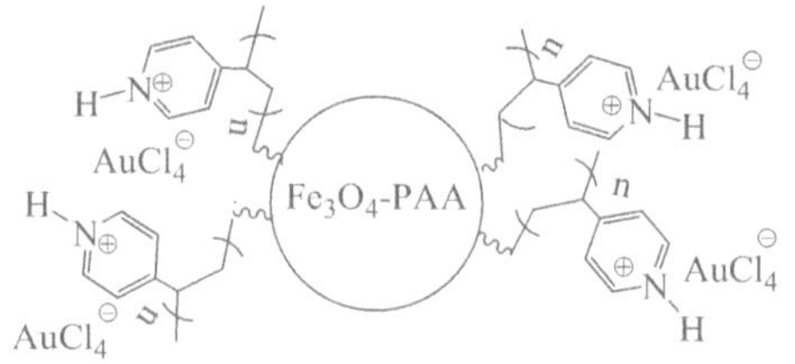

P_4VPy-supported Au nanoparticles immobilized on Fe_3O_4-PAA (Poly(acrylic acid)-modified Fe_3O_4 nanoparticles)

(*Continued*)

TABLE 3.3 (*Continued*)
Polymer-Bound Metal-Complex Catalysts Based on PVPy

Entry | **Polymer Catalysts with Their Applications**

18 Catalytic reduction of 4-nitrophenol [258]

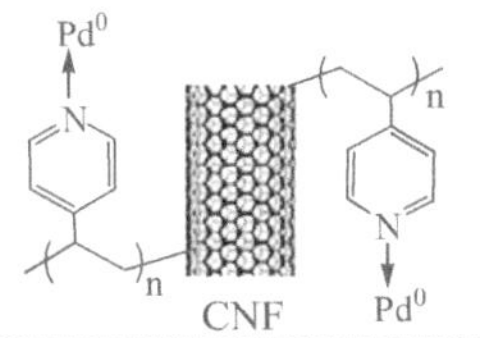

P_4VPy -supported Pd nanoparticles immobilized on carbon-nanofiber

Selective hydrogenation of phenol [259]

19

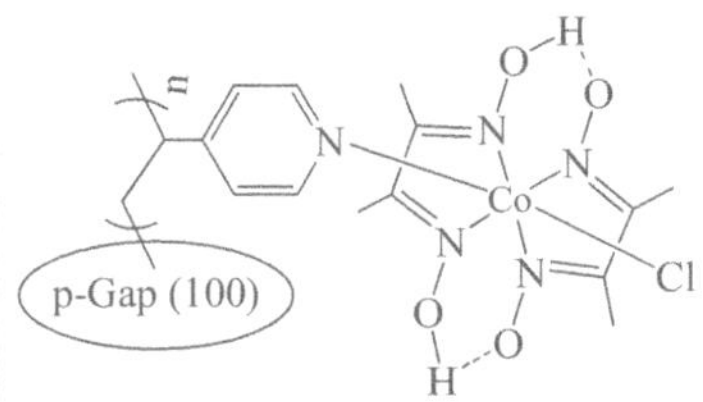

P_4VPy immobilized on p-type (100) gallium phosphide-supported moleular cobalt-containing hydrogen production catalysts

Catalysts for H_2 production to a visible-light absorbing semiconductor [260]

20

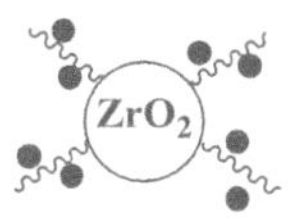

P_4VPy grafted on zirconia particles-supported $H_3PW_{12}O_{40}$

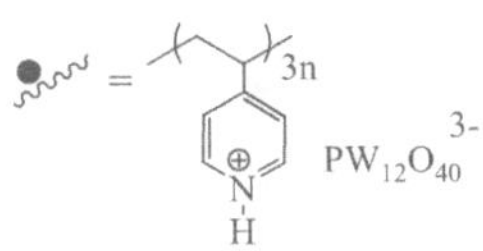

Catalytic oxidation of amines, alkenes, alcohols, and sulfides [261]

21

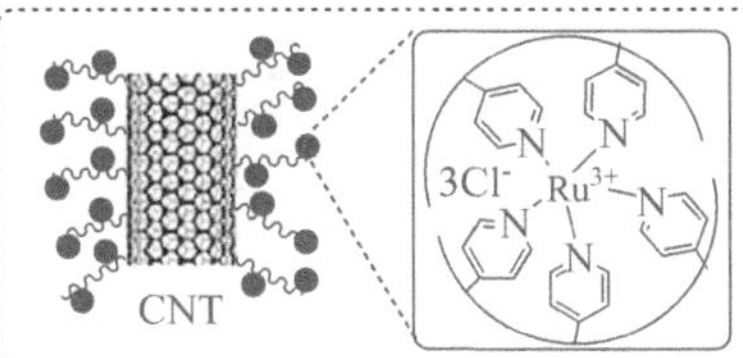

P_4VPy-supported $RuCl_3$ complex immobilized on carbon nanotube

Oxidation of 5-hydroxymethylfurfural [262]

(*Continued*)

TABLE 3.3 (*Continued*)
Polymer-Bound Metal-Complex Catalysts Based on PVPy

Entry	Polymer Catalysts with Their Applications
22	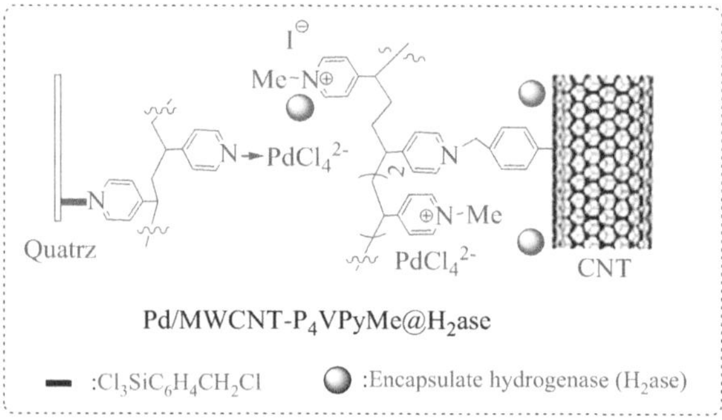 Catalytic oxidation of hydrogen [263]
23	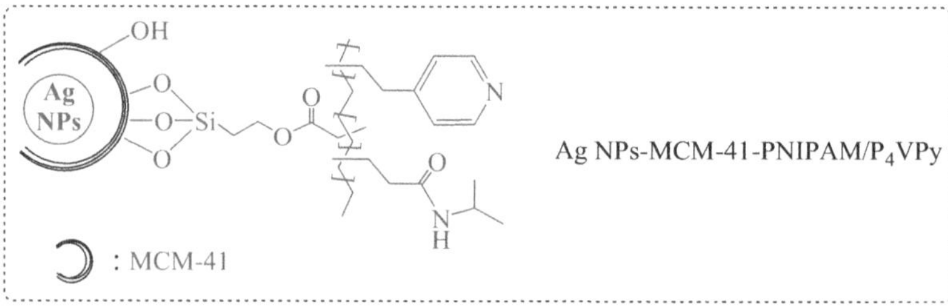 Oxidation of alcohols in aqueous medium [264]
24	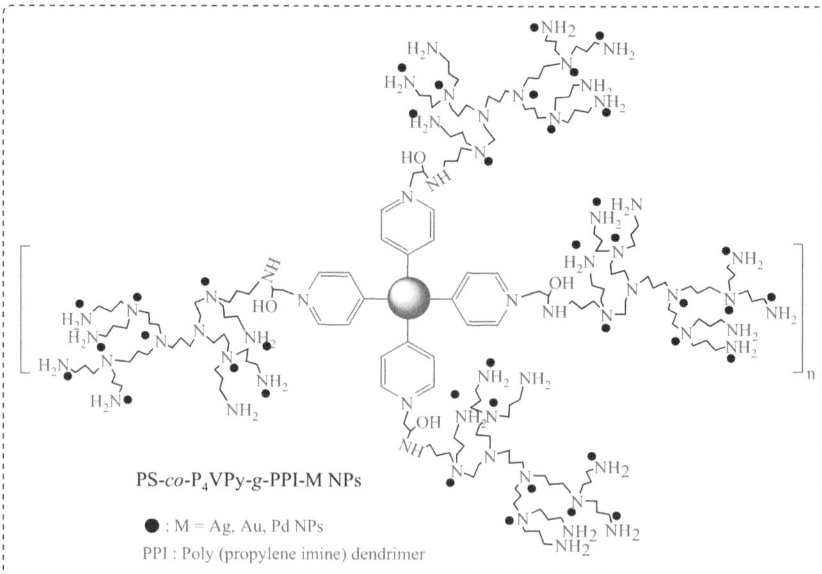 Reduction of trypan blue [265]

(*Continued*)

TABLE 3.3 (*Continued*)
Polymer-Bound Metal-Complex Catalysts Based on PVPy

Entry	Polymer Catalysts with Their Applications
25	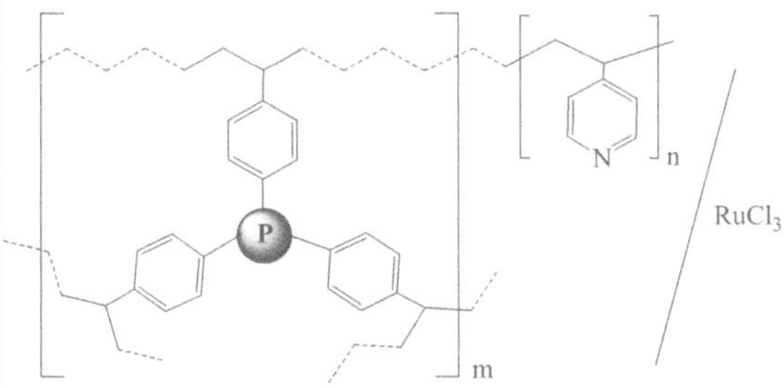 POPs-*co*-P_4VPy/$RuCl_3$ POPs: Porous organic polymers Recyclable catalysts for *N*-formylation of amines with CO_2 and H_2 [266]
26	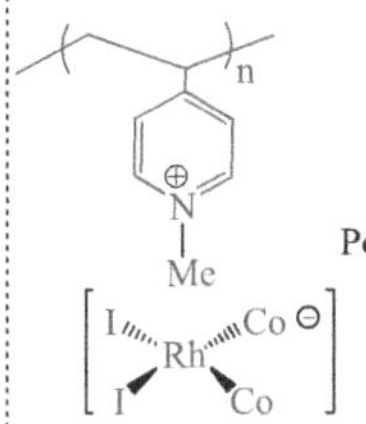 Poly(4-methy-4-vinyl pyridinium) anionic rhodium(I) complex Carbonylation of methyl iodide and methanol to give methyl acetate [267]
27	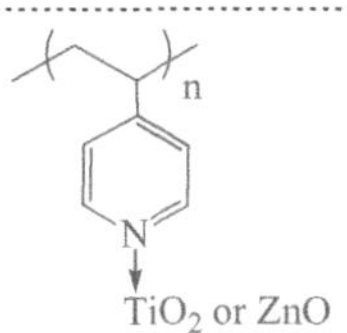 P_4VPy-TiO2 or ZnO spin-coated on glass Photocatalytic degradation of methyl orange and benzoic acid [268]
28	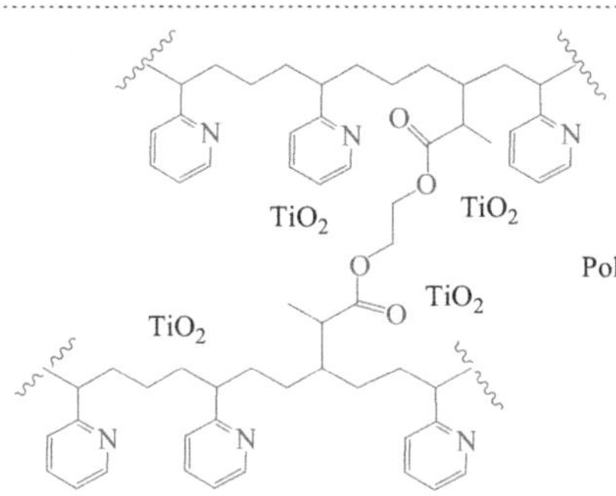 Poly(ethylene glycol dimethacrylate-2-VPy) -TiO_2 Photocatalytic decolorization of textile dyes from aqueous solution [269]

Table 3.3, entry 1 shows P_4VPy-supported metallic complexes or metal nanoparticles (NPs) which have so far been reported in literature as catalysts for organic transformations. The typical applications of some of these catalysts are shown in Scheme 3.14. In most of these reactions, no significant metal leaching into the desired synthesis product was observed. On completion of the reaction, the solid catalyst is filtered off and can be regenerated for reuse. High activity, air and temperature (up to 150°C) stability, ease of separation, stability in many of organic solvents, and high dispersion of catalyst particles are other obvious advantages of such solid green catalysts. Among the P_4VPy-supported metallic complexes, the coupling reaction of CO_2 and propylene oxide attracted attention (Scheme 3.15) [191]. The results of this research have shown that P_4VPy-$ZnBr_2$ exhibits much higher activities than P_4VPy-ZnI_2 and P_4VPy-$ZnCl_2$. $ZnBr_2$ and P_4VPy alone are not able to catalyze the coupling reaction, indicating a significant synergistic effect between them in P_4VPy-$ZnBr_2$ catalyst. A possible mechanism for this reaction was proposed in Scheme 3.15. The solid ^{13}C CP/MAS NMR (carbon-13 cross-polarization/magic-angle spinning/nuclear magnetic resonance) spectrum of the product derived from the reaction of P_4VPy-$ZnBr_2$ with propylene oxide revealed that the pyridinium-propoxy ion-bridged zinc species (complex B) acts as an active species in this process. Also, after the insertion of $^{13}CO_2$ into complex B, the formation of complex C was confirmed.

Another example on PVPy-supported metal complexes is the air oxidation of ethylbenzene in liquid phase at atmospheric pressure catalyzed by nickel acetylacetonate ($Ni(acac)_2$) (Table 3.3, entry 1, Scheme 3.16) [159]. The catalyst was supported on linear and cross-linked (PVPy)s. *N-n*-butylpolyvinylpyridinium hexafluorophosphate was employed as a cocatalyst. Results showed that both the functionalized linear P_2VPy and P_4VPy are not effective catalysts of ethylbenzene oxidation. By contrast, both cross-linked P_2VPy and P_4VPy are very effective catalysts of the ethylbenzene air oxidation. In this case, the ethylbenzene hydroperoxide formation rates are approximately 50-fold that of the linear polymers. The different structure of the homopolymers (linear P_2VPy and P_4VPy) and the copolymers (cross-linked P_2VPy and P_4VPy) might be the reason for the marked differences in their catalytic behavior. The isotactic structure of P_2VPy by X-ray diffraction methods revealed that the polymer consists of threefold helix with three chains passing through a hexagonal unit cell [270]. The results showed that P_4VPy also has an analogous threefold structure with a slight difference. Such structures confine the pendant pyridine groups within the helix and thus, the access of reactants to the active nickel centers is restricted. In contrast, the structures of the copolymers are amorphous. The DVB units keep the active pyridine units apart, resulting in an open structure that facilitates the entry of reactants to the pyridine-coordinated Ni^{2+} centers. As a result, cross-linked P_4VPy-$Ni(acac)_2$ and especially P_2VPy-$Ni(acac)_2$ serve as efficient catalysts for the oxidation of ethylbenzene [159].

It has been reported that various ruthenium(II) salen complexes can effectively serve as catalysts for the olefination of a wide range of aldehydes in a homogeneous phase [271]. However, a major limitation of utilizing metal salens in homogeneous solutions is the generation of μ-oxo dimers and other polymeric species, which can lead to irreversible catalyst deactivation. In order to overcome these issues, two $Ru(II)(salen)(PPh_3)_2$ complexes grafted on P_4VPy have been synthesized and used

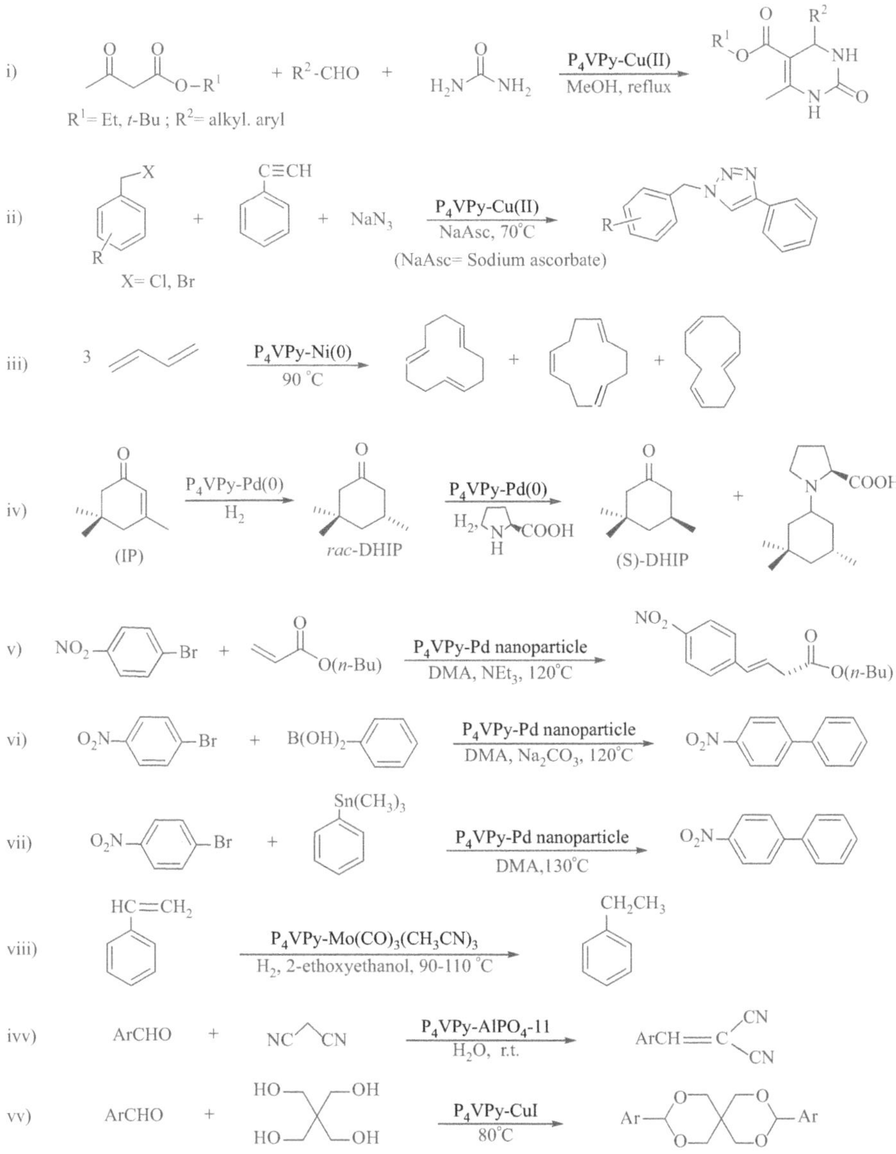

SCHEME 3.14 P_4VPy-supported metallic complexes for the preparation of 3,4-dihydropyrimidin-2(1*H*)ones (i) [152], synthesis of 1,4-disubstituted 1*H*-1,2,3-triazoles (ii) [154], cyclotrimerization of butadiene to cyclododecatriene (iii) [158], the proline-modified asymmetric hydrogenation of isophorone (iv) [165,166], Suzuki, Heck, and Stille-type coupling reactions (v, vi, vii) [181], hydrogenation of styrene to ethylbenzene (viii) [214], Knoevenagel condensation (ivv) [218], and chemoselective synthesis of pentaerythritol diacetals (vv) [219].

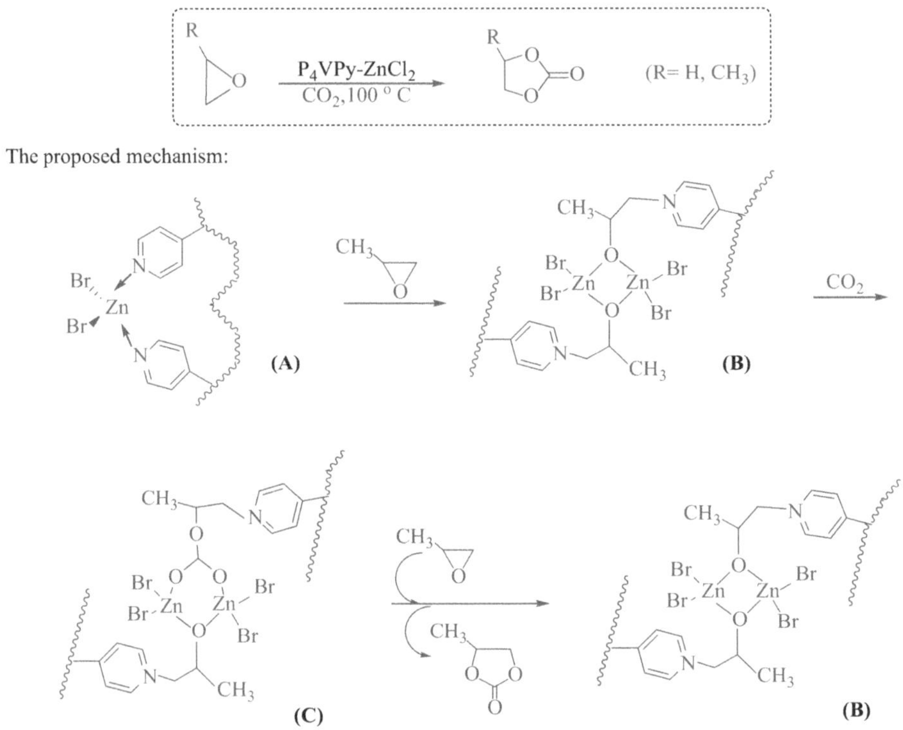

SCHEME 3.15 The synthesis of propylene carbonate from the reaction of propylene oxide with CO_2 using $P_4VPy\text{-}ZnBr_2$ catalyst [191].

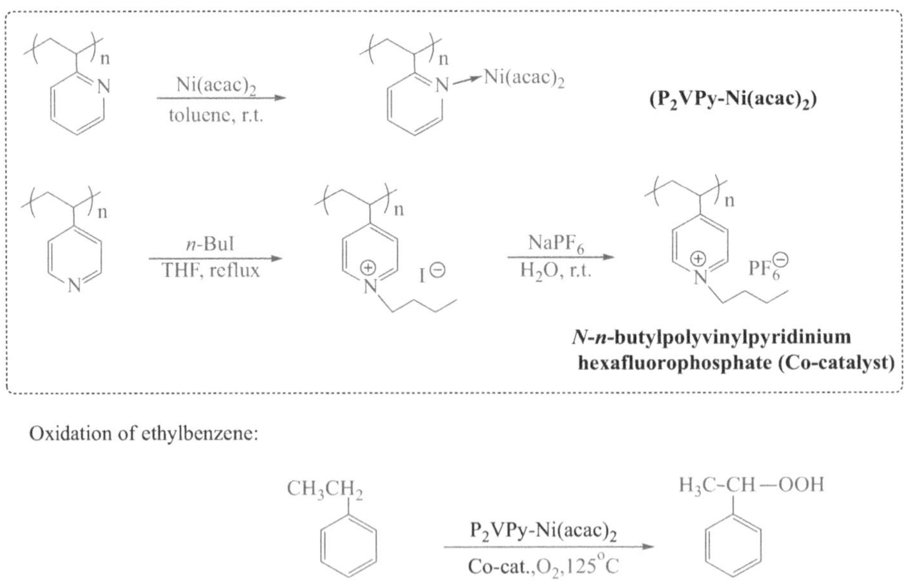

SCHEME 3.16 Catalytic oxidation of ethylbenzene using $P_2VPy\text{-}Ni(acac)_2$ [159].

as recyclable catalysts for the olefination of different aldehydes (Table 3.3, entry 1, Scheme 3.17) [213]. The findings demonstrate that grafting both Ru(II) salen complexes on the polymer enhances their thermal stability, with product yields comparable to those achieved with homogeneous ruthenium(II) salen complexes. Notably, the ruthenium(II) compound with a larger salen ligand exhibits superior recyclability, selectivity, and stability compared to its smaller ligand counterpart. Furthermore, the extent of leaching in all samples is less significant than that observed in silicas (i.e., SBA-15, MCM-41)-grafted ruthenium(II) salen complexes.

In addition to P_4VPy, the copolymers poly(4-vinylpyridine-*co*-styrene) (Table 3.3, entry 3) [235–239], poly(4-vinylpyridine-*co*-propylene) (Table 3.3, entry 2) [234], poly(4-vinylpyridine-*co*-*N*-vinylpyrrolidone) (Table 3.3, entry 6) [242–244], poly(4-vinylpyridine-*co*-acrylic acid) (Table 3.3, entry 8) [247], and poly(4-vinylpyridine-*co*-ethylene) and poly(4-vinylpyridine-*co*-tetrafluoroethylene) (Table 3.3, entry 10) [249] have been used as supports for immobilization of metallic complexes. Typically, Scheme 3.18 shows application of some of these catalysts.

Other types of polymer–metal complexes based on PVPy are P_4VPy-supported bimetallic catalysts. For example, P_4VPy-supported Pd-Co bimetal catalyst has been prepared and employed for the oxidative carbonylation of bisphenol A to polycarbonate (Table 3.3, entry 12, Scheme 3.19) [252].

The immobilization of metal complexes and NPs on the surface of a variety of substrates modified with the neutral or quaternized form of P_4VPy has been investigated (Table 3.3, entries 13, 14, 17–20). These substrates include silica gel, quartz, carbon nanofiber, *p*-type (100) gallium phosphide, and carbon nanotube. For example, P_4VPy-supported Pd^0 immobilized on quartz (Table 3.3, entry 14) [255], P_4VPy immobilized on Fe_3O_4-supported Au NPs (Table 3.3, entry 17) [258], P_4VPy-functionalized carbon nanofiber-supported Pd NPs (Table 3.3, entry 18) [259], and P_4VPy-functionalized carbon nanotube-supported $RuCl_3$ complex (Table 3.3, entry 21) [262] have been prepared and used as efficient catalysts in organic synthesis (Scheme 3.20).

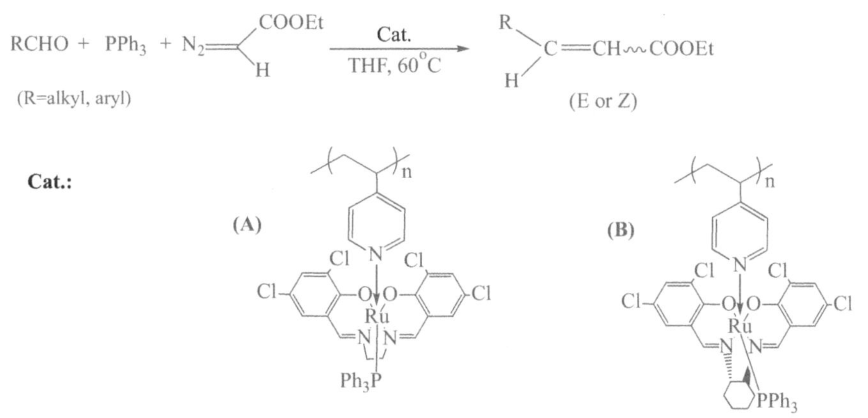

SCHEME 3.17 P_4VPy-supported ruthenium(II) salen complexes for olefination of aldehydes [213].

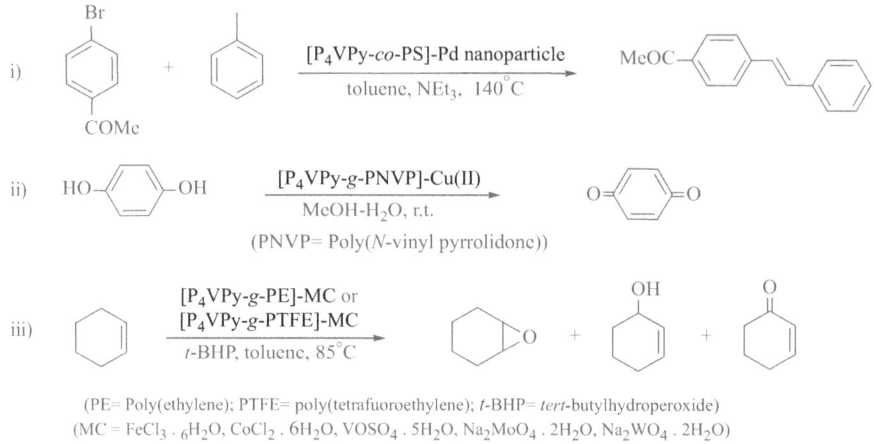

SCHEME 3.18 [P_4VPy-*co*-PS]-supported metallic complexes for Heck reaction (i) [237], oxidation of hydroquinone by [P_4VPy-*co*-PNVP]-$CuSO_4$ (ii) [243], and oxidation of cyclohexene by [P_4VPy-*co*-PE] and [P_4VPy-*co*-PTFE]-metallic complexes (iii) [249].

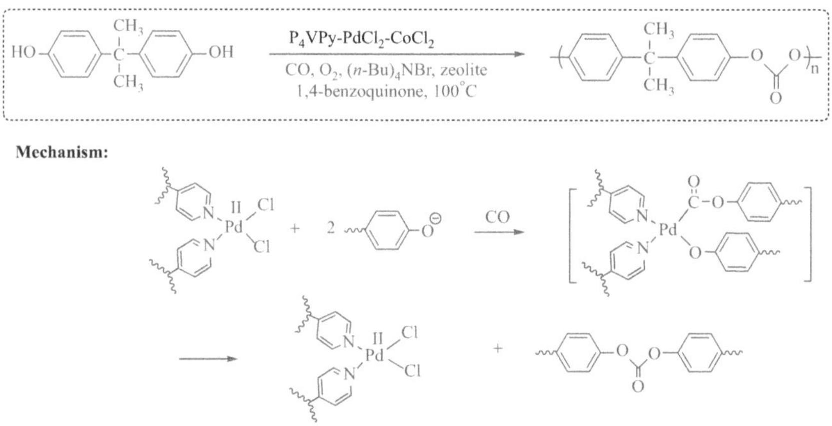

SCHEME 3.19 P_4VPy-supported pd–Co bimetal catalyst for the oxidative carbonylation of bisphenol A to polycarbonate [252].

As mentioned in Chapter 1, core–shell microspheres have diverse morphologies with high performance in different research fields and industries [272,273]. Metal NPs such as palladium, gold, and silver have been successfully incorporated into the shell layer of core–shell poly(styrene-*co*-4-vinylpyridine) microspheres, and the obtained compound was used as a heterogeneous recyclable catalyst for Suzuki reaction (Table 3.3, entry 11, Scheme 3.21) [250,251]. The microspheres which consist of a polystyrene core surrounded by a coordinative shell of P_4VPy were synthesized by one-stage soap-free emulsion polymerization method.

Microencapsulation refers to a technique where tiny solid particles or liquid or gas droplets, ranging from 3 to 800 μm in diameter, are enveloped or coated with a

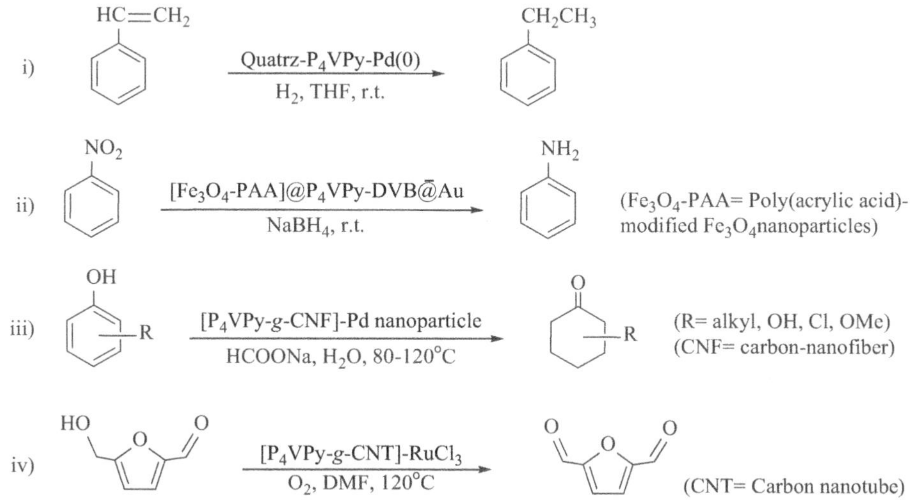

SCHEME 3.20 P_4VPy-supported metal-complex catalysts immobilized on different solid supports for hydrogenation of aromatic conjugated alkenes (i) [255], reduction of 4-nitrophenol (ii) [258], selective hydrogenation of phenol (iii) [259], and oxidation of 5-hydroxymethylfurfural to 2,5-diformylfuran (iv) [262].

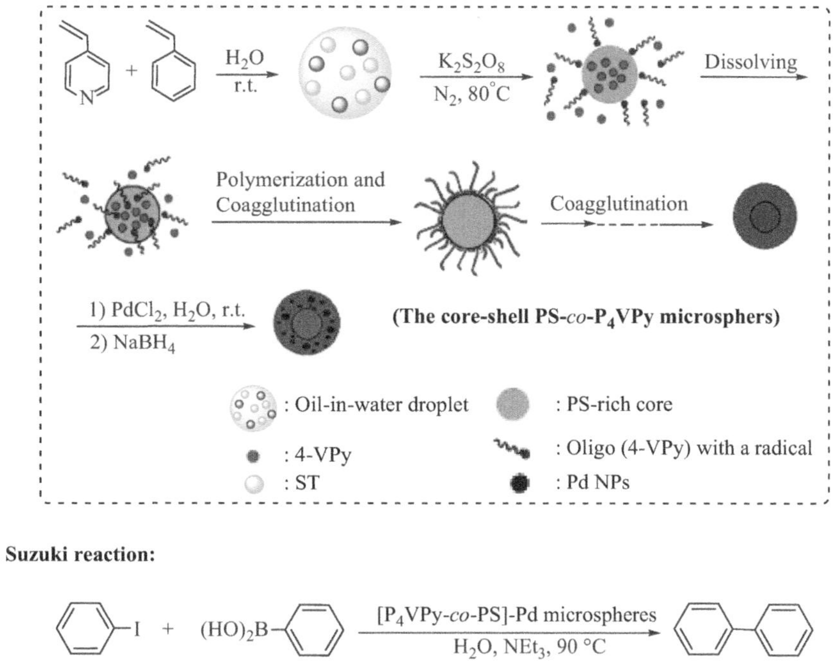

SCHEME 3.21 Synthesis and application of PS-*co*-P_4VP-Pd microspheres as a catalyst for Suzuki reaction [250,251].

continuous layer of polymeric substance, resulting in microencapsulated materials. Microcapsules are attracting increasing interest of medicinal and industrial chemists [274,275]. P_4VPy has been studied for the preparation of microencapsulated catalysts (Table 3.3, entries 15, 16). For example, microencapsulated rhodium/cross-linked P_4VPy (25% DVB) catalyst has been prepared and used for the hydroformylation of 1-hexene (Scheme 3.22) [256]. For the preparation of microcapsules, ethanolic solution of cross-linked P_4VPy (2 or 25% DVB) was exposed to the commercially available reagent acetylacetonatodicarbonyl rhodium(I) under nitrogen at 318 K overnight. The better results were obtained with 25% DVB cross-linked P_4VPy. The major products are aldehydes, attributed to the lower hydrogenation ability of the encapsulated catalyst systems in contrast to the homogeneous ruthenium carbonyls. Additionally, the catalyst can be reused.

Very recently, a dual-responsive heterogeneous catalyst was fabricated by combining P_4VPy and poly(*N*-isopropylacrylamide) (PNIPAM), respectively, as the pH- and temperature-responsive polymers with MCM-41 as integrated support (Table 3.3, entry 23, Scheme 3.23) [264]. The obtained switchable catalyst (Ag NPs-MCM-41-PNIPAM/P_4VPy) was used for selective oxidation of aliphatic and aromatic alcohols to aldehydes and ketones using H_5IO_6 as oxidant. At temperatures <30°C–35°C (the lower critical solution temperature of PNIPAM) and below the pK_a value of P_4VPy, which is 3.2 (pH = 3–5), both PNIPAM and P_4VPy become hydrophilic and consequently the Ag NPs-MCM-41-PNIPAM/P_4VPy catalyst exists in its

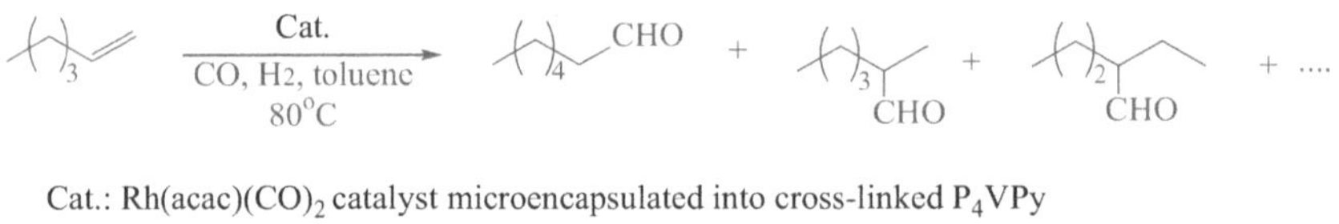

SCHEME 3.22 Rh(acetylacetonate)$(CO)_2$ catalyst microencapsulated into P_4VPy for the hydroformylation of 1-hexene [256].

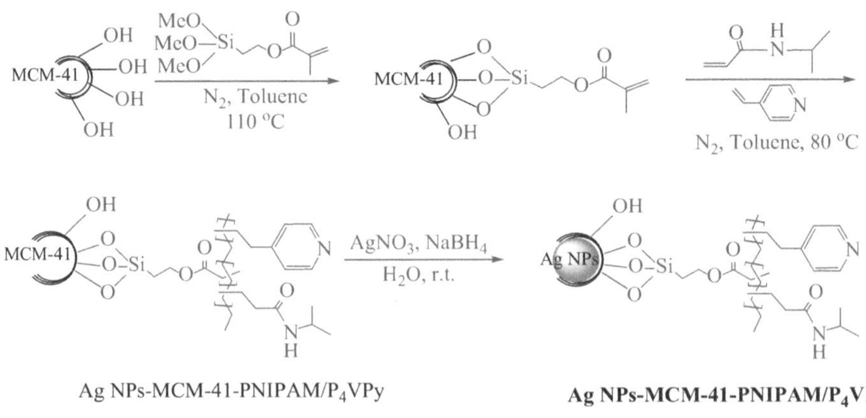

SCHEME 3.23 Synthetic routes for dual-responsive Ag NPs-MCM-41-PNIPAM/P_4VPy catalysts [264].

swollen state, which leads the reactant molecules reach to the active Ag NPs sites for the oxidation reactions. At a high temperature and pH >3.2, the catalyst is in a collapsed state.

3.2.4 Quaternized Forms of PVPy as Polysoaps and Phase Transfer Catalysts

For several decades, polysoaps have been known as a class of compounds that favorably combines the properties of polyelectrolytes and micelles [276,277]. PVPy has been used for the fabrication of some polysoaps. In this regard, a typical polysoap can be formed through quaternization of PVPy with long alkyl chains [278–280]. The resulting polysoap has a structural organization related to micelles formed from simple surfactants. The hydrophobicity of polysoaps is readily adjustable by changing the content and length of the alkyl group.

Also, PVPy has been employed for the preparation of PTCs. The most important PTCs are tetraalkylammonium and tetraalkylphosphonium salts, crown ethers, and cryptands [281]. However, these catalysts are typically not recovered, and this brings issues to the relatively expensive PTCs. Consequently, an increasing amount of research has been directed toward the application of functionalized polymers as PTCs [282,283]. Table 3.4 shows representative examples of polysoaps and PTCs based on P_4VPy.

Typically, the alkaline hydrolysis of *p*-nitrophenyl esters in the presence of polyelectrolytes such as quaternized P_4VPy with ethyl, butyl, and benzyl halides has been investigated (Table 3.4, entry 2, Scheme 3.24 (i)) [285]. The results showed

TABLE 3.4
Polysoaps and Phase Transfer Catalysts Based on PVPy

Entry	Polymer Catalysts with Their Applications
	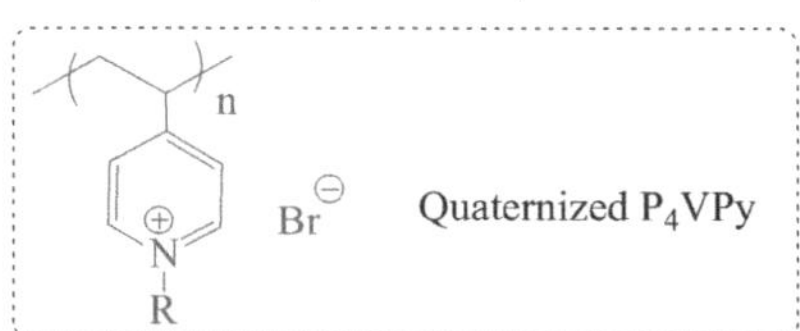
1	R = ethyl, *n*-propyl, *n*-butyl, cetyl, benzyl Alkaline fading reaction of triphenylmethane dyes [284]
2	R = ethyl, *n*-butyl, benzyl Alkaline hydrolysis of *p*-nitrophenyl esters [285,286]
3	R = ethyl, lauryl Hydrolysis of phenyl esters [287]
4	R = $CH_2CH(Me)(Et)$, *n*-butyl Ethylation of phenylacetonitrile [288]
5	R = ethyl, lauryl, dodecyl, octyl Decarboxylation of 6-nitrobenzisoxazole-3-carboxylate anion [289,290]

(Continued)

TABLE 3.4 (*Continued*)
Polysoaps and Phase Transfer Catalysts Based on PVPy

Entry	Polymer Catalysts with Their Applications
6	R=$(CH_2)nCH_3$; n=1_15
	Finkelstein reaction of various alkyl halides [291]
7	R=ethyl
	α, β-Elimination of serine [292,293]
	Cleavage of threonine [292,294]
	α, β-Elimination of tryptophan [295]
8	R= benzyl
	Nucleophilic substitution reactions [296]
9	R= benzyl, n-Bu, n-octyl, $HO(CH_2)_3$, $NC(CH_2)_3$
	Reaction of carbon dioxide and glycidyl phenyl ether [297]
10	R= n-propyl, n-butyl
	Nucleophilic substitution reactions [298,299]

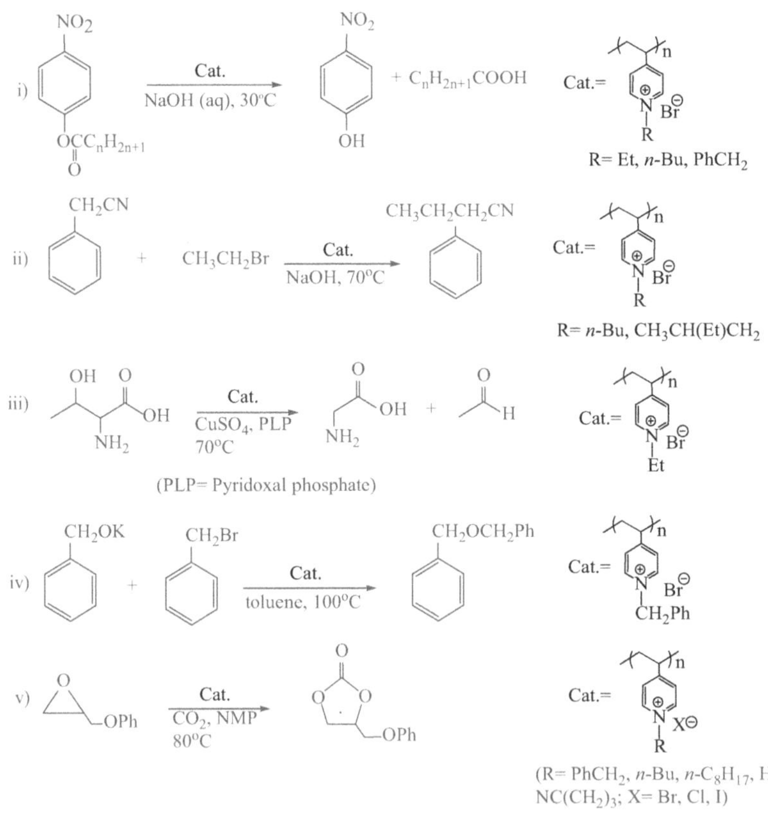

SCHEME 3.24 Quaternized forms of P_4VPy for alkaline hydrolysis of p-nitrophenyl esters (i) [285], ethylation of phenylacetonitrile under phase transfer conditions (ii) [288], cleavage of threonine (iii) [292,294], nucleophilic substitution reactions under phase transfer condition (iv) [296], and the reaction of carbon dioxide and glycidyl phenyl ether (v) [297].

that cationic polysoaps with hydrophobic groups enhanced hydrolysis, while those lacking such groups did not exhibit such effects. Hydroxide ions are attracted to the cationic part of the polymer via the electrostatic interactions, and the ester molecules also gather around the polymer chains due to their hydrophobic nature.

Ethylation of phenylacetonitrile to 2-phenylbutanenitrile using P_4VPy, P_2VPy, and quaternized P_4VPy has been reported (Table 3.4, entry 4, Scheme 3.24 (ii)) [288]. The results showed that in this reaction reactivity of P_4VPy was similar to that of pyridine, while P_2VPy has scarce activity due to its steric effects, and the quaternized P_4VPy was found to be highly active.

REFERENCES

1. Anastas P, Eghbali N. *Chemical Society Reviews.* 2010;39(1):301–12.
2. Anastas PT, Kirchhoff MM. *Accounts of Chemical Research.* 2002;35(9):686–94.
3. Sherrington D. *Chemistry of Waste Minimization.* 1995:141–200.
4. Ley SV, Baxendale IR, Bream RN, Jackson PS, Leach AG, Longbottom DA et al. *Journal of the Chemical Society, Perkin Transactions.* 1. 2000;1(23):3815–4195.
5. Benaglia M, Puglisi A, Cozzi F. *Chemical Reviews.* 2003;103(9):3401–30.
6. Lu J, Toy PH. *Chemical Reviews.* 2009;109(2):815–38.
7. Akiyama R, Kobayashi S. *Chemical Reviews.* 2009;109(2):594–642.
8. Islam M, Mondal S, Manir M, Singha Roy A, Mondal P, Hossain D. *Chinese Journal of Chemistry.* 2010;28(10):1810–20.
9. Deleuze H, Schultze X, Sherrington D. *Polymer.* 1998;39(24):6109–14.
10. Shinde SS, Lee BS, Chi DY. *Tetrahedron Letters.* 2008;49(27):4245–8.
11. Rodrigues Jd S, Gonçalves D, Mangrich A, Soldi V, Bertolino J, Pires A. *Advances in Polymer Technology: Journal of the Polymer Processing Institute.* 2000;19(2):113–9.
12. Rao V, Ashokan P, Shridhar M. *Materials Science and Engineering: A.* 2000;276(1–2):266–8.
13. Baggiani C, Giraudi G, Giovannoli C, Tozzi C, Anfossi L. *Analytica Chimica Acta.* 2004;504(1):43–52.
14. Cha JN, Zhang Y, Wong H-SP, Raoux S, Rettner C, Krupp L, Deline V. *Chemistry of Materials.* 2007;19(4):839–43.
15. Neagu V, Untea I, Tudorache E, Luca C. *Reactive and Functional Polymers.* 2003;57(2–3):119–24.
16. Kumaresan R, Sabharwal K, Srinivasan T, Rao PV, Dhekane G. *Solvent Extraction and Ion Exchange.* 2007;25(4):515–28.
17. Zhai G, Kang E, Neoh K. *Journal of Membrane Science.* 2003;217(1–2):243–59.
18. Khaligh NG, Abbo HS, Johan MR, Titinchi SJ. *Current Organic Chemistry.* 2019;23(4):439–79.
19. Creutz S, Teyssié P, Jérôme R. *Macromolecules.* 1997;30(19):5596–601.
20. Bohrisch J, Wendler U, Jaeger W. *Macromolecular Rapid Communications.* 1997;18(11):975–82.
21. Drelinkiewicz A, Sobczak J, Sobczak E, Krawczyk M, Zięba A, Waksmundzka-Góra A. *Materials Chemistry and Physics.* 2009;114(2–3):763–73.
22. Cao L, Hu Y, Zhang L, Ma C, Wang X, Wang J. *The Journal of Supercritical Fluids.* 2011;58(2):233–8.
23. Groppo E, Uddin MJ, Zavorotynska O, Damin A, Vitillo JG, Spoto G, Zecchina A. *The Journal of Physical Chemistry C.* 2008;112(49):19493–500.
24. Holt P, Tamami B. *Polymer.* 1970;11(10):553–60.
25. Holt P, Tamami B. *Polymer.* 1973;14(12):645–8.

26. Tamami B, Goudarzian N. *Polymer Bulletin*. 1990;23:295–8.
27. Berkowitz JB, Yamin M, Fuoss RM. *Journal of Polymer Science*. 1958;28[116]:69–82.
28. Boyes AG, Strauss UP. *Journal of Polymer Science*. 1956;22[102]:463–76.
29. Onyon P. *Transactions of the Faraday Society*. 1955;51:400–12.
30. Fuoss RM, Maclay WN. *Journal of Polymer Science*. 1951;6(3):305–17.
31. Choukchou-Braham E, Benabadji I, Mansri A, François J. *European Polymer Journal*. 2003;39(2):297–303.
32. Mansri A, Tennouga L, *Desbrieres J. European Polymer Journal*. 2007;43(2):540–9.
33. Morcellet-Sauvage J, Loucheux C. *Die Makromolekulare Chemie: Macromolecular Chemistry and Physics*. 1975;176(2):315–31.
34. Frere Y, Gramain P. *Macromolecules*. 1992;25(12):3184–9.
35. Vorontsov YD, Lukovkin G, Gusev V, Rusak A, Nikolayev N, Golina L, Yevdakov V. *Polymer Science USSR*. 1979;21(4):983–90.
36. Wen T, Wang H-F, Mao Y-C, Chuang W-T, Tsai J-C, Ho R-M. *Polymer*. 2016;107:44–53.
37. Cozzi F. *Advanced Synthesis & Catalysis*. 2006;348(12-13):1367–90.
38. Kosloski-Oh SC, Wood ZA, Manjarrez Y, de Los Rios JP, Fieser ME. *Materials Horizons*. 2021;8(4):1084–129.
39. Letsinger RL, Savereide TJ. *Journal of the American Chemical Society*. 1962;84(16):3122–7.
40. Yoshida J-i, Hashimoto J, Kawabata N. *Bulletin of the Chemical Society of Japan*. 1981;54(1):309–10.
41. Johnston RD, Marston CR, Krieger PE, Goe GL. *Synthesis*. 1988;5(05):393–4.
42. Karimi Zarchi MA, Rahmani F. *Journal of Applied Polymer Science*. 2011;121(1):582–8.
43. Menger F, Chu C. *The Journal of Organic Chemistry*. 1981;46(24):5044–5.
44. Li Z, Ganesan A. *Synthetic Communications*. 1998;28(17):3209–12.
45. Zarchi MAK, Aslani M. *Journal of Applied Polymer Science*. 2012;124(4):3456–62.
46. Sket B, Zupan M. *Journal of Macromolecular Science—Chemistry*. 1983;19(5):643–52.
47. Nomiya K, Murasaki H, Miwa M. *Polyhedron*. 1986;5(4):1031–3.
48. Tamami B, Iranpoor N, Zarchi MK. *Polymer*. 1993;34(9):2011–3.
49. Iranpoor N, Tammi B, Shekarriz M. *Synthetic Communications*. 1999;29(19):3313–21.
50. Tabakovic R, Tabakovic I. *Reactive and Functional Polymers*. 1999;39(3):263–8.
51. Balakrishnan T, Rajendran V. *Journal of Applied Polymer Science*. 2000;78(12):2075–80.
52. Olah GA, Mathew T, Goeppert A, Török B, Bucsi I, Li X-Y et al. *Journal of the American Chemical Society*. 2005;127(16):5964–9.
53. Lee BS, Mahajan S, Janda KD. *Tetrahedron Letters*. 2005;46(5):807–10.
54. Lee BS, Mahajan S, Janda KD. *Tetrahedron*. 2005;61(12):3081–6.
55. Lee S-H, Lee BS. *Bulletin of the Korean Chemical Society*. 2009;30:551–5.
56. Elkholy SS, Khalil KD, Elsabee MZ. *Journal of Applied Polymer Science*. 2006;99(6):3308–17.
57. Khalil KD, Al-Matar HM. *Molecules*. 2013;18(5):5288–305.
58. Olah GA, Mathew T, Panja C, Smith K, Surya Prakash G. *Catalysis Letters*. 2007;114:1–7.
59. Bae HW, Han J-S, Jung S, Cheong M, Kim HS, Lee JS. *Applied Catalysis A: General*. 2007;331:34–8.
60. Jun K-W, Shim E-K, Kim S-B, Lee K-W. *Studies in Surface Science and Catalysis*. 1994;82(1):659–66.
61. Chari MA, Shobha D, Kumar TK, Syamasundar K. *Trends in Applied Sciences Research*. 2007;2(1):80–4.
62. Karuehanon W, Sirathanyarote C, Pattarawarapan M. *Tetrahedron*. 2012;68(46):9423–8.
63. Fan S, Luan Y, Wang J, Gao H, Zhang X, Wang G. *Journal of Molecular Catalysis A: Chemical*. 2015;404:186–92.
64. Borah KJ, Dutta P, Borah R. *Bulletin of the Korean Chemical Society*. 2011;32(1):225–8.
65. Lee SH, Kadam ST. *Applied Organometallic Chemistry*. 2011;25(8):608–15.

66. Borah KJ, Borah R. *Monatshefte für Chemie-Chemical Monthly*. 2011;142:1253–7.
67. Khaligh NG. *Ultrasonics Sonochemistry*. 2013;20(4):1062–8.
68. Khaligh NG, Shirini F. *Journal of Molecular Catalysis A: Chemical*. 2011;348(1–2):20–9.
69. Khaligh NG. *Ultrasonics Sonochemistry*. 2013;20(1):19–25.
70. Khaligh NG. *Tetrahedron Letters*. 2012;53(13):1637–40.
71. Khaligh NG. *Catalysis Science & Technology*. 2012;2(11):2211–5.
72. Khaligh NG. *Ultrasonics Sonochemistry*. 2012;19(4):736–9.
73. Ali M, Zarchi K, Tarabsaz A. *Chinese Journal of Polymer Science*. 2013;31:1660–9.
74. Dutta P, Sarma P, Borah R. *Synthetic Communications*. 2013;43(10):1378–86.
75. Karimi Zarchi MA, Tarabsaz A. *Phosphorus, Sulfur, and Silicon and the Related Elements*. 2015;190(4):550–60.
76. Ghashang M, Mansoor SS, Aswin K. *Chinese Journal of Catalysis*. 2014;35(1):43–8.
77. Prakash GS, Shakhmin A, Glinton KE, Rao S, Mathew T, Olah GA. *Green Chemistry*. 2014;16(7):3616–22.
78. Ghashang M, Mansoor SS, Aswin K, Sudhan S. *Research on Chemical Intermediates*. 2015;41:5239–51.
79. Prakash GS, Gurung L, Glinton KE, Belligund K, Mathew T, Olah GA. *Green Chemistry*. 2015;17(6):3446–51.
80. Khalig NG. *RSC Advances*. 2012;2(8):3321–7.
81. Khaligh NG, Mihankhah T, Johan MR. *Heteroatom Chemistry*. 2018;29(4):e21440.
82. Hajjami M, Ghorbani-Choghamarani A, Norouzi M. *Chinese Journal of Catalysis*. 2012;33(9–10):1661–4.
83. Shirini F, Jolodar OG. *Journal of Molecular Catalysis A: Chemical*. 2012;356:61–9.
84. Shirini F, Abedini M, Pourhasan R. *Dyes and Pigments*. 2013;99(1):250–5.
85. Shirini F, Khaligh NG, Jolodar OG. *Journal of the Iranian Chemical Society*. 2013;10:181–8.
86. Khaligh NG. *Journal of Molecular Catalysis A: Chemical*. 2012;363:90–100.
87. Khaligh NG, Shirini F. *Ultrasonics Sonochemistry*. 2013;20(1):26–31.
88. Khaligh NG. *Chinese Journal of Catalysis*. 2014;35(3):329–34.
89. Shirini F, Esmaeeli-Ranjbar S, Seddighi M. *Chinese Journal of Catalysis*. 2014;35(7):1017–23.
90. Parvanak Boroujeni K, Shojaei P. *Turkish Journal of Chemistry*. 2013;37(5):756–64.
91. Kiasat AR, Mouradzadegun A, Saghanezhad SJ. *Research on Chemical Intermediates*. 2015;41:319–26.
92. Parvanak Boroujeni K, Taheri S, Seyfipour G. *Synthesis and Reactivity in Inorganic, Metal-Organic, and Nano-Metal Chemistry*. 2014;44(1):84–8.
93. Parvanak Boroujeni K, Ghasemi P, Rafienia Z. *Monatshefte für Chemie-Chemical Monthly*. 2014;145:1023–6.
94. Parvanak Boroujeni K, Ghasemi P. *Catalysis Communications*. 2013;37:50–4.
95. Guzmán-Pantoja J, Ornelas-Robles N, Hernández-Cortez JG, Martínez-Palou R. *Journal of Molecular Liquids*. 2025:127651.
96. Khaligh NG, Shirini F. *Ultrasonics Sonochemistry*. 2015;22:397–403.
97. Khaligh NG, Ghasem-Abadi PG. *Chinese Journal of Catalysis*. 2014;35(7):1126–35.
98. Prakash GS, Paknia F, Kulkarni A, Narayanan A, Wang F, Rasul G et al. *Journal of Fluorine Chemistry*. 2015;171:102–12.
99. Li W-Y, Zong Y-X, Wang J-K, Niu Y-Y. *Chinese Chemical Letters*. 2014;25(4):575–8.
100. Wang J, Zong Y, Fu R, Niu Y, Yue G, Quan Z et al. *Ultrasonics Sonochemistry*. 2014;21(1):29–34.
101. Kang J, Kim SY, Jo HH, Zong K. *ChemSusChem*. 2024:e202400216.
102. Marziano NC, Ronchin L, Tortato C, Zingales A, Sheikh-Osman AA. *Journal of Molecular Catalysis A: Chemical*. 2001;174(1–2):265–77.

103. Hyytiäinen M, Appelblad P, Pontén E, Stigbrand M, Irgum K, Jaegfeldt H. *Journal of Chromatography A*. 1996;740(2):279–83.
104. Parvanak Borujeni K, Massah AR. *Reactive and Functional Polymers*. 2006;66(10):1126–31.
105. Parvanak Boroujeni K, Parvanak K. *Journal of the Serbian Chemical Society*. 2011;76(2):155–63.
106. Feller A, Lercher JA. *Advances in Catalysis*. 2004;48:229–95.
107. He Q, Zhang Y, Zhang Z, Qian H-J, Chen Q. *Frontiers in Soft Matter*. 2022;2:980210.
108. Sheldon RA, Arends I, Hanefeld U. *Green Chemistry and Catalysis*: Wiley-VCH, Germany; 2007.
109. Le Goffic F, Sicsic S, Vincent C. *Tetrahedron Letters*. 1976;17(33):2845–7.
110. Fife WK, Zhang Z-D. *Tetrahedron Letters*. 1986;27(Part 3, 41):4937–40.
111. Fife WK, Zhang Z-D. *Tetrahedron Letters*. 1986;27(Part 2, 41):4933–6.
112. Raje VP, Bhat RP, Samant SD. *Synlett*. 2006;2006(16):2676–8.
113. Jagtap SR, Bhor MD, Bhanage BM. *Catalysis Communications*. 2008;9(9):1928–31.
114. Prakash GS, Colmenares JC, Batamack PT, Mathew T, Olah GA. *Tetrahedron Letters*. 2009;50(44):6016–8.
115. Prakash GS, Colmenares JC, Batamack PT, Mathew T, Olah GA. *Journal of Molecular Catalysis A: Chemical*. 2009;310(1–2):180–3.
116. Zhang Y-L, Liu S, Liu S, Liu F, Zhang H, He Y, Xiao F-S. *Catalysis Communications*. 2011;12(13):1212–7.
117. Tai A, Iwaoka Y, Ito H. *Carbohydrate Research*. 2011;346(15):2511–4.
118. Shirini F, Khaligh NG. *Journal of the Iranian Chemical Society*. 2012;9:495–502.
119. Albadi J, Mansournezhad A, Darvishi-Paduk M. *Chinese Chemical Letters*. 2013;24(3):208–10.
120. Albadi J, Fadaeian M, Akbari B-B. *Iranian Journal of Organic Chemistry*. 2013;5:1089–92.
121. Li Q, Tao W, Li A, Zhou Q, Shuang C. *Applied Catalysis A: General*. 2014;484:148–53.
122. Albadi J, Mansournezhad A, Balout-Bangan FA. *Acta Chimica Slovenica*. 2014;61:185–90.
123. Albadi J, Mansournezhad A. *Current Chemistry Letters*. 2014;3(4):221–7.
124. Kashyap B, Phukan P. *Monatshefte für Chemie-Chemical Monthly*. 2016;147:797–800.
125. Nasirmahale LN, Jolodar OG, Shirini F, Tajik H. *Polycyclic Aromatic Compounds*. 2019;41(1):199–210.
126. Jin R, Xu H, Easa J, Chapero-Planell A, O'Brien CP. *The Journal of Physical Chemistry C*. 2023;127(3):1441–54.
127. Jaroonwatana W, Del Gobbo S, D'Elia V, Crespy D. *Chemical Engineering Journal*. 2024;499:155635.
128. Jagtap SR, Raje VP, Samant SD, Bhanage BM. *Journal of Molecular Catalysis A: Chemical*. 2007;266(1–2):69–74.
129. Kalbasi RJ, Kolahdoozan M, Massah A, Shahabian K. *Bulletin of the Korean Chemical Society*. 2010;31(9):2618–26.
130. Kalbasi RJ, Mosaddegh N. *Catalysis Communications*. 2011;12(13):1231–7.
131. Kalbasi RJ, Mosaddegh N. *Journal of Porous Materials*. 2012;19(5):557–65.
132. Kolahdoozan M, Javad R. *Monatshefte für Chemie-Chemical Monthly*. 2012;143:325–34.
133. Kolahdoozan M, Kalbasi RJ, Shahzeidi ZS, Zamani F. *Journal of Chemistry*. 2013:496837.
134. Kalbasi RJ, Kolahdoozan M, Vanani SM. *Journal of Solid State Chemistry*. 2011;184(8):2009–16.
135. Gomha S, Riyadh SM, Mahmmoud EA, Elaasser MM. *Heterocycles*. 2015;91(6):1227–43.
136. Gomha SM, Riyadh SM, Mahmmoud EA, Elaasser MM. *Chemistry of Heterocyclic Compounds*. 2015;51:1030–8.

137. Alshabanah LA, Al-Mutabagani LA, Gomha SM, Ahmed HA. *Frontiers in Chemistry*. 2022;9:762248.
138. Ripoll C, Muller G, Selegny E. *European Polymer Journal*. 1971;7(10):1393–409.
139. Borowiak-Resterna A, Szymanowski J, Voelkel A. *Journal of Radioanalytical and Nuclear Chemistry*. 1996;208(1):75–86.
140. Tsuji J. *Transition Metal Reagents and Catalysts: Innovations in Organic Synthesis*: John Wiley & Sons, New York; 2002.
141. Nasrollahzadeh M, Shafiei N, Maham M, Issaabadi Z, Nezafat Z, Varma RS. *Molecular Catalysis*. 2020;494:111129.
142. Zhang Q, Zhou Y, Fu X, Villarreal E, Sun L, Zou S, Wang H. *The Journal of Physical Chemistry C*. 2019;123(43):26695–704.
143. Panunzi B, Roviello A, Ruffo F, Vivo A. *Inorganica Chimica Acta*. 1998;281(2):141–7.
144. Hayashi H, Mori M, Shigemoto N, Okazaki T. *Industrial & Engineering Chemistry Product Research and Development*. 1978;17(2):128–33.
145. Clear JM, Kelly JM, Peper DC, Vos JG. *Inorganica Chimica Acta*. 1979;33: L139–L40.
146. Haas O, Vos JG. *Journal of Electroanalytical Chemistry and Interfacial Electrochemistry*. 1980;113(1):139–49.
147. Andrieux CP, Haas O, Saveant JM. *Journal of the American Chemical Society*. 1986;108(26):8175–82.
148. Skurlatov YI, Kovner VY, Travin S, Kirsh YE, Purmal A, Kabanov V. *European Polymer Journal*. 1979;15(8):811–5.
149. Chanda M, O'Driscoll K, Rempel G. *Journal of Molecular Catalysis*. 1981;11(1):9–21.
150. Chanda M, O'Driscoll K, Rempel G. *Journal of Molecular Catalysis*. 1980;7(3):389–402.
151. Chanda M, O'Driscoll K, Rempel G. *Journal of Molecular Catalysis*. 1981;12(2):197–211.
152. Yarapathi RV, Kurva S, Tammishetti S. *Catalysis Communications*. 2004;5(9):511–3.
153. Castro IU, Stüber F, Fabregat A, Font J, Fortuny A, Bengoa C. *Journal of Hazardous Materials*. 2009;163(2–3):809–15.
154. Zarchi MAK, Nazem F. *Journal of the Iranian Chemical Society*. 2014;6(11):1731–42.
155. Nishide H, Minakata T, Tsuchida E. *Journal of Molecular Catalysis*. 1982;15(3):327–35.
156. Pardey AJ, Rojas AD, Yánez JE, Betancourt P, Scott C, Urbina C et al. *Polyhedron*. 2005;24(4):511–9.
157. Yánez JE, Rivas AB, Alvarez J, Ortega MC, Pardey AJ, Longo C, Feazell RP. *Journal of Coordination Chemistry*. 2006;59(15):1719–28.
158. Schuchardt U, Santos Dias F. *Journal of Molecular Catalysis*. 1985;29(1):145–9.
159. Alcántara R, Canoira L, Guilherme-Joao P, Rodríguez JG, Alcántara E. *Reaction Kinetics and Catalysis Letters*. 2004;83:105–12.
160. Cainelli G, Contento M, Manescalchi F, Plessi L. *Synthesis*. 1989;1989(01):47–8.
161. Herrmann WA, Kratzer RM, Blümel J, Friedrich HB, Fischer RW, Apperley DC et al. *Journal of Molecular Catalysis A: Chemical*. 1997;120(1–3):197–205.
162. Basavaraju K, Sharma S, Maurya RA, Kim DP. *Angewandte Chemie International Edition*. 2013;52(26):6735–8.
163. Zharmagambetova A, Golodov V, Saltykov YP. *Journal of Molecular Catalysis*. 1989;55(1):406–14.
164. Drelinkiewicz A, Hasik M. *Journal of Molecular Catalysis A: Chemical*. 2001;177(1):149–64.
165. Schäfer C, Mhadgut SC, Kugyela N, Török M, Török B. *Catalysis Science & Technology*. 2015;5(2):716–23.
166. Benkhaled M, Morin S, Pichon C, Thomazeau C, Verdon C, Uzio D. *Applied Catalysis A: General*. 2006;312:1–11.
167. Mathew JP, Srinivasan M. *Polymer International*. 1991;24(4):249–54.
168. Jo Y-D, Park K-S, Ahn J-H, Ihm S-K. *Reactive and Functional Polymers*. 1996;29(2):91–9.

169. Saha CR, Bhattacharya S. *Journal of Chemical Technology & Biotechnology.* 1987;37(4):233–45.
170. Zharmagambetova A, Mukhamedzhanova S, Dusenbina B. *Reactive Polymers.* 1994;24(1):21–5.
171. Li Y, Hong XM, Collard DM, El-Sayed MA. *Organic Letters.* 2000;2(15):2385–8.
172. Jackson JA, Newsham MD, Worsham C, Nocera DG. *Chemistry of Materials.* 1996;8(2):558–64.
173. Friedrich HB, Singh N. *Tetrahedron Letters.* 2000;41(20):3971–4.
174. Friedrich HB, Singh N. *Catalysis Letters.* 2006;110:61–70.
175. Pardey AJ, Mediavilla M, Canestrari M, Urbina C, Moronta D, Lujano E et al. *Catalysis Letters.* 1998;56:231–5.
176. Pardey A, Fernández M, Canestrari M, Baricelli P, Lujano E, Longo C et al. *Reaction Kinetics and Catalysis Letters.* 1999;67:325–31.
177. Pardey AJ, Fernández Ma, Alvarez J, Urbina C, Moronta D, Leon V et al. *Applied Catalysis A: General.* 2000;199(2):275–83.
178. Pardey A, Fernández M, Alvarez J, Urbina C, Moronta D, Leon V et al. *Journal of Molecular Catalysis A: Chemical.* 2000;164(1–2):225–34.
179. Pardey A, Brito J, Rivas A, Ortega M, Longo C, Baricelli P et al. *Journal of the Chilean Chemical Society.* 2003;48(2):57–60.
180. Hung-Low F, Uzcátegui GC, Ortega MC, Rivas AB, Yanez JE, Alvarez J et al. *Catalysis Today.* 2005;107:273–81.
181. Pathak S, Greci MT, Kwong RC, Mercado K, Prakash GS, Olah GA, Thompson ME. *Chemistry of Materials.* 2000;12(7):1985–9.
182. Musikabhumma K, Uozumi T, Sano T, Soga K. *Macromolecular Rapid Communications.* 2000;21(10):675–9.
183. Mugweru A, Rusling JF. *Analytical Chemistry.* 2002;74(16):4044–9.
184. Mugweru A, Rusling JF. *Electrochemistry Communications.* 2001;3(8):406–9.
185. Pardey AJ, Brito J, Fernández M, Rivas AB, Ortega MC, Longo C et al. *Reaction Kinetics and Catalysis Letters.* 2001;74:111–8.
186. Pardey AJ, Hung-Low F, Uzcátegui GC, Ortega MC, Longo C. *Reaction Kinetics and Catalysis Letters.* 2006;88:203–8.
187. Farkas M, Rodríguez E, Longo C, Monasterios M, Ortega M, Rivas A et al. *Journal of the Chilean Chemical Society.* 2006;51(1):829–35.
188. Pardey AJ, Suárez JD, Gascón G, Ortega MC, Longo C, Moya SA. *Catalysis Letters.* 2008;126:112–8.
189. Fang M, Machalaba N, Sánchez-Delgado RA. *Dalton Transactions.* 2011; 40(40):10621–32.
190. Sánchez-Delgado RA, Machalaba N. *Catalysis Communications.* 2007;8(12):2115–8.
191. Kim HS, Kim JJ, Kwon HN, Chung MJ, Lee BG, Jang HG. *Journal of Catalysis.* 2002;205(1):226–9.
192. Saladino R, Neri V, Pelliccia AR, Caminiti R, Sadun C. *The Journal of Organic Chemistry.* 2002;67(4):1323–32.
193. Verma S, Joshi A, De SR, Jat JL. *New Journal of Chemistry.* 2022;46(5):2005–27.
194. Saladino R, Neri V, Mincione E, Filippone P. *Tetrahedron.* 2002;58(42):8493–500.
195. Goti A, Cardona F, Soldaini G, Crestini C, Fiani C, Saladino R. *Advanced Synthesis & Catalysis.* 2006;348(4-5):476–86.
196. Bianchini G, Crucianelli M, Crestini C, Saladino R. *Topics in Catalysis.* 2006;40:221–7.
197. Saladino R, Bernini R, Neri V, Crestini C. *Applied Catalysis A: General.* 2009;360(2):171–6.
198. Drelinkiewicza A, Waksmundzka A, Makowski W, Sobczak J, Król A, Zieba A. *Catalysis Letters.* 2004;94:143–56.

199. Yamada YM, Watanabe T, Torii K, Uozumi Y. *Chemical Communications.* 2009(37):5594–6.
200. Ohno A, Sato T, Mase T, Uozumi Y, Yamada YM. *Advanced Synthesis & Catalysis.* 2020;362(21):4687–98.
201. Bulatov E, Lahtinen E, Kivijärvi L, Hey-Hawkins E, Haukka M. *ChemCatChem.* 2020;12(19):4831–8.
202. Fusini G, Rizzo F, Angelici G, Pitzalis E, Evangelisti C, Carpita A. *Catalysts.* 2020;10(3):330.
203. Albano G, Interlandi S, Evangelisti C, Aronica LA. *Catalysis Letters.* 2020;150:652–9.
204. Pardey A, Morillo B, Alvarez J, Yanez J, Ortega M, Longo C. *Catalysis Letters.* 2005;104:141–50.
205. Sabitha G, Reddy KB, Yadav J, Shailaja D, Sivudu KS. *Tetrahedron Letters.* 2005;46(47):8221–4.
206. Centomo P, Zecca M, Lora S, Vitulli G, Caporusso A, Tropeano M et al. *Journal of Catalysis.* 2005;229(2):283–97.
207. Lončarević D, Čupić Ž. *Materials Science Forum.* 2005;494:363–8.
208. Efendiev A. *Macromolecular Symposia.* 2006;235(1):89–97.
209. Hung-Low F, Uzcátegui GC, Alvarez J, Ortega MC, Pardey AJ, Longo C. *Reaction Kinetics and Catalysis Letters.* 2006;88:143–8.
210. Caporusso A, Innocenti P, Aronica LA, Vitulli G, Gallina R, Biffis A et al. *Journal of Catalysis.* 2005;234(1):1–13.
211. Evangelisti C, Panziera N, Pertici P, Vitulli G, Salvadori P, Battocchio C, *Polzonetti G. Journal of Catalysis.* 2009;262(2):287–93.
212. Syukri S, Hijazi AK, Sakthivel A, Al-Hmaideen AI, Kühn FE. *Inorganica Chimica Acta.* 2007;360(1):197–202.
213. Syukri S, Sun W, Kühn FE. *Tetrahedron Letters.* 2007;48(9):1613–7.
214. Pardey AJ, Manosalva N, Bartolini M, Molina J, Ortega MC, D'Ornelas L et al. *Catalysis Letters.* 2008;122:274–80.
215. Cao LJ, Wang G, Shi L, Yang M, Sun DB. *Journal of Applied Polymer Science.* 2010;116(6):3178–83.
216. Cao L, Yang M, Wang G, Wei Y, Sun D. *Journal of Polymer Science Part A: Polymer Chemistry.* 2010;48(3):558–62.
217. Lončarević D, Krstić J, Dostanić J, Manojlović D, Čupić Ž, Jovanović DM. *Chemical Engineering Journal.* 2010;157(1):181–8.
218. Kalbasi RJ, Izadi E. *Comptes Rendus Chimie.* 2011;14(11):1002–13.
219. Albadi J, Iravani N, Shirini F, Dehghan F. *Journal of Chemical Research.* 2012;36(10):610–1.
220. Albadi J, Shirini F, Abasi J, Armand N, Motaharizadeh T. *Comptes Rendus Chimie.* 2013;16(5):407–11.
221. Albadi J, Keshavarz M, Abedini M, Khoshakhlagh M. *Journal of Chemical Sciences.* 2013;125:295–8.
222. Albadi J, Shiran JA, Mansournezhad A. *Acta Chimica Slovenica.* 2014;61(4):900–3.
223. Albadi J, Keshavarz M, Abedini M, Vafaie-nezhad M. *Chinese Chemical Letters.* 2012;23(7):797–800.
224. Albadi J, Keshavarz M, Shirini F, Vafaienezhad M. *Catalysis Communications.* 2012;27:17–20.
225. Albadi J, Iravani N, Khoshakhlagh M. *Iranian Journal of Catalysis.* 2012;2(2):85–9.
226. Albadi J, Mansournezhad A, Baghernehad M, Frozan N. *Journal of the Korean Chemical Society.* 2013;57:169–71.
227. Albadi J, Mansournezhad A. *Research on Chemical Intermediates.* 2016;42(6):5739–52.
228. Najafpour MM, Haghighi B, Ghobadi MZ, Sedigh DJ. *Chemical Communications.* 2013;49(78):8824–6.

229. Li J, Huang Z, Yang M, Tan L, Zhang X, Gao H et al. *New Journal of Chemistry.* 2015;39(4):2949–55.
230. Qin Q-H, Na H, Zhang C, Yu Q, Zhang X-Q, Zhang H-X. *Journal of Nanoscience and Nanotechnology.* 2015;15(5):3909–12.
231. Matsuo A, Hasegawa S, Takano S, Tsukuda T. *Langmuir.* 2020;36(27):7844–9.
232. Sen A, Sato T, Ohno A, Baek H, Muranaka A, Yamada YM. *Journal of the American Chemical Society Au.* 2021;1(11):2080–7.
233. Isazade AF, Mammadova UA, Tagiev DB, Zeynalov NA. *New Materials, Compounds and Applications.* 2021;5(3):198–204.
234. Hartley F, McCaffrey D, Murray S, Nicholson P. *Journal of Organometallic Chemistry.* 1981;206(3):347–59.
235. Verlaan J, Bootsma J, Challa G. *Journal of Molecular Catalysis.* 1982;14(2):211–8.
236. Seregina MV, Bronstein LM, Platonova OA, Chernyshov DM, Valetsky PM, Hartmann J et al. *Chemistry of Materials.* 1997;9(4):923–31.
237. Klingelhöfer S, Heitz W, Greiner A, Oestreich S, Förster S, Antonietti M. *Journal of the American Chemical Society.* 1997;119(42):10116–20.
238. Liu Y, Lor C, Fu Q, Pan D, Ding L, Liu J, Lu J. *The Journal of Physical Chemistry C.* 2010;114(13):5767–72.
239. Shi P, Gao C, He X, Sun P, Zhang W. *Macromolecules.* 2015;48(5):1380–9.
240. Piccinino D, Abdalghani I, Botta G, Crucianelli M, Passacantando M, Di Vacri ML, Saladino R. *Applied Catalysis B: Environmental.* 2017;200:392–401.
241. Mathew JP, Srinivasan M. *Polymer International.* 1992;29(3):179–84.
242. Yamashita K, Kanamori T, Tsuda K. *Journal of Macromolecular Science: Part A - Chemistry.* 1990;27(7):897–909.
243. Yamashita K, Kanamori T, Nango M, Tsuda K. *Polymer.* 1993;34(12):2638–43.
244. Yamashita K, Okada I, Suzuki Y, Tsuda K. *Die Makromolekulare Chemie, Rapid Communications.* 1988;9(10):705–8.
245. Bronstein LM, Chernyshov DM, Volkov IO, Ezernitskaya MG, Valetsky PM, Matveeva VG, Sulman EM. *Journal of Catalysis.* 2000;196(2):302–14.
246. Sulman EM, Matveeva VG, Sulman MG, Demidenko GN, Valetsky PM, Stein B et al. *Journal of Catalysis.* 2009;262(1):150–8.
247. Xi X, Liu Y, Shi J, Cao S. *Journal of Molecular Catalysis A: Chemical.* 2003;192(1–2):1–7.
248. Gao B, Kong D, Zhang Y. *Journal of Molecular Catalysis A: Chemical.* 2008;286(1–2):143–8.
249. Turmanova S, Vassilev K, Boneva S. Reactive and *Functional Polymers.* 2008;68(3):759–67.
250. Wen F, Zhang W, Wei G, Wang Y, Zhang J, Zhang M, Shi L. *Chemistry of Materials.* 2008;20(6):2144–50.
251. Wen F, Zhang W, Zheng P, Zhang X, Yang X, Wang Y et al. *Journal of Polymer Science Part A: Polymer Chemistry.* 2008;46(4):1192–202.
252. Okamoto M, Sugiyama Ji. *Journal of Applied Polymer Science.* 2009;114(4):2226–34.
253. Tamami B, Farjadian F. *Journal of the Iranian Chemical Society.* 2011;8:S77–S88.
254. Farjadian F, Tamami B. *ChemPlusChem.* 2014;79(12):1767–73.
255. Ren Z, Wang H-L, Cai Y-Q, Chen M, Qian D-J. *Materials Chemistry and Physics.* 2011;127(1–2):310–5.
256. Kontkanen M-L, Vlasova L, Suvanto S, Haukka M. *Applied Catalysis A: General.* 2011;401(1–2):141–6.
257. Kontkanen M-L, Haukka M. *Catalysis Communications.* 2012;23:25–9.
258. Guo W, Wang Q, Wang G, Yang M, Dong W, Yu J. *Chemistry–An Asian Journal.* 2013;8(6):1160–7.
259. Chen A, Li Y, Chen J, Zhao G, Ma L, Yu Y. *ChemPlusChem.* 2013;78(11):1370–8.

260. Krawicz A, Yang J, Anzenberg E, Yano J, Sharp ID, Moore GF. *Journal of the American Chemical Society*. 2013;135(32):11861–8.
261. Yang J, Zhang N, Li H, Xu B, Tian W, Dong D. *Polymer International*. 2015;64(6):804–10.
262. Chen J, Zhong J, Guo Y, Chen L. *RSC Advances*. 2015;5(8):5933–40.
263. Liu J, Zorin NA, Chen M, Qian D-J. *Langmuir*. 2015;31(23):6546–53.
264. Mohan A, Peter J, Rout L, Thomas AM, Nagappan S, Parambadath S et al. *Colloids and Surfaces A: Physicochemical and Engineering Aspects*. 2021;611:125846.
265. Shanmugam P, Rajakumar K, Boddula R, Ngullie RC, Wei W, Xie J, Murugan E. *Materials Science for Energy Technologies*. 2019;2(3):532–42.
266. Zhang K, Zong L, Jia X. *Advanced Synthesis & Catalysis*. 2021;363(5):1335–40.
267. Ivko SA, James AM, Derry MJ, Dawson R, Haynes A. *Catalysis Science & Technology*. 2022;12(2):664–73.
268. Martínez I, Santillán R, Fuentes Camargo I, Rodríguez JL, Andraca Adame JA, Gutiérrez HM. *Polymers*. 2022;14(21):4666.
269. Yalçin Ş, Ali K. *Journal of Natural and Applied Sciences*. 2022;26(1):115–21.
270. Puterman M, Kolpak F, Blackwell J, Lando J. *Journal of Polymer Science: Polymer Physics Edition*. 1977;15(5):805–19.
271. Sun W, Yu B, Kühn FE. *Tetrahedron Letters*. 2006;47(12):1993–6.
272. Galogahi FM, Zhu Y, An H, Nguyen N-T. *Journal of Science: Advanced Materials and Devices*. 2020;5(4):417–35.
273. Li W-H, Stöver HD. *Macromolecules*. 2000;33(12):4354–60.
274. Donbrow M. *Microcapsules and Nanoparticles in Medicine and Pharmacy*: CRC Press, Boca Raton; 1991.
275. Benita S. *Microencapsulation: Methods and Industrial Applications*: Taylor and Francis, New York; 2006.
276. Afolabi F, Mahmood SM, Yekeen N, Akbari S, Sharifigaliuk H. *Journal of Petroleum Science and Engineering*. 2022;208(A):109358.
277. Maclay WN, Fuoss RM. *Journal of Polymer Science*. 1951;6(5):511–21.
278. Strauss UP, Jackson EG. *Journal of polymer Science*. 1951;6(5):649–59.
279. Jackson EG, Strauss UP. *Journal of Polymer Science*. 1951;7(5):473–84.
280. Strauss UP, Assony SJ, Jackson EG, Layton LH. *Journal of Polymer Science*. 1952;9(6):509–18.
281. Dehmlow EV, Dehmlow SS. *Phase Transfer Catalysis*: Wiley, Michigan; 1993.
282. Keller WE. *Phase-Transfer Reactions*: Thieme, New York; 1986.
283. Tamami B, Mahdavi H. *Tetrahedron Letters*. 2002;43(35):6225–8.
284. Okubo T, Ise N. *Journal of the American Chemical Society*. 1973;95(7):2293–7.
285. Okubo T, Ise N. *The Journal of Organic Chemistry*. 1973;38(18):3120–2.
286. Rodulfo T, Hamilton J, Cordes E. *The Journal of Organic Chemistry*. 1974;39(15):2281–4.
287. Kunitake T, Shinkai S, Hirotsu S. *Journal of Polymer Science: Polymer Letters Edition*. 1975;13(6):377–81.
288. Chiellini E, Solaro R, D'Antone S. *Die Makromolekulare Chemie: Macromolecular Chemistry and Physics*. 1977;178(11):3165–70.
289. Kunitake T, Shinkai S, Hirotsu S. *The Journal of Organic Chemistry*. 1977;42(2):306–12.
290. Shinkai S, Hirakawa SI, Shimomura M, Kunitake T. *Journal of Organic Chemistry*. 1981;46(5):868–72.
291. Fréchet JM, de Meftahi MV. *British Polymer Journal*. 1984;16(4):193–8.
292. Nakano H, Goto T, Yamamoto Y. *Journal of Polymer Science: Polymer Letters Edition*. 1980;18(5):381–3.
293. Nakano H, Nishioka M, Sangen O, Yamamoto Y. *Journal of Polymer Science: Polymer Chemistry Edition*. 1981;19(11):2919–28.
294. Nakano H, Yamane R, Sangen O, Yamamoto Y. *Journal of Polymer Science: Polymer Chemistry Edition*. 1982;20(8):2335–9.

295. Nakano H, Yagi T, Sangen O, Yamamoto Y. *Journal of Polymer Science: Polymer Letters Edition*. 1982;20(1):23–6.
296. MacKenzie W, Sherrington D. *Polymer*. 1981;22(4):431–3.
297. Ochiai B, Endo T. *Journal of Polymer Science Part A: Polymer Chemistry*. 2007;45(23):5673–8.
298. Jia Z, Zhen T, Zhang X, Gu Q. *Journal of Membrane Science*. 2013;448:74–80.
299. Jia Z, Zhen T, Zhang X. *Journal of Membrane Science*. 2014;454:316–21.

4 Polymeric Reagents Based on Poly(vinylpyridine)

4.1 POLYMERIC REAGENTS

As stated in Chapter 2, polymer-supported reagents have become regularly used species in various fields because the attaching of reagents to the polymer matrix can eliminate the complications of product separation, instability, toxicity, or limited recovery and reusability usually associated with their unsupported counterparts [1–11]. They may exhibit greater compliance with continuous flow methods compared to batch processes [12]. The growing interest in polymeric reagents is also attributed to their remarkable selectivity and versatility in various synthetic transformations. They are often used in stoichiometric quantities or more to drive chemical transformation to complete conversion. However, the most important shortcomings of polymeric reagents may be reduced reactivity because of diffusional constraints, steric hindrance, site isolation effects, and challenges associated with their structural characterization. Furthermore, sometimes they are expensive and somewhat difficult to prepare. For example, Scheme 4.1 shows a typical polymeric reagent used in chemical reactions. The active functional group of the support would lead to the required chemical modification of the substrate. Such groups may be attached to the polymeric carriers by chemical bonding or by physical adsorption. Usually, physically adsorbed species are unsatisfactory due to leaching of active groups in the reaction mixture, and they are therefore unreusable and also inadequate for column applications.

The support used may be either hydrophobic polymers (e.g., polystyrene and its derivatives) or hydrophilic polymers (e.g., polyacrylamide, poly(vinylpyrrolidone)). Along with polystyrene that has frequently been employed for the preparation of polymeric reagents, PVPy has also gained increasing attention due to ease of functionalization through the nitrogen atom of the pyridine ring, availability, a more favorable balance between its hydrophobic/hydrophilic nature compared to polystyrene, the feasibility of quaternization of pyridine rings, and possibility of copolymerization with other monomers, resulting in a diverse array of block copolymers with unique properties.

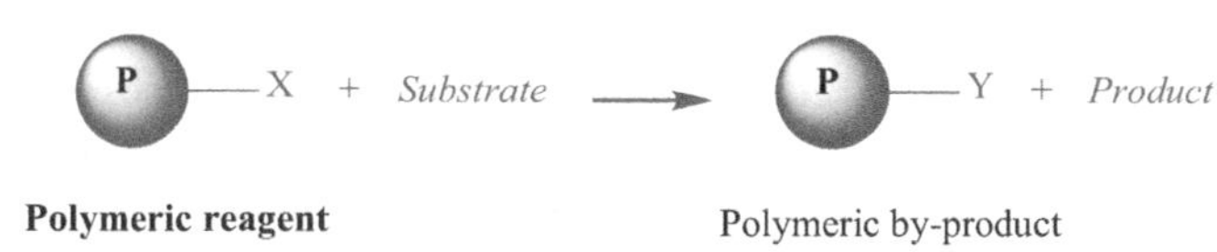

SCHEME 4.1 Polymer-supported reagent in a chemical process.

DOI: 10.1201/9781003656104-4

In 2009, we reported a review on PVPy-supported reagents [13]. Herein, I want to update that task. The discussions are concentrated on the introduction of PVPy-supported reagents, in three categories (oxidizing, reducing, and halogenating reagents), and a relatively complete overview of the existing literature on such reagents.

4.2 POLYMERIC OXIDIZING AGENTS BASED ON PVPy

The oxidation of organic compounds plays a crucial role in the production of valuable oxygen-containing compounds. In the recent decades, with the progress of the industry, there has been a demand for the development of new methods of controlled oxidation. So far, many oxidizing agents such as chromium-based reagents, e.g., Jones reagent (CrO_3/H_2SO_4), Sarett and Collins reagents (complexes of CrO_3 with pyridine), PDC (pyridinium dichromate), PCC (pyridinium chlorochromate), $Na_2Cr_2O_7$ (sodium dichromate), $K_2Cr_2O_7$ (potassium dichromate), $ZnCr_2O_7{\cdot}3H_2O$ (hydrated zinc dichromate), $Fe_2(Cr_2O_7)_3$ (ferric dichromate), $KMnO_4$ (potassium permanganate), MnO_2 (manganese dioxide), $BaMnO_4$ (barium manganate), H_2O_2 (hydrogen peroxide), t-BuOOH (tert-butyl hydroperoxide), I_2/Li_2CO_3 (iodine/lithium carbonate), IBX (o-iodoxybenzoic acid), and NaOCl (sodium hypochlorite) have been introduced [14–17]. However, these methods are plagued by the limitations such as the requirement for stoichiometric quantities of oxidants, the occurrence of over-oxidation or side reactions, and generation of waste. In addition, some of them are corrosive, difficult to handle, and unreusable reagents. Thus, these shortcomings made them a prime target for heterogenization. In this context, numerous supported oxidants have been documented and utilized as substitutes for traditional oxidants. They often exhibit higher selectivity, and require easier handling and simpler experimental setup in eco-friendly conditions [13,18–27].

Like other polymers, PVPy has been used as an ideal support for the immobilization of oxidizing agents. Table 4.1 shows PVPy-supported oxidizing agents with their applications. Representative examples of applications of these reagents are listed in Scheme 4.2. Often, anchoring of oxidizing reagents on PVPy has been achieved through electrostatic interactions and somewhat by coordination bonds.

The oxidation process facilitated by water-soluble dichromate salts is influenced by pH levels, and these oxidants mostly work at low pH, damaging acid-sensitive substrates [15]. To solve this problem, several pyridinium salts of dichromate (e.g., PCC, PDC) as well as the analogous heterogeneous reagents were introduced, which were less acidic and are therefore more appropriate for the oxidation of acid-sensitive compounds [13,14]. These alternatives also have other advantages including high selectivity, simple work-up, and safe handling. PVPy has been used as support for chromium-based oxidizing reagents (Table 4.1). P_4VPy- and P_2VPy-supported zinc and ferric dichromate have been prepared from their starting materials and used for oxidation of alcohols to the corresponding aldehydes and ketones (Table 4.1, entry 9, Schemes 4.3 and 4.4) [45,63]. As shown in Scheme 4.4, they were as active as the corresponding low-molecular weight analog, ferric dichromate and zinc dichromate while having the advantages of being less acidic and hygroscopic [63,64]. PVPy-supported $Fe_2(Cr_2O_7)_3$ was more efficient than supported $ZnCr_2O_7$ ones.

TABLE 4.1
Polymeric Oxidizing Agents Based on PVPy

Entry	Polymer Reagents with Their Applications
1	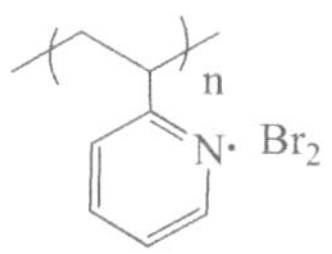 Poly(2-vinylpyridine)-bromine Preparation of disulfides from thiols [28]
2	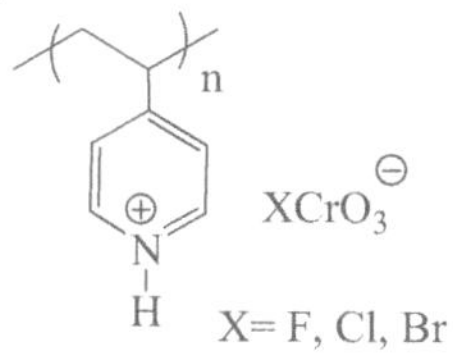 Poly(4-vinylpyridinium fluorochromate) Poly(4-vinylpyridinium chlorochromate) Poly(4-vinylpyridinium bromochromate) Oxidation of oximes and alcohols [29–32]
3	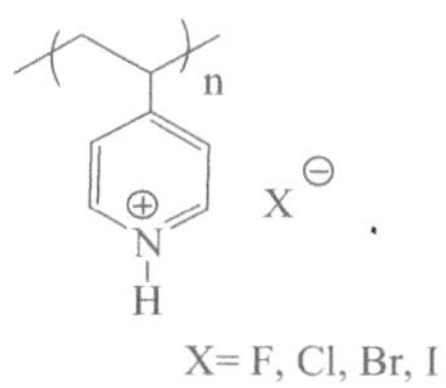 Poly(4-vinylpyridinium)fluoride Poly(4-vinylpyridinium)chloride Poly(4-vinylpyridinium)bromide Poly(4-vinylpyridinium)iodide Electrochemical oxidation of alcohols, alkenes, and sulfides [33–37]
4	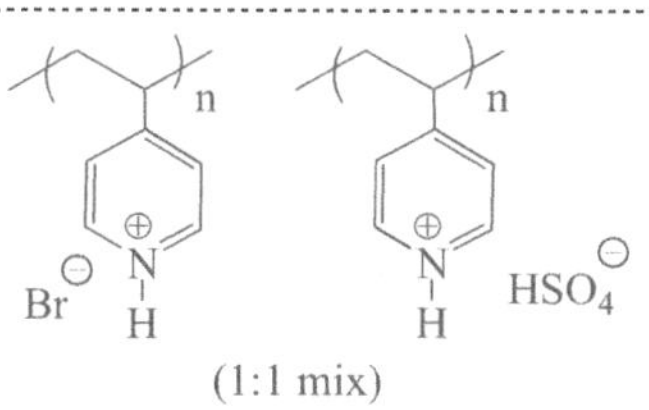 Poly(4-vinylpyridinium)bromide Poly(4-vinylpyridinium)hydrogen sulfate Electrochemical oxidation of alkylbenzenes, alkenes, and sulfides [36]
5	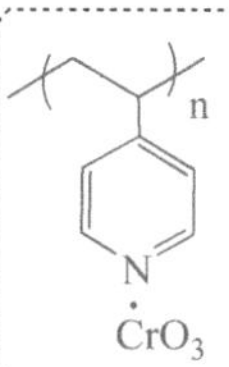 Poly(4-vinylpyridine)-chromium trioxide Oxidation of alcohols [38]

(Continued)

TABLE 4.1 (*Continued*)
Polymeric Oxidizing Agents Based on PVPy

Entry	Polymer Reagents with Their Applications
6	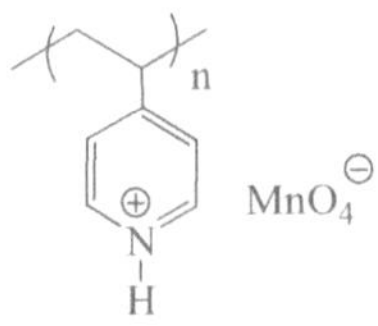 Poly(4-vinylpyridinium)permanganate Oxidation of alcohols [39,40]
7	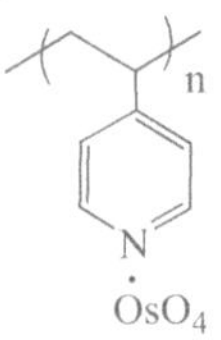 Poly(4-vinylpyridine)-osmium tetroxide Conversion of alkenes to aldehydes and ketones [41] Oxidative cleavage of olefins [42]
8	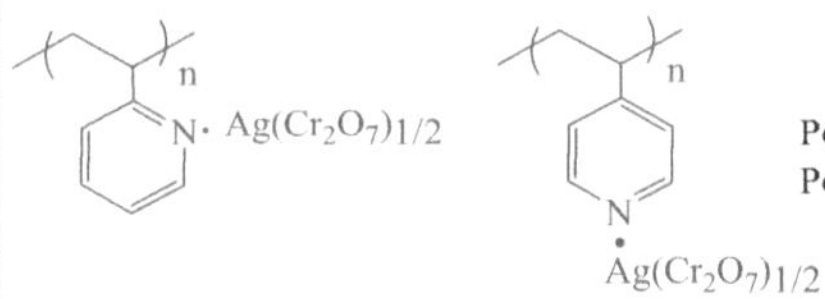 Poly(4-vinylpyridine)-silver dichromate Poly(2-vinylpyridine)-silver dichromate Oxidation of oximes, sulfide, amines, and arenes [43,44]
9	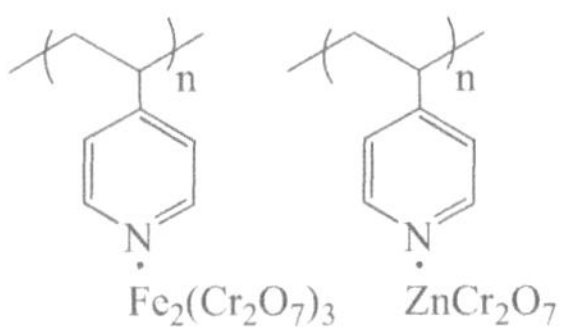 Poly(4-vinylpyridine)-ferric dichromate Poly(4-vinylpyridine)-zinc dichromate Oxidation of alcohols and aromatic compounds [45]
10	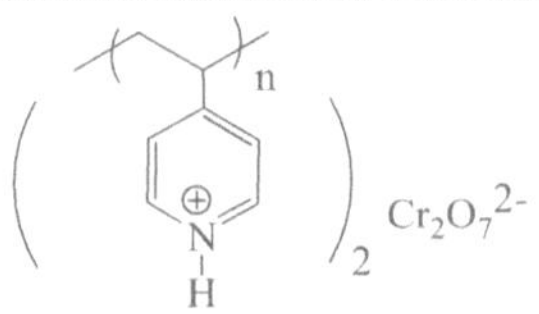 Poly(4-vinylpyridinium)dichromate Oxidation of alcohols [46,47]

(*Continued*)

TABLE 4.1 (*Continued*)
Polymeric Oxidizing Agents Based on PVPy

Entry	Polymer Reagents with Their Applications
11	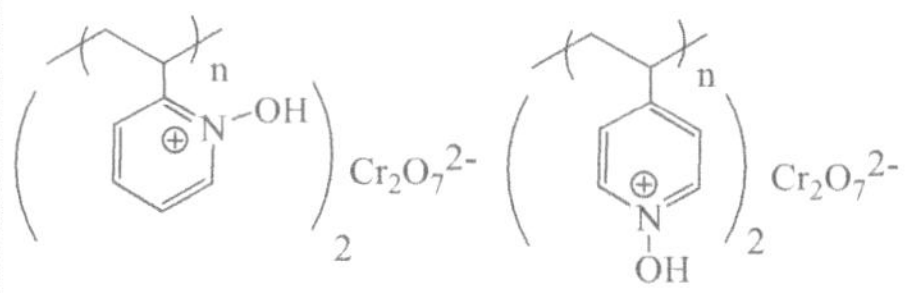 Poly(2-vinylpyridine *N*-oxide)dichromate Poly(4-vinylpyridine *N*-oxide)dichromate Oxidation of alcohols, amines, sulfide, and acyloins [48]
12	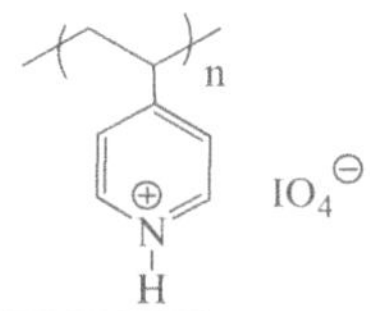 Poly(4-vinylpyridinium)periodate Oxidation of alcohols, acyloins, azines, decarboxylation of carboxylic acids, and C–C cleavage of epoxides [13]
13	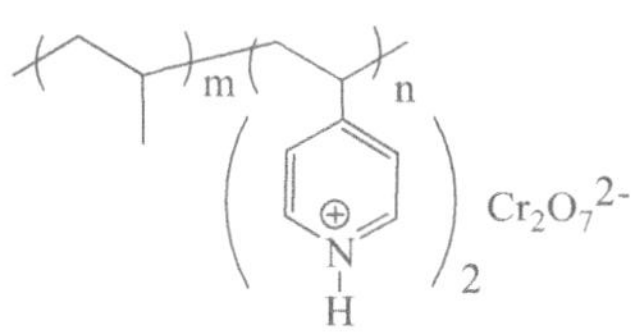 Poly[propylene-g-(4-vinylpyridinium dichromate)] Oxidation of alcohols [49]
14	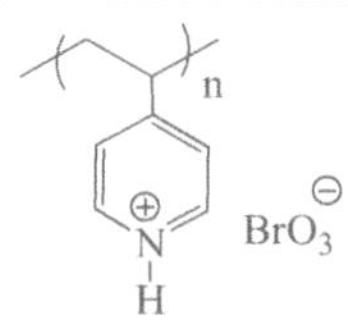 Poly(4-vinylpyridinium)bromate Oxidation of alcohols, sulfides, and alkyl benzenes [50]
15	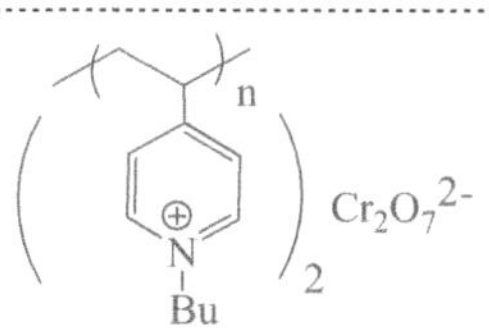 Poly(*n*-butyl-4-vinylpyridinium)dichromate Oxidation of alcohols and acyloins [51]

(*Continued*)

TABLE 4.1 (*Continued*)
Polymeric Oxidizing Agents Based on PVPy

Entry	Polymer Reagents with Their Applications
16	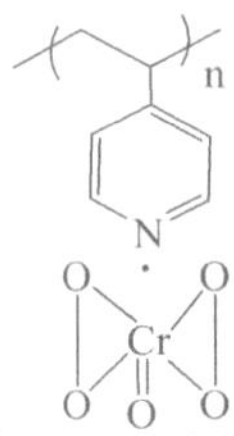 Poly(4-vinylpyridine)-chromium peroxide Oxidation of alcohols [52]
17	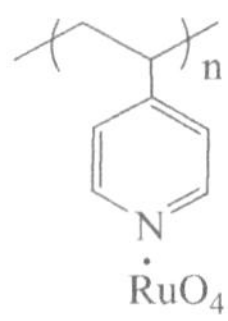 Poly(4-vinylpyridine)-ruthenate Oxidation of alcohols [53]
18	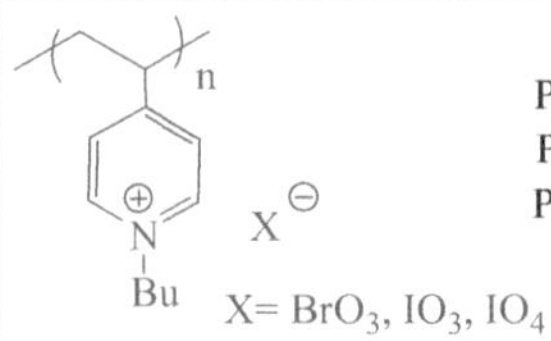 Poly(*n*-butyl-4-vinylpyridinium)bromate Poly(*n*-butyl-4-vinylpyridinium)iodate Poly(*n*-butyl-4-vinylpyridinium)periodate Oxidation of alcohols, sulfides, and diols [54]
19	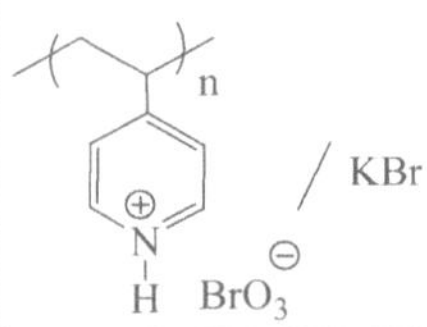 Poly(4-vinylpyridinium)bromate Bromination of aromatic compounds [55]
20	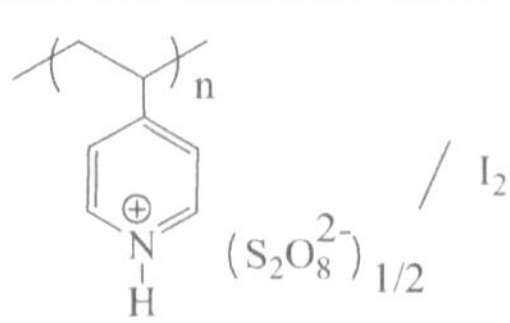 Poly(4-vinylpyridinium)peroxodisulfate Iodination of aromatic compounds [56]

(*Continued*)

TABLE 4.1 (*Continued*)
Polymeric Oxidizing Agents Based on PVPy

Entry **Polymer Reagents with Their Applications**

21

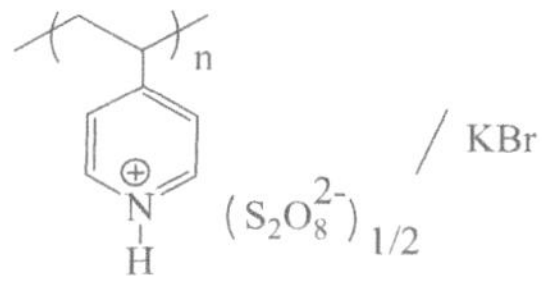

Poly(4-vinylpyridinium)peroxodisulfate

Bromination of aromatic compounds [57]

22

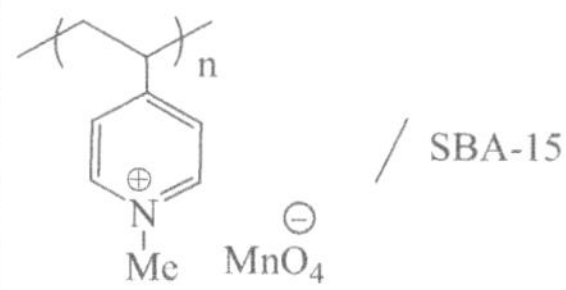

Poly(*N*-methyl-4-vinylpyridinium)permanganate

Oxidation of aromatic alcohols [58]

23

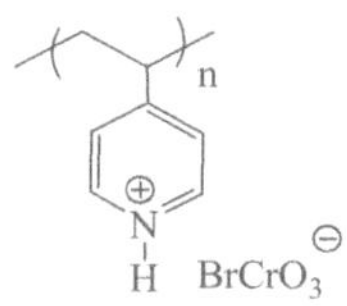

Poly(4-vinylpyridinium)bromochromate

Bromination of aromatic compounds [59]

24

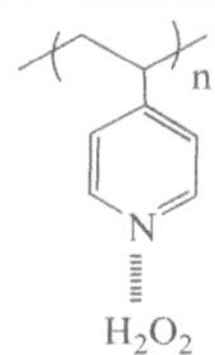

Poly(4-vinylpyridine)-hydrogen peroxide

Synthesis of phenols from arylboronic acids [60]

Conversion of sulfides to sulfoxides and ketones to gem-dihydroperoxides [61]

25

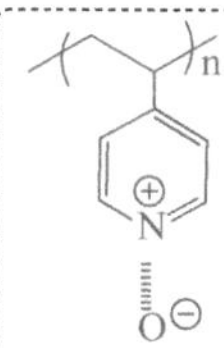

Poly(4-vinylpyridine-N-oxide) (P_4VPyNO)

Oxidation of phosphorus trichloride [62]

(*Continued*)

TABLE 4.1 (*Continued*)
Polymeric Oxidizing Agents Based on PVPy

Entry	Polymer Reagents with Their Applications
26	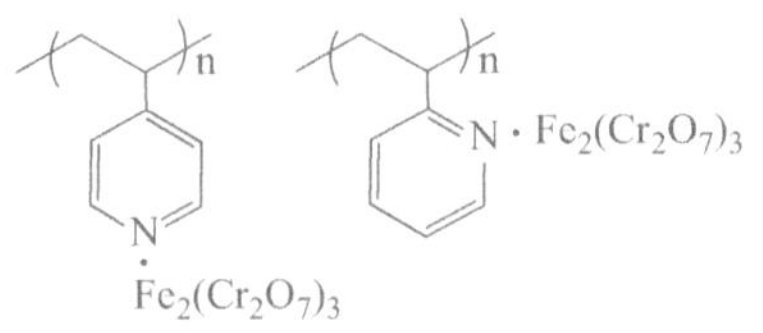 Poly(4-vinylpyridine)-ferric dichromate Poly(2-vinylpyridine)-ferric dichromate Oxidation of alcohols, hydrocarbons, olefinic and acetylenic double bonds [63]

They showed high selectivity in oxidizing reactions. For example, the oxidation of benzyl alcohol with different amounts of PVPy-..$ZnCr_2O_7$ produces benzaldehyde as the sole product, while using $ZnCr_2O_7$benzaldehyde product is generated when the molar ratio of oxidant to alcohol is 0.5, and with increasing molar amounts of oxidant to 1.5, over-oxidation occurs, resulting in the formation of benzoic acid product. It is worth noting that the oxidizing power of $P_2VPy\text{-}Fe_2(Cr_2O_7)_3$ and $P_2VPy\text{-}ZnCr_2O_7$ was more than their corresponding poly-4-isomers, as reported elsewhere [43,44,63]. It can be attributed to the fact that the proximity of the nitrogen atoms of the pyridine rings to the polymer backbone in poly 2-isomes enhances the binding of chromate ions on polymer reagent. In addition, due to slight interactions between hydrogens in the polymer chain and the nitrogen of the rings in poly 2-isomes, it has a more compact structure [65]. Tamami et al. prepared atactic, isotactic, and syndiotactic forms of poly(vinylpyridinium)dichromate (PVPy-DC) and investigated their activity for the oxidation of different alcohols [65]. In all cases of oxidation reactions, poly 2-isomer was more efficient than poly 4-isomer and the activity of the reagents was in the following order: syndiotactic P_2VPy-DC > atactic P_2VPy-DC > isotactic P_2VPy-DC. Considering that these polymeric oxidizing agents have almost similar molecular weights, the reagent loadings, and swelling properties, the differences in their activities can be related to the fact that syndiotactic and isotactic P_2VPy have different conformations, zigzag and spiral, respectively; therefore, nitrogen atom in the pyridine rings is spatially located differently in these two stereoregular polymers which affects their reactivity.

Tamami et al. described poly(vinylpyridine N-oxide) (PVPyNO)-supported dichromate as an oxidizing agent (Table 4.1, entry 11) [48]. Scheme 4.5 shows synthetic pathways for these reagents [48,66]. They are able to oxidize different hydroxy compounds, amines, and thiols (Scheme 4.6). In contrast to dichromate of pyridine N-oxide ($[VPyNO]_2Cr_2O_7$) which is a combustible and difficult to handle reagent, PVPyNO-supported dichromates are the environmentally acceptable, safe, easy-to-handle, and regenerable reagents. They are much cheaper than their silver and zinc analogues [43–45]. Unlike PVPy-DC [46], these compounds are nonacidic

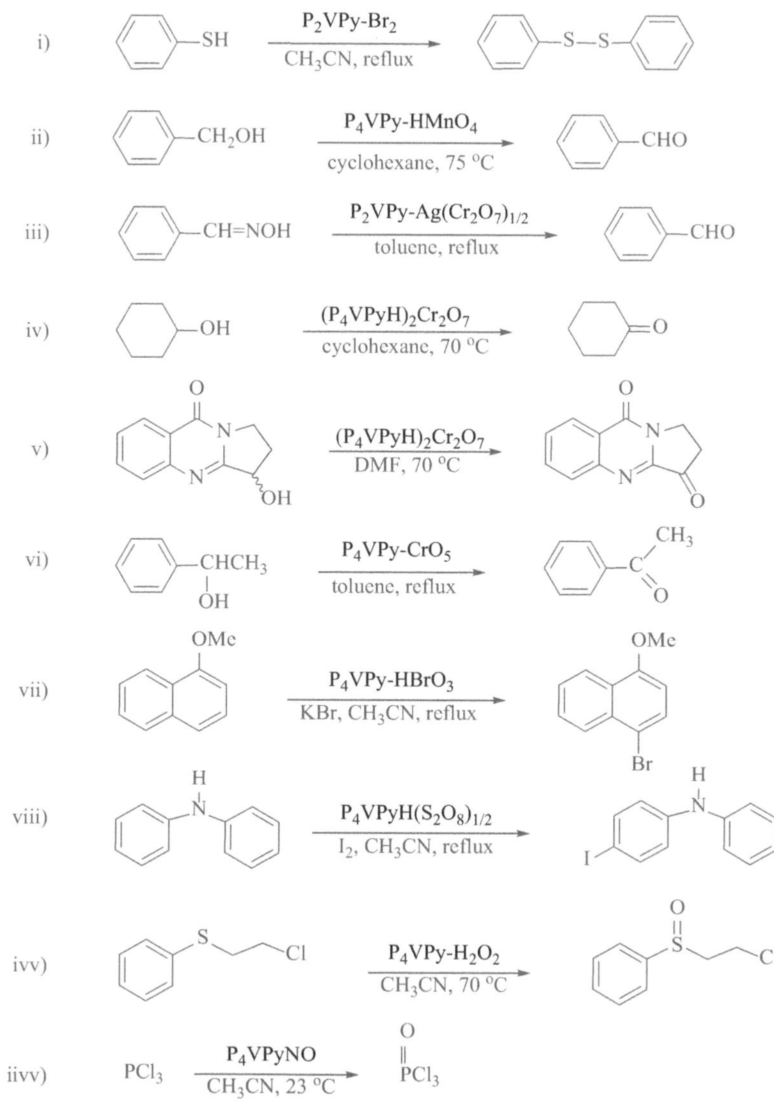

SCHEME 4.2 Oxidizing agents based on PVPy polymers for conversion of sulfides to disulfides (i) [28], oxidation of alcohols to aldehyde and ketones (ii, iv, v, vi) [39,46,47,52], oxidation of oximes (iii) [44], halogenation of aromatic compounds (vii, viii) [55,56], preparation of sulfoxides from sulfides (ivv) [61], and oxidation of phosphorus trichloride (iivv) [62].

and do not require the wetting of reagents for oxidation reactions. It should be noticed that $[P_2VPyNO]_2Cr_2O_7$ is more efficient than $[P_4VPyNO]_2Cr_2O_7$, similar to previous reports [43,44]. As mentioned before, this can be related to the higher local concentration and better availability of chromate ions on poly 2-isomes [65].

P_4VPy-supported osmium tetroxide ($P_4VPy-OsO_4$) and P_4VPy-supported sodium ruthenate ($P_4VPy-RuO_4$) have been reported as efficient oxidizing agents for

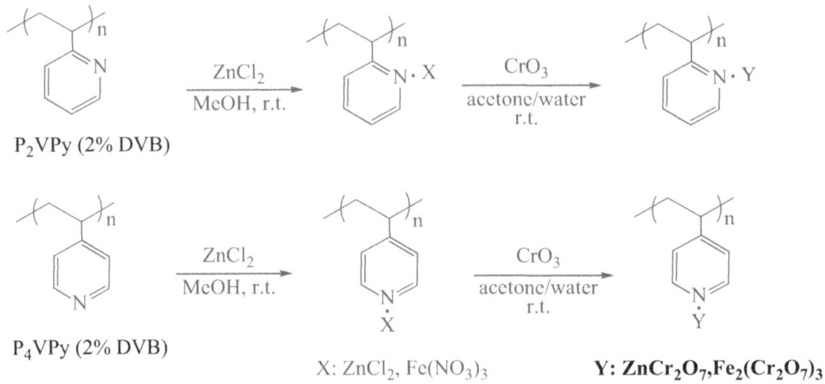

SCHEME 4.3 Preparation of $P_2VPy\text{-}ZnCr_2O_7$, $P_4VPy\text{-}ZnCr_2O_7$ [45], $P_2VPy\text{-}Fe_2(Cr_2O_7)_3$, and $P_4VPy\text{-}Fe_2(Cr_2O_7)_3$ [63].

Reaction condition	Yield (%)
$P_2VPy\text{-}ZnCr_2O_7$, toluene, reflux, 0.5h	95%
$P_4VPy\text{-}ZnCr_2O_7$, toluene, reflux, 1h	95%
$P_2VPy\text{-}Fe_2(Cr_2O_7)_3$, toluene, reflux, 0.15h	97%
$P_4VPy\text{-}Fe_2(Cr_2O_7)_3$, toluene, reflux, 0.25h	98%
$ZnCr_2O_7$, CH_2Cl_2, r.t., 0.1h	95%
$Fe_2(Cr_2O_7)_3$, MeCN, reflux, 1.5h	75%

SCHEME 4.4 Oxidation of benzyl alcohol in the presence of PVPy-supported zinc and ferric dichromate and unsupported counterparts [45,63,64].

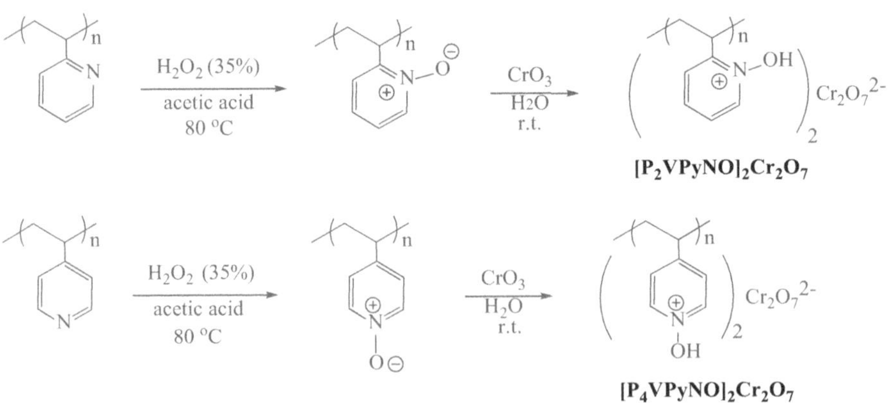

SCHEME 4.5 Preparation of PVPyNO-supported dichromate [48,66].

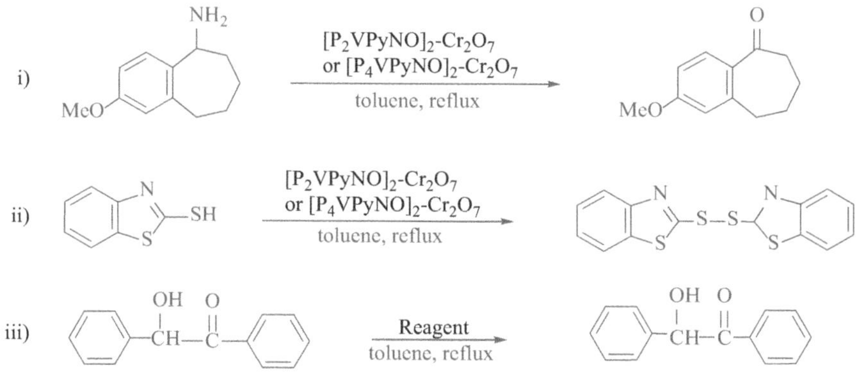

Reagent	Yield (%)
$[P_2VPyNO]_2Cr_2O_7$,150 min	80%
$[P_4VPyNO]_2Cr_2O_7$, 240 min	70%
$[VPyNO]_2Cr_2O_7$, 60 min	85%

SCHEME 4.6 PVPyNO-supported dichromate for oxidation of amines (i) and thiols (ii) and comparison of its oxidative reactivity with $[VPyNO]_2Cr_2O_7$ (iii) [48].

oxidative cleavage of olefins and oxidation of alcohols (Table 4.1, entries 7 and 17) [41,42,53]. For both reagents, a complex structure has been proposed. The synthetic routes for these reagents are presented in Scheme 4.7. As illustrated in Scheme 4.7, P_4VPy-RuO_4 selectively oxidizes alcohols to their corresponding aldehydes, whereas

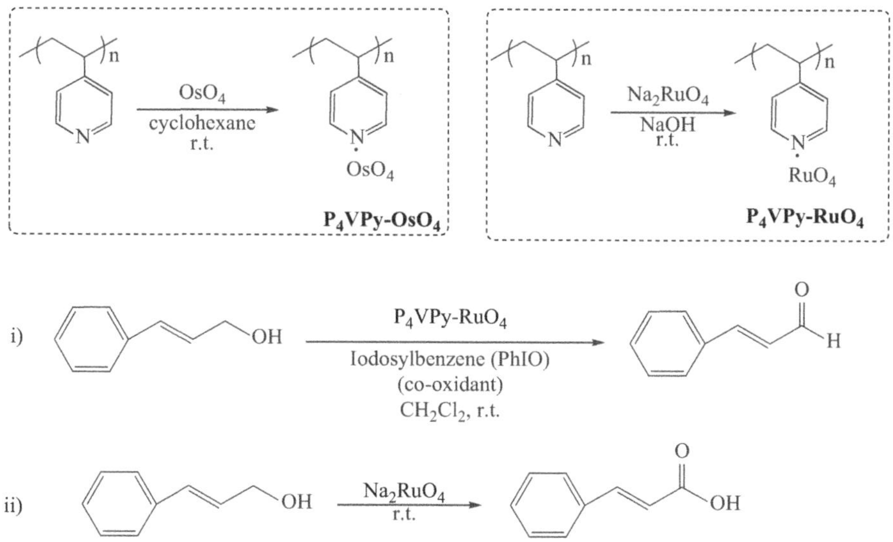

SCHEME 4.7 Preparation of P_4VPy-OsO_4 and P_4VPy-RuO_4 [41,42,53] and selective oxidation of alcohols to the corresponding aldehydes with P_4VPy-RuO_4 (i) and oxidation of alcohols to carboxylic acids with Na_2RuO_4 (ii) [53,67].

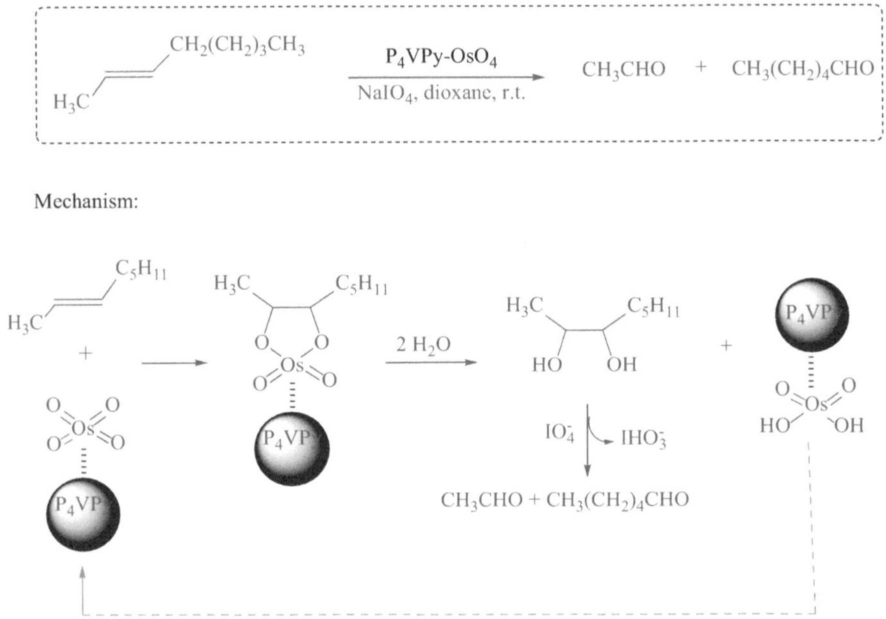

SCHEME 4.8 Oxidative cleavage of alkenes using $P_4VPy\text{-}OsO_4$ [41].

Na_2RuO_4 oxidizes alcohols to carboxylic acids [67]. Unlike OsO_4, which is expensive and toxic, $P_4VPy\text{-}OsO_4$ is stable (as a benchtop reagent) and nontoxic, and its reusability is important in reducing costs. The probable mechanism of cleavage of alkenes using $P_4VPy\text{-}OsO_4$ in the presence of sodium periodate is depicted in Scheme 4.8.

Hydrogen peroxide (H_2O_2) is one of the most frequently used oxidizing reagents due to its high oxygen content, superior efficiency relative to other peroxides, and the fact that it generates water as the only by-product. However, this reagent has several drawbacks, such as its corrosive nature, difficulties in handling, stability that is dependent on pH levels, a propensity to over-oxidize substrates, and interactions with bases due to its high reactivity. To solve the above-mentioned problems, concepts of H_2O_2 heterogenization have been established. In view of these, recently PVPy-supported H_2O_2 was introduced as solid alternative H_2O_2 and its utility for oxidation of sulfides to sulfoxides and conversion of arylboronic acids to phenols and halogenated phenols was evaluated (Table 4.1, entry 24, Scheme 4.9) [60,61]. In comparison to H_2O_2, PVPy-H_2O_2 complex is much safer and easier to handle with simpler work-up while reducing the amounts of by-products and wastes.

4.3 POLYMERIC REDUCING AGENTS BASED ON PVPy

The reduction process is a widely recognized fundamental transformation that holds significant practical relevance in both laboratories and industrial settings [68]. A variety of reducing agents that include lithium aluminum hydride ($LiAlH_4$), sodium borohydride ($NaBH_4$), sodium hydride (NaH), hydrazine (NH_2NH_2), zinc amalgam (Zn(Hg)), sodium amalgam (Na(Hg)), aluminum amalgam (Al(Hg)), samarium (II)

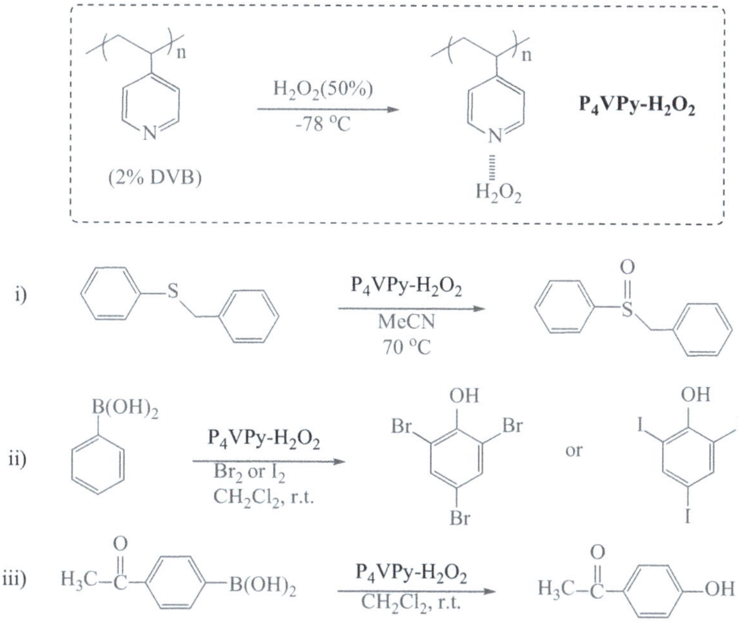

SCHEME 4.9 Preparation of PVPy-supported H_2O_2 and its use for the oxidation of sulfides to sulfoxides (i) and arylboronic acids to phenols (ii, iii) [60,61].

iodide (SmI_2), Sn/HCl, triethylsilane (Et_3SiH), hydrogen without or with a catalyst, borane-tetrahydrofuran (BH_3-THF), diisobutylaluminum hydride (i-Bu_2AlH), and Raney nickel (Ni) have been reported for this purpose [69]. Functional polymers are leading to the development of new heterogeneous reducing agents as environmentally benign alternatives to chemically homogeneous reducing reagents. A diverse range of the synthesized polymeric reducing agents, mostly based on borohydrides, is well documented [70–76]. Alongside the application of PVPy as a support for oxidizing agents, several polymeric reducing agents based on PVPy have also been introduced, which showed good reproducibility and activity. Table 4.2 shows polymeric supported reducing reagents based on PVPy that have so far been reported. Scheme 4.10 represents examples of applications of these reagents.

P_4VPy-supported zirconium borohydride ($P_4VPy-Zr(BH_4)_4$) is one of the polymer-supported reducing reagents which has been shown to possess remarkable benefits in oxidation reactions (Table 4.2, entry 4) [81]. In contrast to $Zr(BH_4)_4$, which is a volatile, unstable, and explosive material, $P_4VPy-Zr(BH_4)_4$ is a bench-top and nonhygroscopic reagent with ease of handling, recovery, and regeneration. Scheme 4.11 shows the preparation of $P_4VPy-Zr(BH_4)_4$ and its reducing activity compared to the previously reported $P_4VPy-Zn(BH_4)_2$ [79,80] and $P_4VPy-AlCl_2BH_4$ [82] in the reduction of cinnamaldehyde to cinnamyl alcohol. In most oxidation reactions, the efficiency of these polymeric reducing agents is in the following order: $P_4VPy-AlCl_2BH_4 \approx P_4VPy-Zr(BH_4)_4 > P_4VPy-Zn(BH_4)_2$. Capacity of these reagents is in the order of $P_4VPy-Zr(BH_4)_4 > P_4VPy-Zn(BH_4)_2 > P_4VPy-AlCl_2BH_4$. It is worth noting that all three of the above polymeric reagents exhibited more efficiency than monomeric

TABLE 4.2
Polymeric Reducing Agents based on PVPy

Entry	Polymer Reagents with Their Applications
1	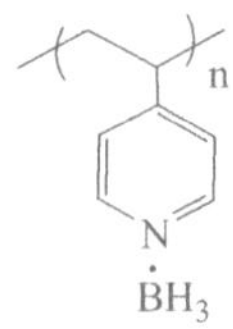 Poly(4-vinylpyridine borane) Reduction of carbonyl compounds [77]
2	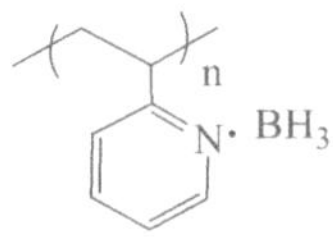 Poly(2-vinylpyridine borane) Reduction of carbonyl compounds [78]
3	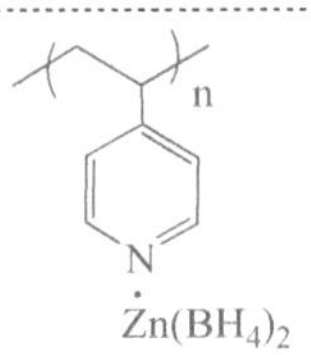 Poly(4-vinylpyridine)-zinc borohydride Reduction of aldehydes, ketones, azides, and disulfides [79,80]
4	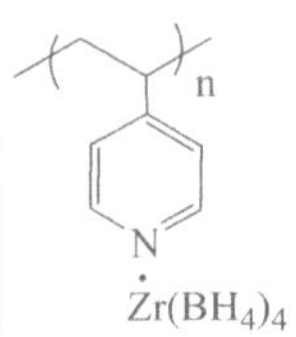 Poly(4-vinylpyridine)-Zirconium borohydride Reduction of carbonyl compounds [81]
5	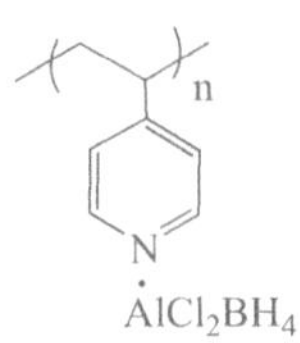 Poly(4-vinylpyridine)-chloroaluminium borohydride Reduction of carbonyl compounds, oxiranes, and azides [82] Reduction of oxiranes to alcohols [83]

(Continued)

TABLE 4.2 (*Continued*)
Polymeric Reducing Agents based on PVPy

Entry	Polymer Reagents with Their Applications
6	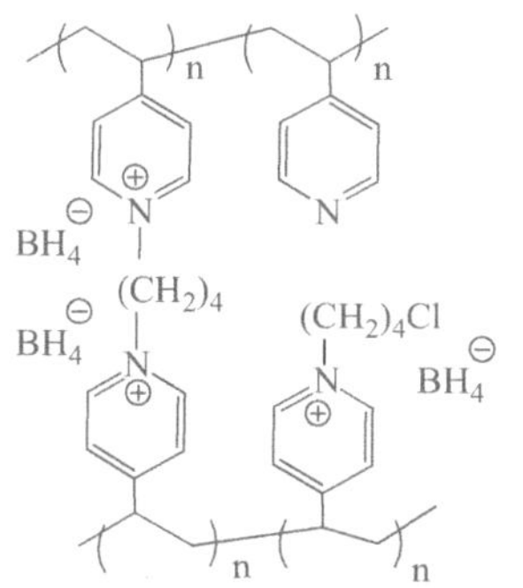 Poly(1,4-butyl-bis-vinylpyridinium)borohydride Reduction of carbonyl compounds [84]
7	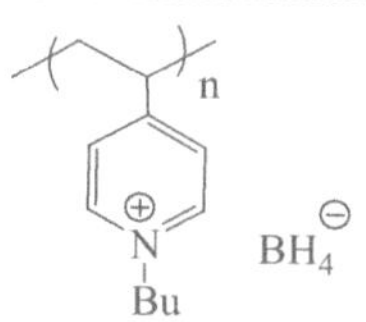 Poly(*n*-butyl-4-vinylpyridinium)borohydride Reduction of carbonyl compounds [85]

reducing agent 4-VPy-BH_3 [77] and cross-linked P_4VPy-BH_3, in terms of higher yields (Scheme 4.12).

Borohydride ion can also bind electrostatically to quaternized PVPy (Table 4.2) [84,85]. Scheme 4.13 presents the preparation and mechanism of action of one of these polymeric reducing agents called poly(n-butyl-4-vinylpyridinium)borohydride (P_4VPy-BuBH_4) for the reduction of carbonyl compounds. In addition to the common advantages of polymeric reagents, P_4VPy-BuBH_4 exhibited regioselectivity in the reduction of α,β-unsaturated ketones. For example, the reduction of 2-cyclohexen-1-one with $NaBH_4$ yields a mixture of 2-cyclohexen-1-ol and cyclohexanol in similar amounts, whereas P_4VPy-BuBH_4 exclusively produces 2-cyclohexen-1-ol (Scheme 4.14).

4.4 POLYMERIC HALOGENATING AGENTS BASED ON PVPy

Halogenated compounds hold significant promise for various applications, including pharmaceutical syntheses and materials science [86]. They serve as crucial intermediates in organic chemistry, e.g., for the preparation of organometallic compounds,

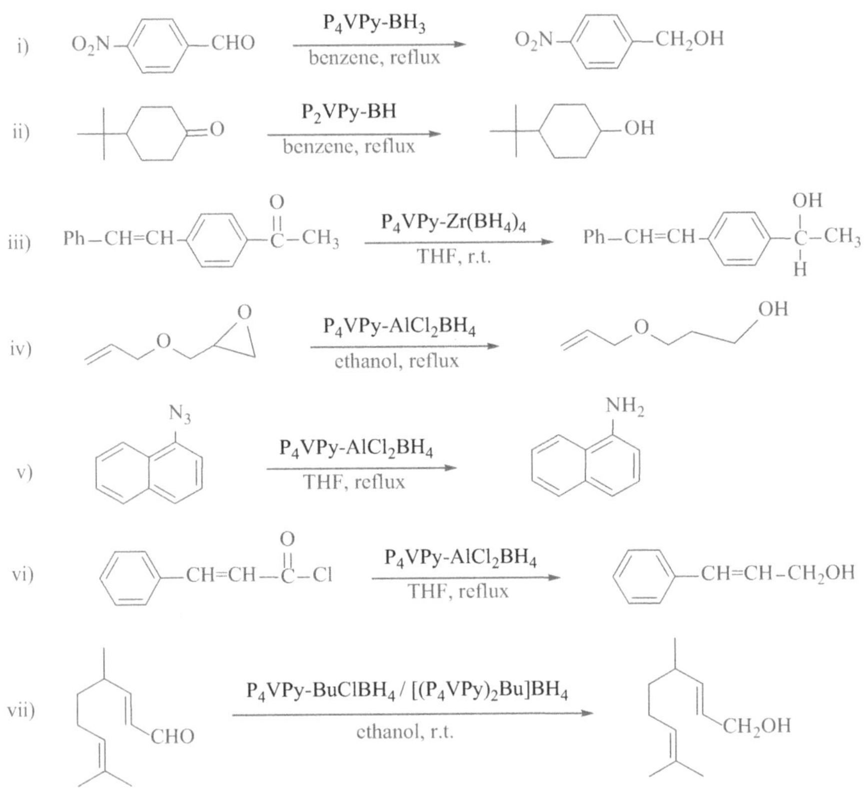

SCHEME 4.10 Representative examples of applications of PVPy-supported reducing agents for the reduction of aldehydes and ketones (i, ii, iii, vii) [77,78,81,84], epoxides (iv) [83], azides (v) [82], and acid chlorides (vi) [82].

as well as the starting materials for Suzuki–Miyaura and Mizoroki–Heck reactions [87,88]. However, the toxicity, volatility, and corrosiveness of halogens and halogenating agents pose challenges in their handling [89]. Consequently, there has been considerable interest in developing clean methods that employ environmentally friendly and green halogenating reagents. Among these, polymer-bound halogenating reagents are regarded as practical substitutes for traditional halogenating agents due to their nontoxic, noncorrosive, recyclable nature and ease of handling [90–96]. PVPy has been used as support for molecular halogens or halogen ions to produce efficient heterogeneous halogenating agents [13]. As discussed in Chapter 1, sometimes, the synthesis of PVPy bearing halogens is of fundamental and technological interest because of its applications in different fields of material science like optics and electronics [97]. The immobilization is often accomplished through both complexation of the pyridine rings with halogens and electrostatic interactions between pyridinium rings and halogen ions. Table 4.3 shows various halogenating agents based on PVPy. A schematic representation of applications of these reagents for a variety of organic transformations is depicted in Scheme 4.15.

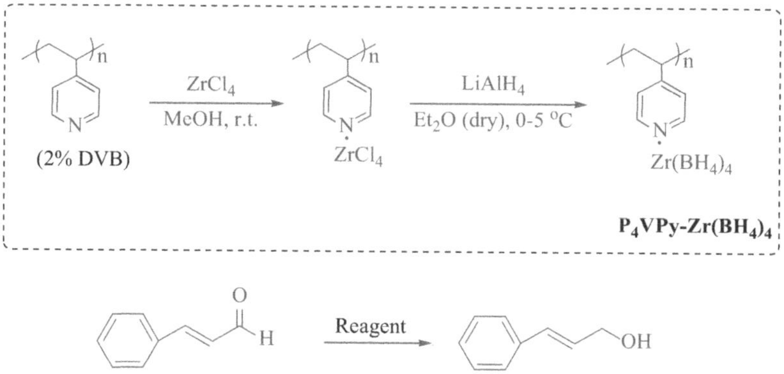

Reaction condition	Yield (%)
$P_4VPy\text{-}Zr(BH_4)_4$, THF, r.t., 6h	95%
$P_4VPy\text{-}Zn(BH_4)_2$, EtOH, reflux, 9h	90%
$P_4VPy\text{-}AlCl_2BH_4$, EtOH, reflux, 4h	98%

SCHEME 4.11 Preparation of $P_4VPy\text{-}Zr(BH_4)_4$ [81] and comparison of its oxidative reactivity with $P_4VPy\text{-}Zn(BH_4)_2$ [79] and $P_4VPy\text{-}AlCl_2BH_4$ [82].

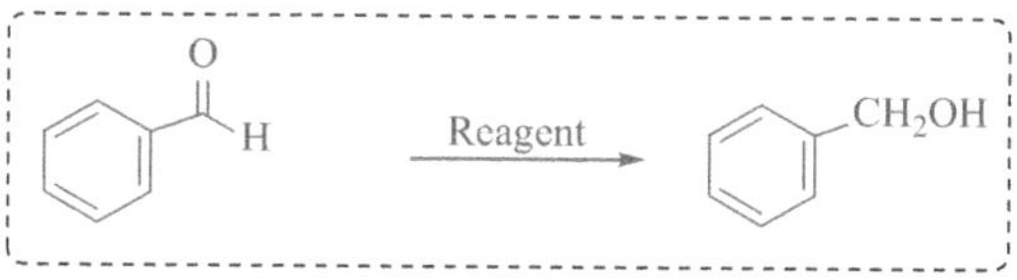

Reaction condition	Yield (%)
$P_4VPy\text{-}Zr(BH_4)_4$, THF, r.t., 1h	88%
$P_4VPy\text{-}Zn(BH_4)_2$, EtOH, reflux, 8h	80%
$P_4VPy\text{-}AlCl_2BH_4$, EtOH, reflux, 5h	88%
$4\text{-}VPy\text{-}BH_3$, benzene, reflux, 4h	24%
$P_4VPy\text{-}BH_3$, benzene, reflux, 4h	14%

SCHEME 4.12 Comparison of oxidation activity of $P_4VPy\text{-}Zr(BH_4)_4$ [81], $P_4VPy\text{-}Zn(BH_4)_2$ [79], and $P_4VPy\text{-}AlCl_2BH_4$ [82] with monomeric reducing agent $4\text{-}VPy\text{-}BH_3$ and cross-linked $P_4VPy\text{-}BH_3$ [77].

Among the various polymeric halogenating agents utilized in organic transformations, brominating agents have gained heightened interest from the organic chemistry community owing to their distinct benefits compared to bromine and its soluble complexes, which present numerous issues including toxicity, corrosiveness, overbromination, etc. Along this line, PVPy has been used as support for bromine and its derivatives (Table 4.3). Cross-linked and linear copolymers of (−)-*p*-(3-menthyl)

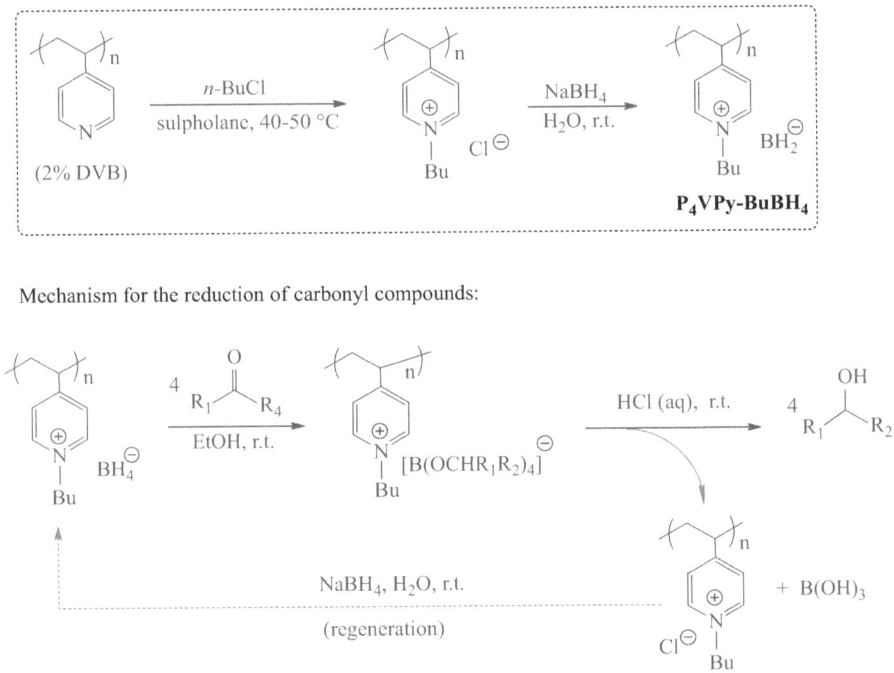

SCHEME 4.13 Preparation of poly(n-butyl-4-vinylpyridinium)borohydride as a polymer reagent for the reduction of aldehydes and ketones [85].

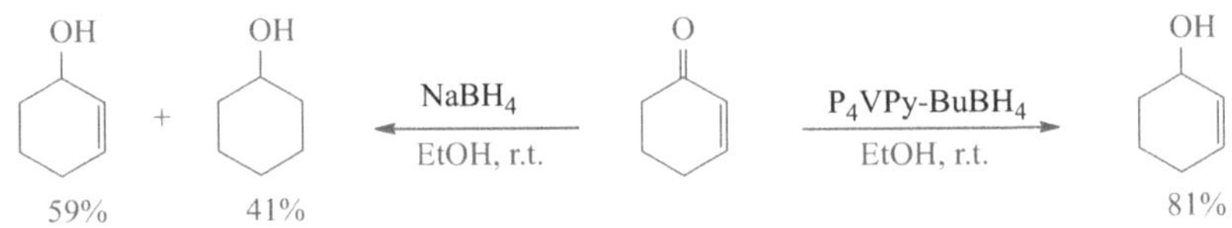

SCHEME 4.14 The successful regioselective reduction of α,β-unsaturated ketones using $P_4VPy\text{-}BuBH_4$ [85].

acrylate or *p*-(3-menthyl)methacrylate with 4-VPy monomers (denoted as P_4VPy-co-PMA and P_4VPy-co-PMMA, respectively) have been used as carriers for the immobilization of bromine molecule (Table 4.3, entry 4) [106,120]. They are optically active copolymers which are prepared by free radical polymerization of the corresponding monomers (Scheme 4.16). It is worth mentioning that spectroscopic methods (e.g., ultraviolet–visible, Fourier transform infrared, and coupled conductivity) predict two types of charge-transfer complex structures for the interaction of bromine with pyridine, "external" and "internal" complexes, that are in equilibrium with each other (Scheme 4.17) [106]. Investigations revealed the prevalence of external complexes in $P_4VPy\text{-}Br_2$ and internal complexes in P_4VPy-co-PMA or P_4VPy-co-PMMA [106]. All three polymers and copolymers mentioned above were successfully used in the bromination of alkenes (Scheme 4.18). None of the above brominating reagents showed a tendency to continue reacting with alkyl dibromide products to give pyridinium salts.

TABLE 4.3
Polymeric Halogenating Agents Based on PVPy

Entry	Polymer Reagents with Their Applications
1	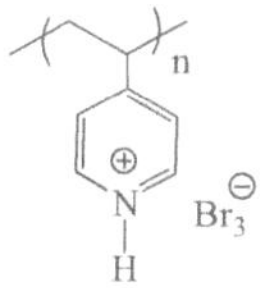 Poly(4-vinylpyridinium)hydrobromideperbromide Bromination of ketones [13,98–102]
2	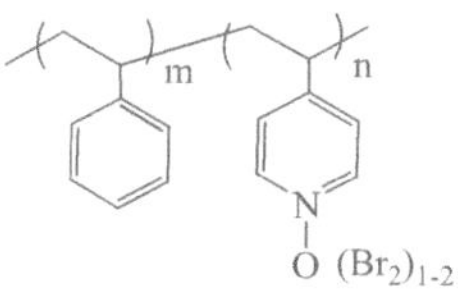 Poly(styrene-*co*-4-vinylpyridine *N*-oxide)bromine Bromination of alkenes [103]
3	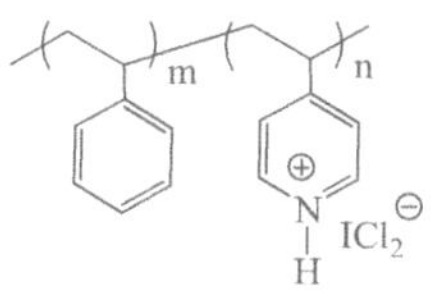 Poly(styrene-[4-vinylpyridinium dichloroiodate]) Iodination of aromatic compounds [104,105]
4	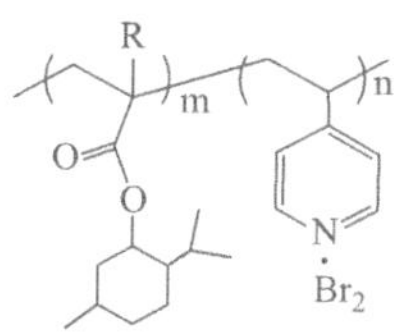 P_4VPy-*co*-PMA-bromine, R= H P_4VPy-*co*-PMMA-bromine, R= Me Bromination of alkenes [106]
5	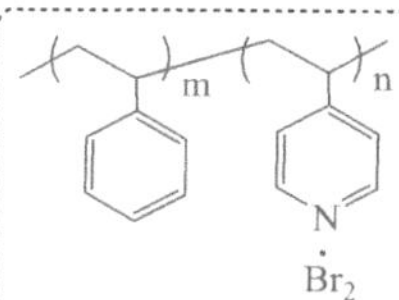 Poly(styrene-*co*-4-vinylpyridine)-bromine Side chain bromination of arenes [107] α-Bromination of ketones [108] Bromination alkenes [109]

(*Continued*)

TABLE 4.3 (*Continued*)
Polymeric Halogenating Agents Based on PVPy

Entry	Polymer Reagents with Their Applications
6	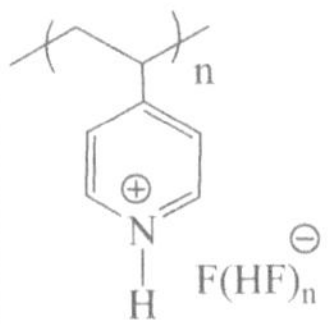 Poly(4-vinylpyridinium)poly(hydrogen fluoride) (n= 5-7) Iodofluorination of olefines using N-iodosuccinimide [110] Fluorination of olefins and alcohols [111] Preparation of gem-difluoro compounds [112]
7	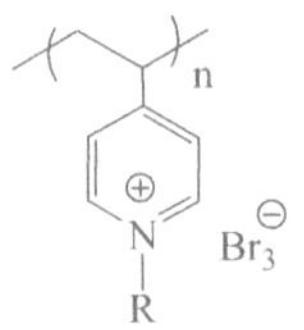 Poly(*n*-alkyl-4-vinylpyridinium) hydrobromide perbromide R= n-C_6H_{13}: Bromination of aromatic molecules [113] R= n-C_6H_9: α-Bromination of ketones and dibromination of alkenes [13]
8	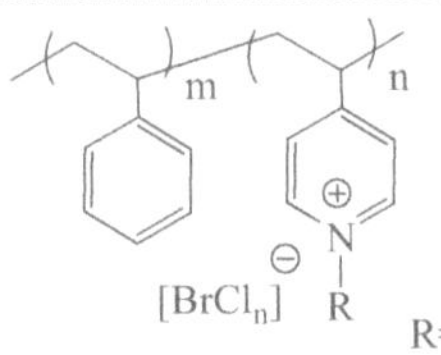 Poly(styrene-*co*-[1-alkyl-4-vinylpyridinium bromochloride]) R= H, C_2H_5, *n*-C_4H_5, *n*-C_6H_{13}, *n*-$C_{12}H_{25}$ Chlorobromination of olefins [114]
9	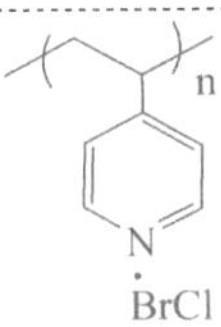 Poly(4-vinylpyridine)-bromochloride Bromochlorination of alkenes [115]

(*Continued*)

TABLE 4.3 (*Continued*)
Polymeric Halogenating Agents Based on PVPy

Entry	Polymer Reagents with Their Applications
10	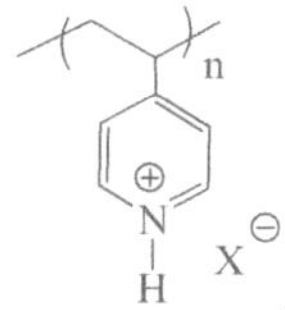 Poly(4-vinylpyridinium)bromide Poly(4-vinylpyridinium)chloride X= Br,Cl Preparation of β-bromohydrins and β-chlorohydrins [116]
11	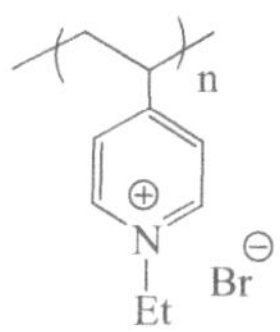 Poly(*N*-ethyl-4-vinylpyridinium)bromide Synthesis of arylbromide [117]
12	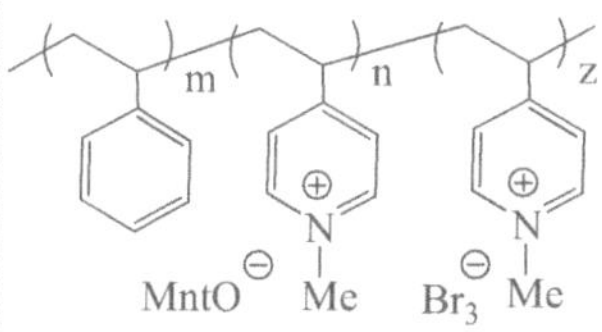 Montmorillonite clay-poly(styrene-*co*-(*N*-methyl-4-vinylpyridinium)bromide hydrobromide Bromination of ketones, alkenes, and aromatic compounds [118]
13	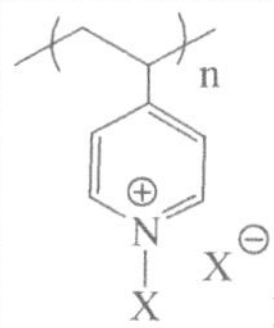 Poly(4-vinylpyridine)-iodine Poly(4-vinylpyridine)-bromine X= Br,I ipso-Iodination and bromination of arylboronic acids [119]

The above brominated polymers were examined in the dibromination of conjugated dienes (both racemic and prochiral substrates), and they exhibited good regioselectivity compared to free bromine, although no stereo-election was observed (Scheme 4.19). The regioselective dibromination of alkenes with both endocyclic and exocyclic double bonds represents another intriguing characteristic of $P_4VPy\text{-}Br_2$. Scheme 4.20 shows the high affinity of $P_4VPy\text{-}Br_2$ for dibromination of endocyclic double bond in 4-vinylcyclohexene. As it can be seen from the scheme, both bromine solution and

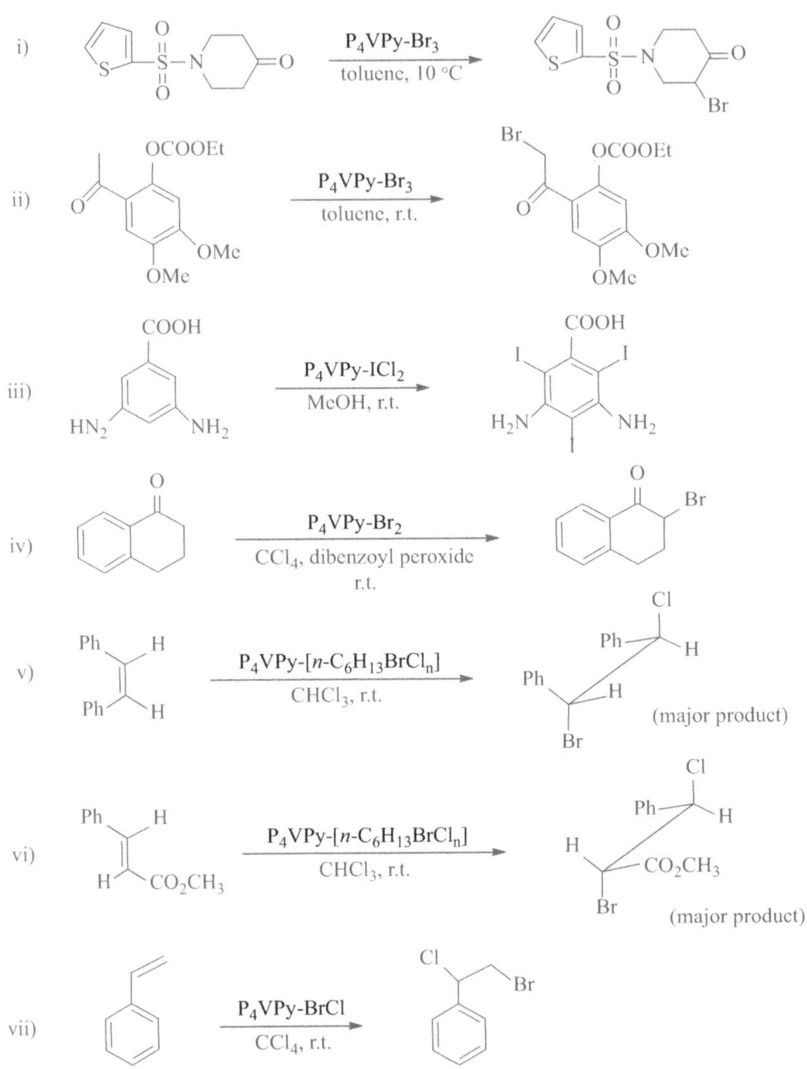

SCHEME 4.15 Schematic representation of applications of halogenated polymers based on PVPy in α-bromination of ketones (i, ii, iv) [99,102,108], iodination of aromatic compounds (iii) [105], and bromochlorination of alkenes (v, vi, vii) [114,115].

pyridinium hydrobromide perbromide in bromination of 4-vinylcyclohexene produced the corresponding tetrabromoalkane [121].

Poly(N-ethyl-4-vinylpyridinium)bromide (P_4VPy-EtBr) is another brominating reagent, which has been used for the conversion of aromatic amines to aryl bromides (Table 4.3, entry 11) [117]. It is easily prepared from the reaction of cross-linked P_4VPy (2% DVB) with ethyl bromide (Scheme 4.21). Following the transformation of primary aromatic amines into their respective diazonium salts via routine

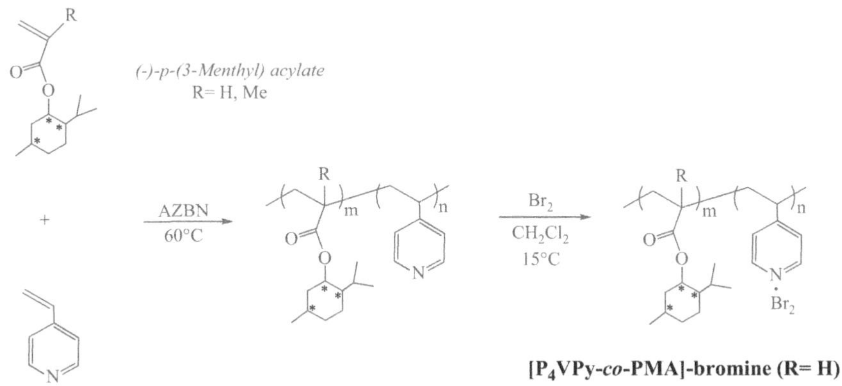

[P_4VPy-*co*-PMA]-bromine (R= H)
[P_4VPy-*co*-PMMA]-bromine (R= Me)

SCHEME 4.16 Synthesis of P_4VPy-co-PMA or P_4VPy-co-PMMA [106,120].

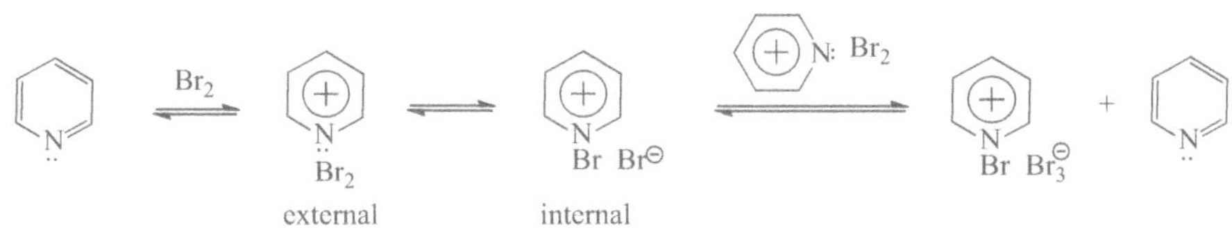

SCHEME 4.17 Types of charge-transfer complexes from the interaction of pyridine with bromine molecule [106].

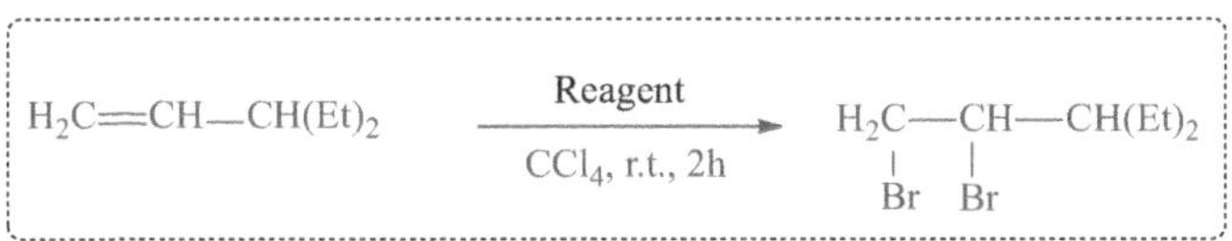

Reagent	Yield (%)
P_4VPy-bromine (linear)	74%
[P_4VPy-*co*-PMA]-bromine (linear)	87%
[P_4VPy-*co*-PMA]-bromine (crosslinked)	87%
[P_4VPy-*co*-PMMA]-bromine (linear)	87%

SCHEME 4.18 Bromination of alkenes using bromine complexes of linear P_4VPy and linear and cross-linked P_4VPy-co-PMA and P_4VPy-co-PMMA copolymers [106].

sequential reactions, the obtained diazonium salts are converted to aryl bromides using P_4VPy-EtBr reagent in the presence of a catalytic amount of CuBr.

Notwithstanding the distinctive characteristics of PVPy reagents, they also suffer from some problems. One of the most important disadvantages of PVPy reagents is their limited thermal and mechanical resistance. One way to overcome this shortcoming is anchoring polymeric reagents onto solid carriers such as clay, zeolite, silica, carbon nanotube, and graphene oxide [122–125]. Recently, a copolymer nanocomposite consisting

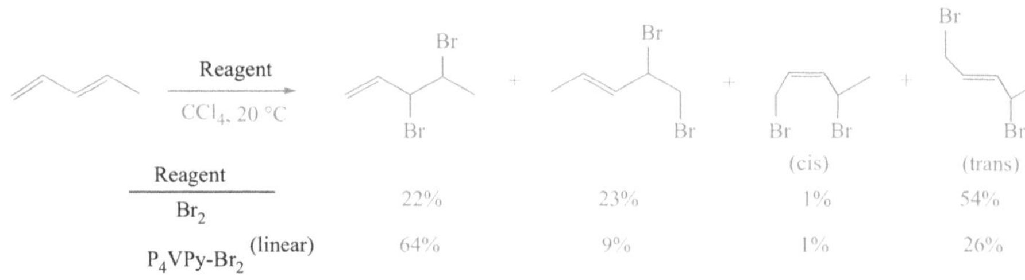

SCHEME 4.19 Regioselective dibromination of conjugated dienes using $P_4VPy-Br_2$ complex versus bromine solution [106].

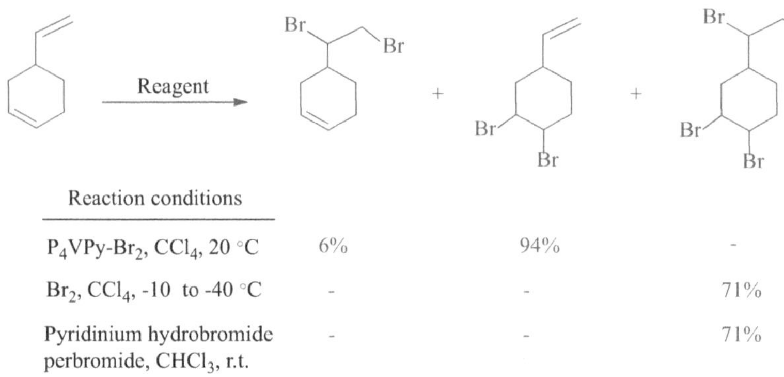

SCHEME 4.20 Regioselective dibromination of 4-vinylcyclohexene using $P_4VPy-Br_2$ complex versus bromine solution and pyridinium hydrobromide perbromide [106,121].

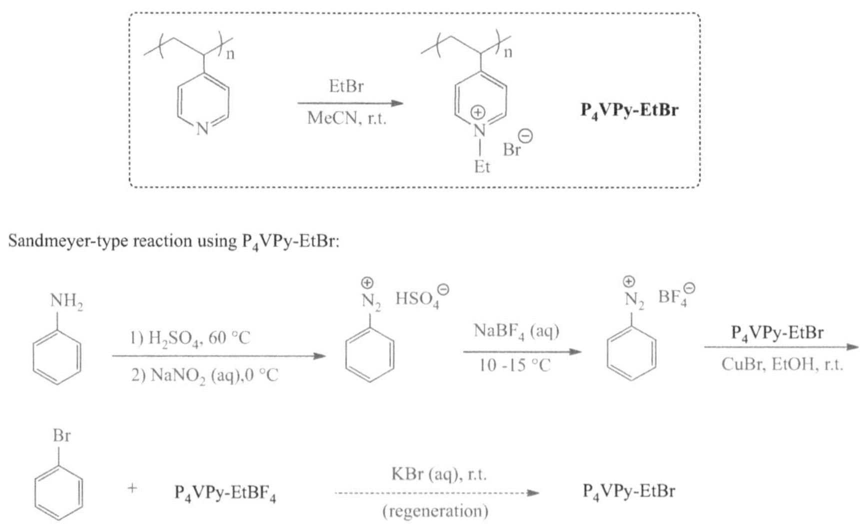

SCHEME 4.21 Preparation of arylbromides from their corresponding amines using P_4VPy-EtBr reagent through Sandmeyer-type reaction [117].

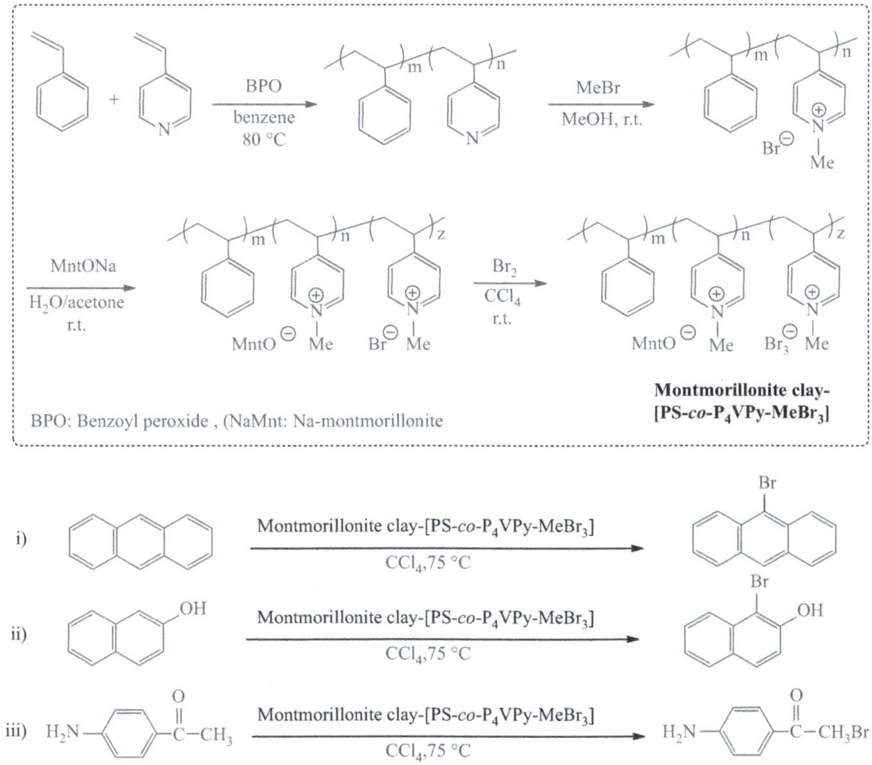

SCHEME 4.22 Preparation and applications of montmorillonite clay-poly(styrene-co-(N-m ethyl-4-vinylpyridinium)bromide hydrobromide for the bromination of aromatic compounds (i, ii) and α-bromination of ketones (iii) [118].

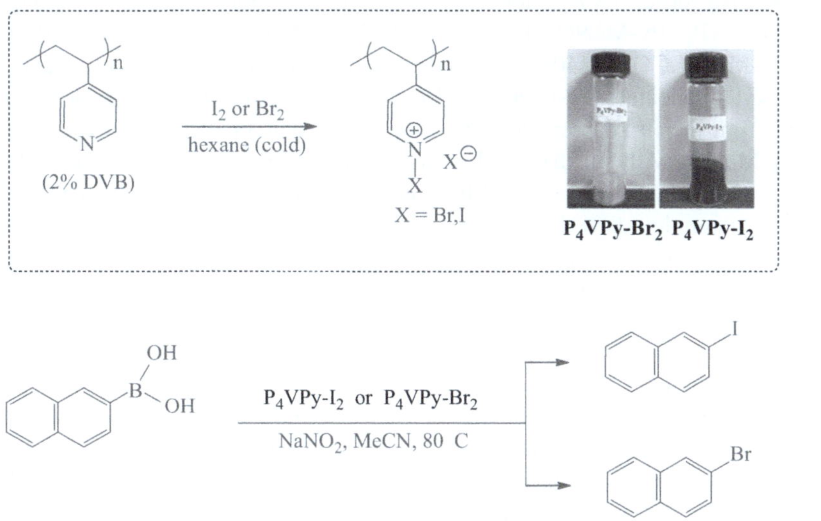

SCHEME 4.23 P_4VPy-iodine and bromine for ipso-halogenation of arylboronic acids [119].

of montmorillonite clay and poly(styrene-co-(N-methyl-4-vinylpyridinium)bromide hydrobromide reagent was synthesized from Na^+ exchange of Na-montmorillonite (NaMnt) with pyridinium cations in poly(styrene-co-(N-methyl-4-vinylpyridinium)bromide (Table 4.3, entry 12, Scheme 4.22) [118]. The obtained reagent was used for the bromination of aromatic compounds and α-bromination of ketones. The results showed that this reagent has superior brominating ability compared to poly(styrene-co-(N-methyl-4-vinylpyridinium)bromide hydrobromide alone in terms of yields, presumably due to high surface area and porosity of clay.

PVPy-supported iodine and bromine were surveyed for the preparation of aryl iodides and bromides from boronic acid derivatives in the presence of $NaNO_2$ as catalyst (Table 4.3, entry 13, Scheme 4.23) [119]. The reagents are more convenient and safer than Br_2 and I_2. They are regenerable reagents, which could be stored in closed vessel for several months.

In addition to their applications as oxidizing, reducing, and halogenating gents, PVPy-supported reagents have also proven to have other roles in organic transformations as shown in Table 4.4. Most of the reported procedures for the preparation of PVPy-supported reagents listed in Table 4.4 are classified based on a general type of process. This includes the quaternization of PVPy with alkyl or aryl halides, followed by a straightforward anion exchange of halide counterions with the desired anion groups.

The utilizations of some polymeric reagents based on PVPy in several organic transformations are shown in Scheme 4.24. A representative selection of these polymeric reagents is poly(N-methy-4-vinylpyridinium)benzoate (P_4VPy-$MePhCO_2$, Table 4.4, entry 7) [139]. As shown in Scheme 4.25, this reagent was synthesized via anion exchange of iodide counter-anions in quaternized poly(N-methyl-4-vinylpyridinium)iodide (P_4VPy-MeI) with benzoate ion. P_4VPy-$MePhCO_2$ was stable for many months, and it was used as a regenerable polymeric reagent for the preparation of aryl benzoates from in situ generated diazonium salts of primary aromatic amines.

Likewise, other PVPy-supported reagents, e.g., poly(N-methyl-4-vinylpyridinium)nitrite (P_4VPy-$MeNO_2$) [133] and poly(N-methyl-4-vinylpyridinium)thioacetate (P_4VPy-$MeCH_3COS$) [140], have been explored for the preparation of aryl azides and alkyl thioacetate, respectively. In the former example, aryl azides were produced from the in-situ reaction of nitrous acid (generated through the reaction of P_4VPy-$MeNO_2$ with sulfuric acid) with an ammonium hydrogen sulfate salt of primary aromatic amines, followed by reaction with sodium azide (Scheme 4.26). In the later example, alkyl thioacetates were obtained from the nucleophilic attack of thioacetate counter-anions in P_4VPy-$MeCH_3COS$ on alkyl halides (Scheme 4.27). Both P_4VPy-$MeNO_2$ and P_4VPy-$MeCH_3COS$ enjoy the general advantages of the solid-supported reagents.

Organoazides are important intermediates in various organic synthesis processes and possess biological significance [151]. Hydrogen azide and sodium azide are the two most commonly used azide agents in the preparation of organoazides, but have problems of toxicity and explosiveness. To solve these shortcomings, polymeric azide reagents have gained increasing attention as possible "green" alternatives to hazardous and unrecyclable azide agents [152,153]. In this context, azide anion has been already immobilized on PVPy and used as heterogeneous reagent for several organic

TABLE 4.4
Miscellaneous Polymeric Reagents Based on PVPy

Entry **Polymer Reagents with Their Applications**

1

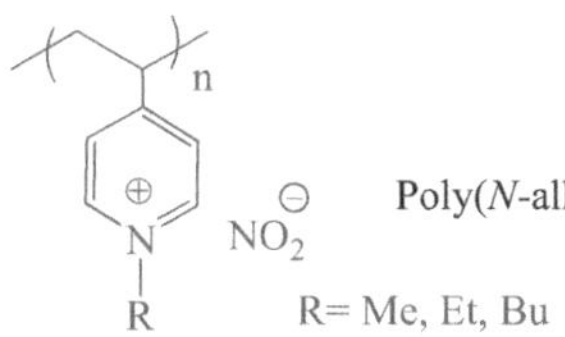

Poly(*N*-alkyl-4-vinylpyridinium)nitrite

R= Et, Bu; Conversion of alkyl halides to nitroalkanes [126]

R= Me; N-Nitrosation of secondary amines [127,128]

R= Me; Synthesis of azo chromophores [129,130]

R= Me; Preparation of aryl iodides [131]

R= Me; Nitration of aromatic compounds [132]

R= Me; Preparation of aryl azides [133]

2

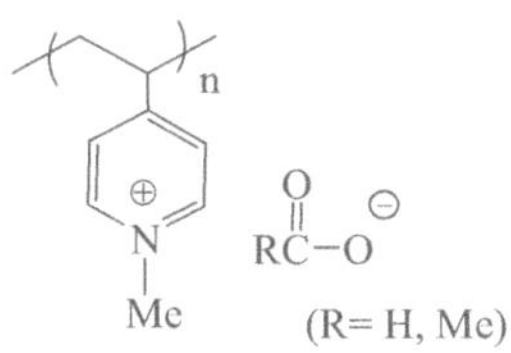

Poly(*N*-methyl-4-vinylpyridinium)carboxylate

Preparation of benzyl acetate or benzyl formats from benzyl halides [134]

3

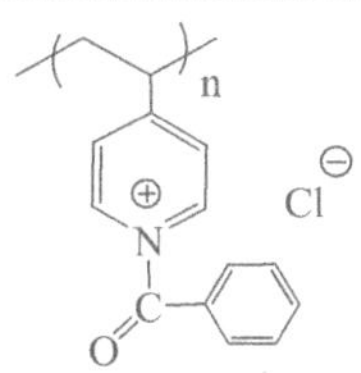

Poly(*N*-benzoyl-4-vinylpyridinium)chloride

Preparation of benzoate esters [135]

4

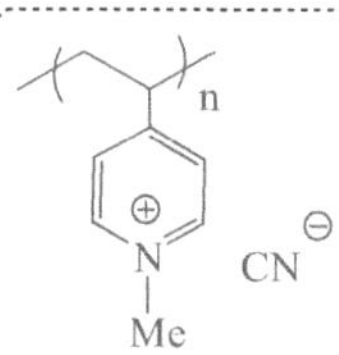

Poly(*N*-methyl-4-vinylpyridinium)cyanide

Cyanation of arenediazonium salts [136]

(Continued)

TABLE 4.4 (*Continued*)
Miscellaneous Polymeric Reagents Based on PVPy

Entry	Polymer Reagents with Their Applications
5	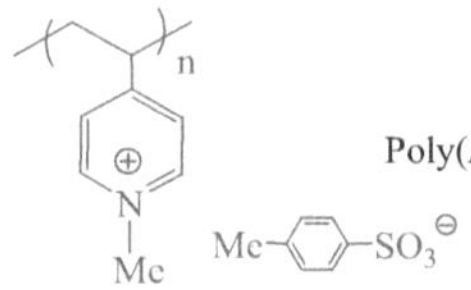 Poly(*N*-methyl-4-vinylpyridinium)*p*-toluene sulfonate Synthesis of oligonucleotides [137]
6	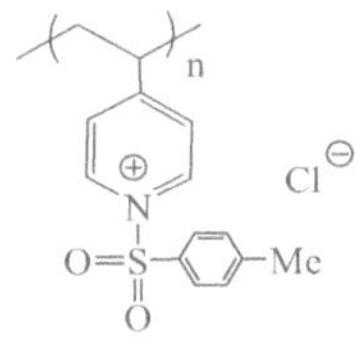 Poly(*N*-tosyl-4-vinylpyridinium)chloride Synthesis of carboxylic anhydrides [138]
7	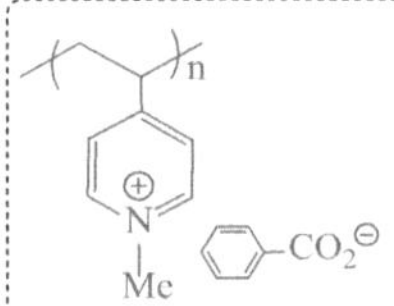 Poly(*N*-methyl-4-vinylpyridinium)benzoate Preparation of aryl benzoates [139]
8	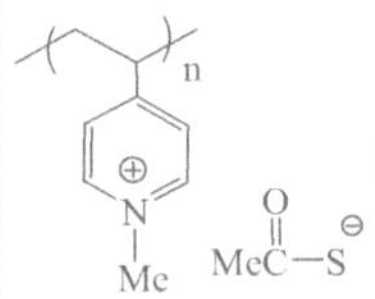 Poly(*N*-methyl-4-vinylpyridinium)thioacetate Preparation of alkyl thioacetate [140]
9	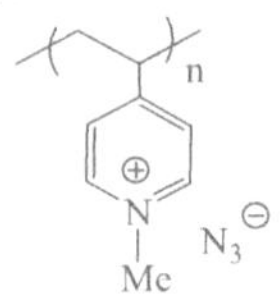 Poly(*N*-methyl-4-vinylpyridinium)azide Preparation of alkyl azides from alkyl halides [141] Diazotization–azidation of amines [142,143] Preparation of azidohydrins from oxiranes [144] Synthesis of tetrazoles [145] Acyl azides synthesis from acyl chlorides [146]

(*Continued*)

TABLE 4.4 (*Continued*)
Miscellaneous Polymeric Reagents Based on PVPy

Entry	Polymer Reagents with Their Applications
10	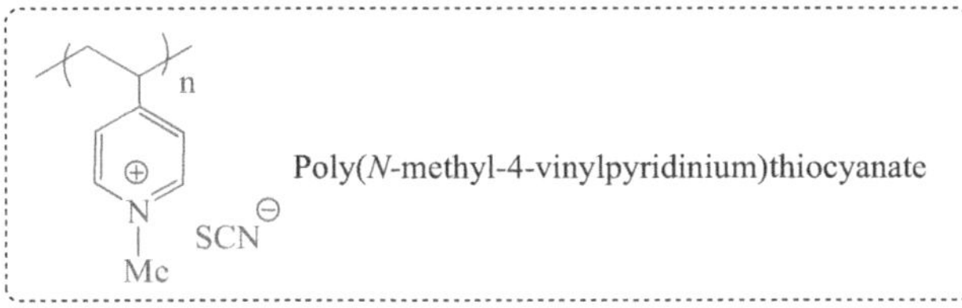
	Thiocyanation of aromatic compounds [147]
11	
	Aromatization of urazoles and 1,4-dihydropyridines [148]
	Conversion of thiols to disulfides [149]
	Preparation of sulfoxides from sulfides [150]

transformations including preparation of alkyl azides from alkyl halides [141], diazotization–azidation of amines [142,143], conversion of epoxides to azidohydrins [144], and acyl azides synthesis from acyl chlorides [146]. An additional noteworthy application of PVPy-supported azide ion lies in the synthesis of 1-substituted tetrazoles [145], which have traditionally been synthesized using hazardous hydrogen azide, trimethylsilyl azide, or other less environmentally friendly alternatives [151]. Scheme 4.28 shows the synthetic pathway for this reagent and its use as a stable, regenerable, and nonexplosive material for the preparation of 1-substituted tetrazoles in the presence of aromatic amines and triethyl orthoformate.

Finally, another interesting example of PVP-supported reagents is poly(4-vinylpyridinium)nitrate ($P_4VPy\text{-}HNO_3$, Table 4.4, entry 11) [148–150]. Scheme 4.29 shows synthetic pathways for $P_4VPy\text{-}HNO_3$ and the plausible mechanism of oxidation of urazoles and 1,4-dihydropyridine promoted by $P_4VPy\text{-}HNO_3$ and silica sulfuric acid ($SiO_2\text{-}OSO_3H$). The nitronium ions (NO_2^+) generated in situ from $P_4VPy\text{-}HNO_3$ and $SiO_2\text{-}OSO_3H$ are the active species that result in oxidation of substrates.

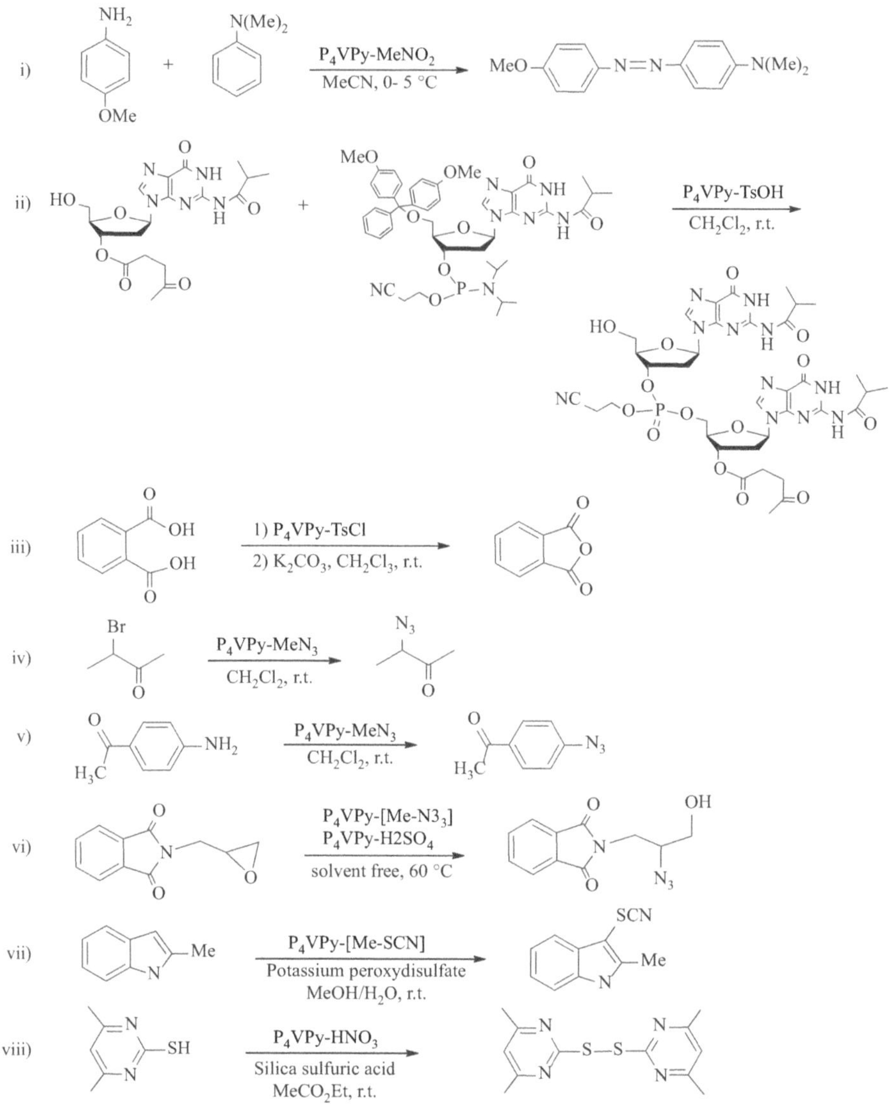

SCHEME 4.24 The use of some PVPy-supported reagents for synthesis of azo chromophores (i) [129], phosphoramidite coupling in the preparation of oligonucleotides (ii) [137], carboxylic anhydrides synthesis (iii) [138], synthesis of alkyl azides (iv) [141], preparation of aromatic azides (v) [143], the preparation of azidohydrin from epoxides (vi) [144], thiocyanation of compounds (vii) [147], and conversion of sulfides to disulfides (viii) [149].

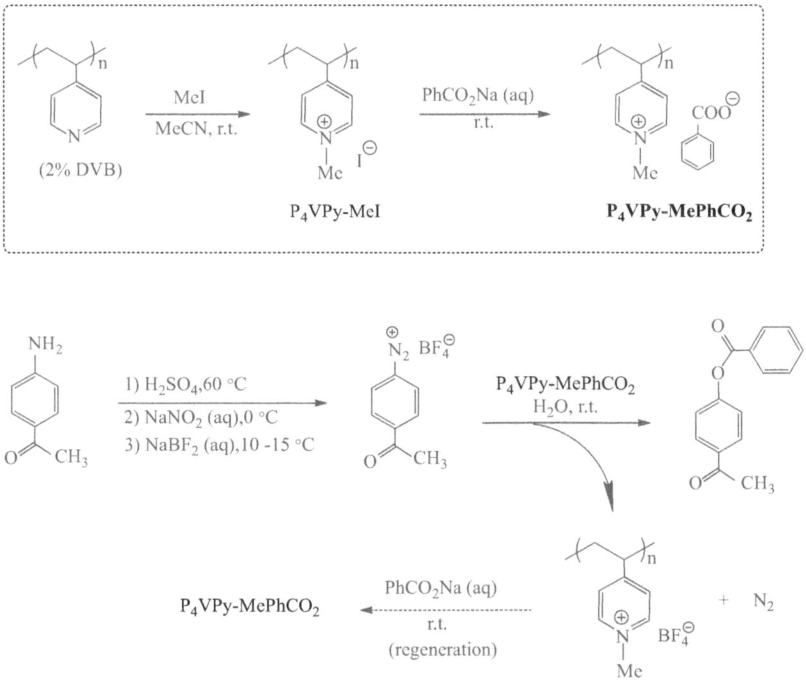

SCHEME 4.25 Synthesis and application of poly(N-methy-4-vinylpyridinium)benzoate for the preparation of aryl benzoates from stabilized arenediazonium tetrafluoroborate [139].

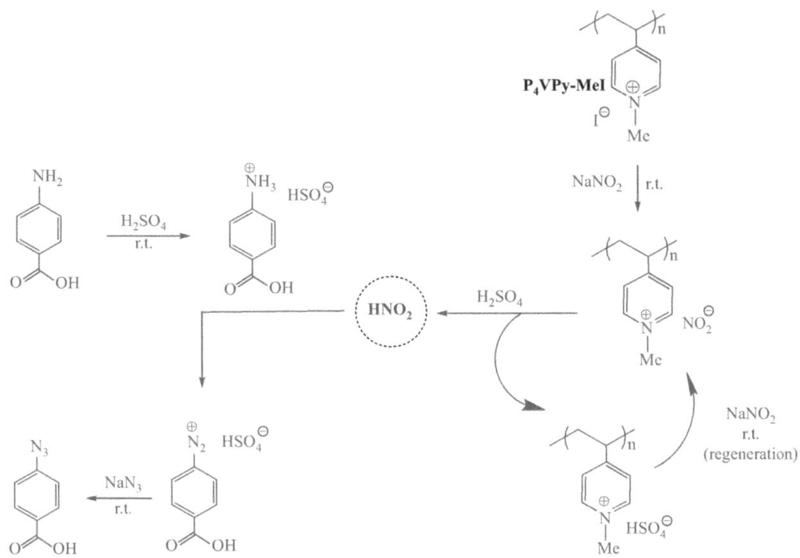

SCHEME 4.26 Poly(N-methyl-4-vinylpyridinium)nitrite as a reagent for the diazotization of aromatic amines [133].

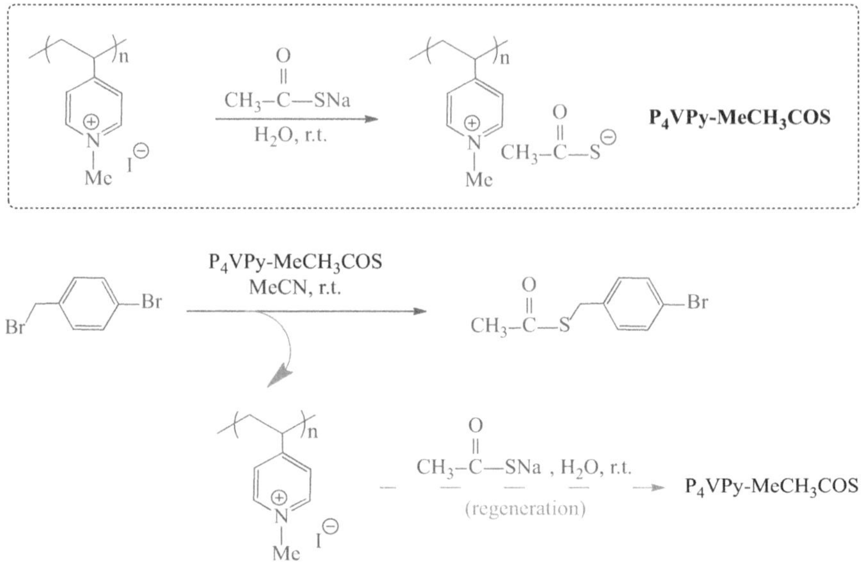

SCHEME 4.27 Poly(N-methyl-4-vinylpyridinium)thioacetate as a reagent for the preparation of alkyl thioacetate [140].

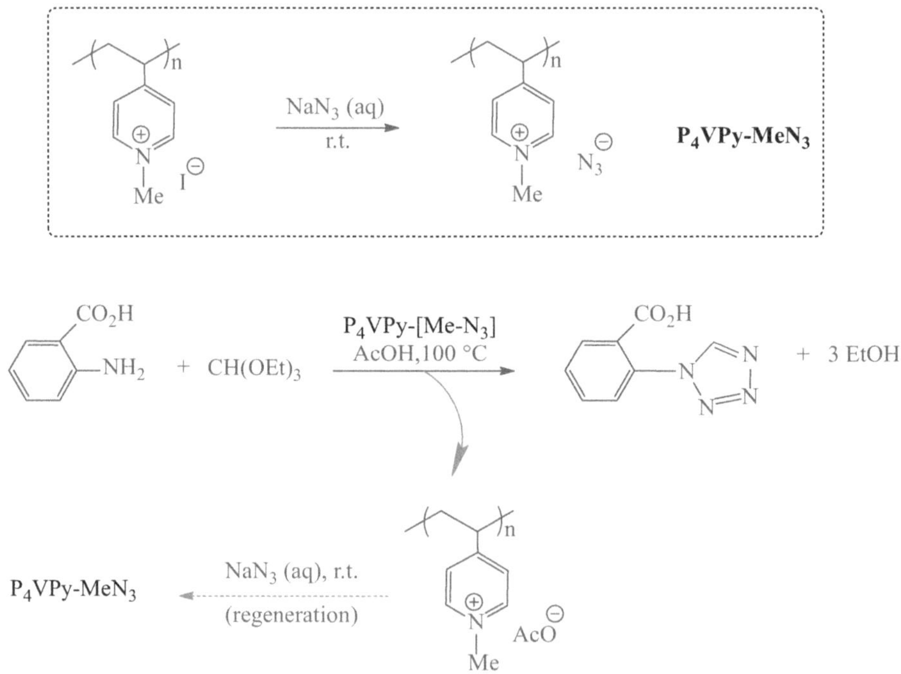

SCHEME 4.28 Illustration of the preparation and application of poly(N-methyl-4-vinylpy ridinium)azide in 1-substituted tetrazoles synthesis [145].

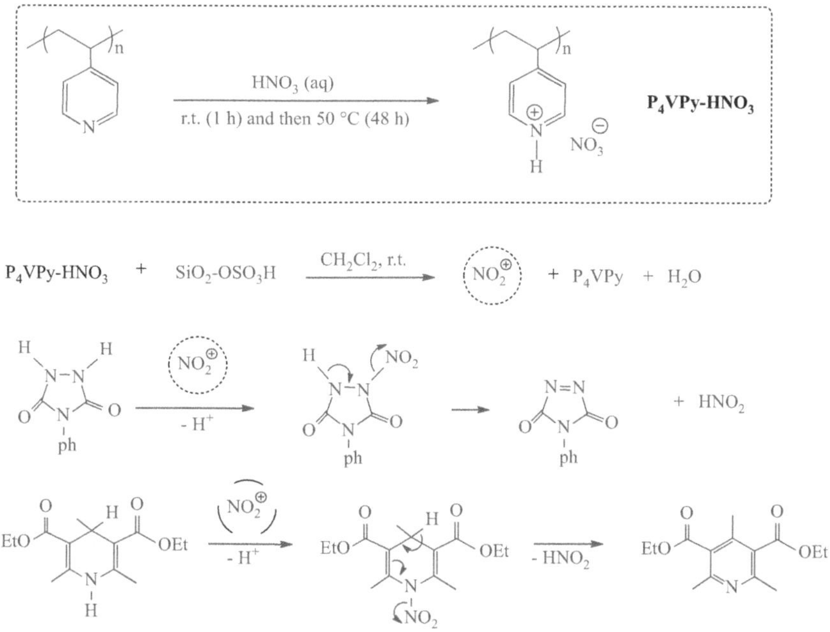

SCHEME 4.29 Oxidation of 4-substituted urazoles (i) and 1,4-dihydropyridines (ii) in the presence of poly(4-vinylpyridinium)nitrate and silica sulfuric acid [148].

REFERENCES

1. Salimi H, Rahimi A, Pourjavadi A. *Monatshefte für Chemie-Chemical Monthly.* 2007;138:363–79.
2. Kraus M, Patchornik A. *Macromolecular Reviews Part D-Journal of Polymer Science.* 1980;15:55–106.
3. Ford WT. *ACS Symposium Series.* 1986;308:1–16.
4. Akelah A, Sherrington D. *Polymer.* 1983;24(11):1369–86.
5. Heitz W. Polymeric reagents polymer design, scope, and limitations. In: Heitz W, Batz HG, Shimidzu T, Goethals EJ (eds.). *Reactivities, Advances in Polymer Science*: Springer, Berlin, Heidelberg; 1977, vol. 23, pp. 1–23.
6. Akelah A. *Reactive Polymers, Ion Exchangers, Sorbents.* 1988;8(3):273–84.
7. Hinzen B. *Methods and Principles in Medicinal Chemistry.* 2000;9:209–37.
8. Bartolozzi A, Grogan MJ, Kates SA. *Methods in Enzymology.* 2003; 369:347–66.
9. Solinas A, Taddei M. *Synthesis.* 2007:2409–53.
10. Hodge P. *Industrial & Engineering Chemistry Research.* 2005;44(23):8542–53.
11. Gelbard G. *Polymer-Supported Reagents.* Blackwell Science Ltd, Oxford; 2002.
12. Baxendale IR, Ley SV, Mansfield AC, Smith CD. *Angewandte Chemie.* 2009;121(22):4077–81.
13. Tamami B, Parvanak Boroujeni K. *Iranian Polymer Journal.* 2009;18(3):191–206.
14. Tojo G, Fernández MI. *Oxidation of Alcohols to Aldehydes and Ketones: A Guide to Current Common Practice*: Springer, New York; 2006.
15. Ėmanuėl′ NM, Zaikov GE, Maĭzus ZK. *Oxidation of Organic Compounds: Medium Effects in Radical Reactions*: Pergamon Press, New York; 1984.

16. Shirini F, Abedini M, Atghia SV. Ionic liquids as green solvents for the oxidation of alcohols. In: Mohammad A, Inamuddin, D (eds.). *Green Solvents II: Properties and Applications of Ionic Liquids*: Springer, Dordrecht; 2012. pp. 363–98.
17. Goswami S, Kar A. *Synthetic Communications*. 2011;41(17):2500–4.
18. Mizuno N. *Modern Heterogeneous Oxidation Catalysis: Design, Reactions and Characterization*: Wiley, Germany; 2009.
19. Yang H, Li B. *Synthetic Communications*. 1991;21(14):1521–6.
20. Zakavi S, Abasi A, Pourali AR, Rayati S. *ChemInform*. 2008;39(4):866–8.
21. Pourali AR, Tabaean M, Nazifi SMR. *Chinese Chemical Letters*. 2012;23(1):21–4.
22. Tamami B, Heiran R, Montazer RE. *Journal of the Serbian Chemical Society*. 2012;77(5):685–97.
23. Ghorbani-Choghamarani A, Nikoorazm M, Tahmasbi B, Norouzi M. *Journal of Saudi Chemical Society*. 2017;21(4):408–14.
24. Samunual P, Bergbreiter DE. *Tetrahedron Letters*. 2016;57(30):3272–6.
25. Chung W-J, Kim D-K, Lee Y-S. *Tetrahedron Letters*. 2003;44(52):9251–4.
26. Harrison CR, Hodge P. *Journal of the Chemical Society, Perkin Transactions 1*. 1982:509–11.
27. Geethakumari K, Sreekumar K. *Reactive and Functional Polymers*. 1999;42(1):11–9.
28. Christensen L, Heacock D. *Synthesis*. 1978;1978(01):50–1.
29. Frechet JM, Warnock J, Farrall MJ. *The Journal of Organic Chemistry*. 1978;43(13):2618–21.
30. Narayanan N, Balasubramanian T. *Journal of Chemical Research Synopses*. 1992(4):132–3.
31. Frechet J, Darling P, MJ F. *Polymer Preprints* 1980;21(2):272–3.
32. Brunelet T, Gelbard G, Guyot A. *Polymer Bulletin*. 1981;5(3):145–9.
33. Yoshida J, Nakai R, Kawabata N. *The Journal of Organic Chemistry*. 1980;45(26):5269–73.
34. Zupan M, Dolenc D. *Tetrahedron*. 1991;47(27):5025–8.
35. Yoshida J, Hashimoto J, Kawabata N. *The Journal of Organic Chemistry*. 1982;47(18):3575–7.
36. Yoshida J, Ogura K, Kawabata N. *The Journal of Organic Chemistry*. 1984;49(18):3419–21.
37. Yoshida J-i, Sofuku H, Kawabata N. *Bulletin of the Chemical Society of Japan*. 1983;56(4):1243–4.
38. Brunelet T, Gelbard G. *Nouveau Journal de Chimie*. 1983;7:483–90 (CAN 100:52121t, 1984).
39. John K, Pillai VR. *Journal of Polymer Science Part A: Polymer Chemistry*. 1989;27(9):2897–906.
40. Chacko A, Philip C, Mathew B. *The Chemist*. 2018;91(2):63.
41. Cainelli G, Contento M, Manescalchi F, Plessi L. *Synthesis*. 1989;1989(01):47–8.
42. Basavaraju K, Sharma S, Maurya RA, Kim DP. *Angewandte Chemie International Edition*. 2013;52(26).
43. Tamami B, Hatam M, Mohadjer D. *Polymer Bulletin*. 1989;21:531–3.
44. Tamami B, Hatam M, Mohadjer D. *Polymer*. 1991;32(14):2666–70.
45. Firouzabadi H, Tamami B, Goudarzian N, Hatam M, Lakouraj MM. *Synthetic Communications*. 1991;21(20):2077–88.
46. Frechet JM, Darling P, Farrall MJ. *The Journal of Organic Chemistry*. 1981;46(8):1728–30.
47. Kamal A, Devaiah V, Shankaraiah N, Reddy KL. *Synlett*. 2006;2006(16):2609–12.
48. Tamami B, Goudarzian N. *European Polymer Journal*. 1992;28(9):1035–8.
49. Peltonen RT, Ekman KB, Nasman JH. *Industrial & Engineering Chemistry Research*. 1994;33(2):235–8.

50. Tamami B, Zarchi MK. *European Polymer Journal.* 1995;31(8):715–7.
51. Tamami B, Kiasat AR. *Iranian Polymer Journal.* 1997;6:273–80.
52. Mansour Lakouraj M. *Journal of Chemical Research, Synopses.* 1999(3):206–7.
53. Friedrich HB, Singh N. *Tetrahedron Letters.* 2000;41(20):3971–4.
54. Tamami B, Parvanak Boroujeni K, Khakzad MM. *Iranian Polymer Journal.* 2003;12(4):331–8.
55. Tajik H, Shirini F, Hassan-zadeh P, Rafiee Rashtabadi H. *Synthetic Communications.* 2005;35(14):1947–52.
56. Tajik H, Mohammadpoor-Baltork I, Rasht-Abadi HR. *Synthetic Communications.* 2004;34(19):3579–85.
57. Tajik H, Mohammadpoor-Baltork I, Hassan-Zadeh P, Rafiee Rashtabadi H. *Russian Journal of Organic Chemistry.* 2007;43:1282–4.
58. Kalbasi RJ, Kolahdoozan M, Shahabian K, Zamani F. *Catalysis Communications.* 2010;11(14):1109–15.
59. Albadi J, Tajik H, Keshavarz M, Abedini M. *Monatshefte für Chemie-Chemical Monthly.* 2013;144:179–81.
60. Prakash GS, Chacko S, Panja C, Thomas TE, Gurung L, Rasul G, et al. *Advanced Synthesis & Catalysis.* 2009;351(10):1567–74.
61. Prakash GS, Shakhmin A, Glinton KE, Rao S, Mathew T, Olah GA. *Green Chemistry.* 2014;16(7):3616–22.
62. Bauer AM, Ramey EE, Oberle KG, Fata GA, Hutchison CD, Turlington CR. *Tetrahedron Letters.* 2019;60(43):151193.
63. Tamami B, Firouzabadi H, Lakouraj MM, Mahdavian A-R. *Iranian Journal of Polymer Science and Technology.* 1994;3(2):82–7.
64. Firouzabadi H, Sardarian A, Moosavipour H, Afshari G. *Synthesis.* 1986;1986(04):285–8.
65. Tamami B, Goudarzian N. *Polymer Bulletin.* 1990;23:295–8.
66. Holt P, Nasrallah E. *Journal of the Chemical Society B: Physical Organic.* 1968:233–7.
67. Lee DG, Hall DT, Cleland JH. *Canadian Journal of Chemistry.* 1972;50(22):3741–3.
68. Gribble GW. *Journal of Natural Products.* 2000;63(5):735–6.
69. Burke SD. Handbook of reagents for organic synthesis. In: Danheiser R (ed.). *Oxidizing and Reducing Agents*: Wiley, Germany; 200, vol. 2.
70. Gibson HW, Bailey FC. *Journal of the Chemical Society, Chemical Communications.* 1977(22):815a.
71. Goudgaon NM, Wadgaonkar PP, Kabalka GW. *Synthetic Communications.* 1989;19(5–6):805–11.
72. Hutchins RO, Natale NR, Taffer IM. *Journal of the Chemical Society, Chemical Communications.* 1978(24):1088–9.
73. Kabalka GW, Wadgaonkar PP, Chatla N. *Synthetic Communications.* 1990;20(2):293–9.
74. Lakouraj MM, Tajbakhsh M, Mahalli MS. *Monatshefte für Chemie-Chemical Monthly.* 2008;139:117–23.
75. Devaky K, Leelamma P. *Journal of Chemical and Pharmaceutical Research.* 2016;8(4):704–13.
76. Tajbakhsh M, Lakouraj MM, Mahalli MS. *Letters in Organic Chemistry.* 2008;5(2):148–52.
77. Hallensleben ML. *Journal of Polymer Science: Polymer Symposia.* 1974;47(1):1–9.
78. Menger FM, Shinozaki H, Lee H-C. *The Journal of Organic Chemistry.* 1980;45(13):2724–5.
79. Firouzabadi H, Tamami B, Goudarzian N. *Synthetic Communications.* 1991;21(22):2275–85.
80. Tamami B, Goudarzian M. *Iranian Journal of Chemistry and Chemical Engineering.* 1996;15(2):63–71.

81. Tamami B, Goudarzian N. *Journal of the Chemical Society, Chemical Communications.* 1994(9):1079.
82. Tamami B, Mansour Lakoraj M, Yeganeh H. *Iranian Polymer Journal* 1997;6:159–68.
83. Mansour Lakouraj M. *Journal of Chemical Research, Synopses.* 1997(9):330–1.
84. Khaligh NG. *Comptes Rendus Chimie.* 2013;16(8):721–7.
85. Khaligh NG, Ghasem-Abadi PG, Mihankhah T. *Comptes Rendus Chimie.* 2014;17(1):23–9.
86. Cantillo D, Kappe CO. *Reaction Chemistry & Engineering.* 2017;2(1):7–19.
87. Sasson Y. Formation of carbon-halogen bonds (Cl, Br, I). In: Patai S, Rappoprt Z (eds.). Supplement *D2: The Chemistry of Halides, Pseudo-Halides and Azides*: Wiley, New York; 1995.
88. Castellanos A, Fletcher SP. *Chemistry– A European Journal.* 2011;17(21):5766–76.
89. Sequeira CA. *High Temperature Corrosion: Fundamentals and Engineering.* Wiley, New York; 2018.
90. Han B, Zheng Z, Zheng D, Zhang L, Cui P, Shi J, Li C. *Synthetic Communications.* 2019;49(19):2512–20.
91. Bongini A, Cainelli G, Contento M, Manescalchi F. *Journal of the Chemical Society, Chemical Communications.* 1980(24):1278–9.
92. Bongini A, Cainelli G, Contento M, Manescalchi F. *Synthesis.* 1980;1980(02):143–6.
93. Afonso CA, Vieira NM, Motherwell WB. *Synlett.* 2000;2000(03):382–4.
94. Chrétien J-M, Zammattio F, Le Grognec E, Paris M, Cahingt B, Montavon G, Quintard J-P. *The Journal of Organic Chemistry.* 2005;70(7):2870–3.
95. Koshy EP, Zacharias J, Pillai VR. *Reactive and Functional Polymers.* 2006;66(8):845–50.
96. Biswas S, Das A. *ChemNanoMat.* 2021;7(7):748–72.
97. Huo N, Rivkin J, Jia R, Zhao Y, Tenhaeff WE. *Advanced Optical Materials.* 2024;12(10):2302201.
98. Frechet J, Farrall And M, Nuyens L. *Journal of Macromolecular Science—Chemistry.* 1977;11(3):507–14.
99. Habermann J, Ley SV, Scott JS. *Journal of the Chemical Society, Perkin Transactions 1.* 1998(19):3127–30.
100. Habermann J, Ley SV, Smits R. *Journal of the Chemical Society, Perkin Transactions 1.* 1999(17):2421–3.
101. Habermann J, Ley SV, Scicinski JJ, Scott JS, Smits R, Thomas AW. *Journal of the Chemical Society, Perkin Transactions 1.* 1999(17):2425–7.
102. Ploypradith P, Kagan RK, Ruchirawat S. *The Journal of Organic Chemistry.* 2005;70(13):5119–25.
103. Johar Y, Zupan M, Šket B. *Journal of the Chemical Society, Perkin Transactions 1.* 1982:2059–62.
104. Zupan M, Šket B, Johar Y. *Journal of Macromolecular Science—Chemistry.* 1982;17(5):759–69.
105. Šket B, Zupet P, Zupan M. *Journal of the Chemical Society, Perkin Transactions 1.* 1989(12):2279–81.
106. Chiellini E, Callaioli A, Solaro R. *Reactive Polymers, Ion Exchangers, Sorbents.* 1985;3(4):357–68.
107. Sket B, Zupan M. *The Journal of Organic Chemistry.* 1986;51(6):929–31.
108. Sket B, Zupan M. *Synthetic Communications.* 1989;19(13–14):2481–7.
109. Zupan M, Šegatin N. *Synthetic Communications.* 1994;24(18):2617–26.
110. Gregorcic A, Zupan M. *Bulletin of the Chemical Society of Japan.* 1987;60(8):3083–5.
111. Olah GA, Li X-Y. *Synlett.* 1990;1990(05):267–9.
112. Prakash GS, Reddy VP, Li X-Y, Olah GA. *Synlett.* 1990;1990(10):594–6.
113. Zajc B, Zupan M. *Tetrahedron.* 1989;45(24):7869–78.
114. Zajc B, Zupan M. *Tetrahedron.* 1990;46(17):6161–6.

115. Zabicky J, Mhasalkar M, Oren I. *Macromolecules*. 1990;23(16):3755–62.
116. Karimi Zarchi MA, Tarabsaz A. *Journal of Polymer Research*. 2013;20:1–9.
117. Karimi Zarchi MA, Mousavi SZ. *Journal of Polymer Research*. 2014;21:1–9.
118. El-Hamshary H, Selim AI, Salahuddin NA, Mandour HS. *Clays and Clay Minerals*. 2015;63(4):328–36.
119. Fu F, Gurung L, Czaun M, Mathew T, Prakash GS. *Tetrahedron Letters*. 2019;60(38):151020.
120. Majumdar RN, Carlini C, Bertucci C. *Die Makromolekulare Chemie: Macromolecular Chemistry and Physics*. 1982;183(9):2047–57.
121. Husstedt U, Schäfer HJ. *Synthesis*. 1979;1979(12):964–6.
122. Salahuddin N, Akelah A. *Polymers for Advanced Technologies*. 2002;13(5):339–45.
123. Han J, Gao C. *Nano-Micro Letters*. 2010;2:213–26.
124. Parvanak Boroujeni K, Asadi F, Kazemi R, Fadavi A. *Journal of Nanoparticle Research*. 2019;21:1–13.
125. Parvanak Boroujeni K, Tohidiyan Z, Fadavi A, Eskandari MM, Shahsanaei HA. *ChemistrySelect*. 2019;4(26):7734–44.
126. Zarchi MAK, Zarei A. *Journal of the Chinese Chemical Society*. 2005;52(2):309–11.
127. Zarchi MK, Noei J. *Journal of Applied Polymer Science*. 2007;104(2):1064–7.
128. Karimi Zarchi MA, Noei J. *Journal of Applied Polymer Science*. 2009;114(4):2134–8.
129. Karimi Zarchi MA, Karimi M. *Journal of Applied Polymer Science*. 2011;120(1):538–43.
130. Karimi Zarchi MA, Karimi M. *Journal of Applied Polymer Science*. 2012;123(5):2762–7.
131. Karimi Zarchi MA, Ebrahimi N. *Journal of Applied Polymer Science*. 2011;121(5):2621–5.
132. Karimi Zarchi MA, Rahmani F. *Journal of Applied Polymer Science*. 2011;121(1):582–8.
133. Karimi Zarchi MA, Ebrahimi N. *Iranian Polymer Journal*. 2012;21:591–9.
134. Zarchi MAK, Mirjalili BF, Ebrahimi N. *Bulletin-Korean Chemical Society*. 2008;29(5):1079.
135. Karimi Zarchi MA, Mirjalili BF, Aval AK. *Journal of Applied Polymer Science*. 2010;115(1):237–41.
136. Karimi Zarchi MA, Ebrahimi N. *Journal of Applied Polymer Science*. 2012;125(3):2163–9.
137. Dueymes C, Schönberger A, Adamo I, Navarro A-E, Meyer A, Lange M, et al. *Organic Letters*. 2005;7(16):3485–8.
138. Karimi Zarchi M, Mirjalili B, Kahrizsangi ZS, Tayefi M. *Journal of the Iranian Chemical Society*. 2010;7:455–60.
139. Karimi Zarchi MA, Ebrahimi N. *Journal of Polymer Research*. 2012;19:1–7.
140. Karimi Zarchi MA, Nejabat M. *Journal of the Iranian Chemical Society*. 2012;9:767–74.
141. Karimi Zarchi MA, Escandari Z. *Journal of Applied Polymer Science*. 2011;121(4):1916–20.
142. Karimi Zarchi MA, Nabaei R. *Journal of Applied Polymer Science*. 2012;124(3):2362–9.
143. Karimi Zarchi MA, Nabaei R, Barani S. *Journal of Applied Polymer Science*. 2012;123(2):788–95.
144. Ali M, Zarchi K, Tarabsaz A. *Chinese Journal of Polymer Science*. 2013;31:1660–9.
145. Karimi Zarchi MA, Nazem F. *Journal of the Iranian Chemical Society*. 2014;11:91–9.
146. Zarchi MAK, Barani S. *Chinese Journal of Polymer Science*. 2013;31:1002–10.
147. Zarchi MK, Banihashemi R. *Journal of Sulfur Chemistry*. 2014;35(4):458–69.
148. Ghorbani-Choghamarani A, Hajjami M, Norouzi M, Amiri A. *Bulgarian Chemical Communications*. 2014;46:384–8.
149. Ghorbani-Choghamarani A, Sardari S. *Journal of Sulfur Chemistry*. 2011;32(1):63–9.
150. Ghorbani-Choghamarani A, Sardari S. *Chinese Journal of Catalysis*. 2010;31(11–12):1347–50.

151. Sarreshtehdar Aslaheh H, Payamifar S, Poursattar Marjani A. *Applied Organometallic Chemistry*. 2024;38(12):e7692.
152. Marinescu LG, Pedersen CM, Bols M. *Tetrahedron*. 2005;61(1):123–7.
153. Lu Y, Taylor RT. *Tetrahedron Letters*. 2003;44(52):9267–9.

Index

O

P

Q

R

S

T

For Product Safety Concerns and Information please contact our
EU representative GPSR@taylorandfrancis.com Taylor & Francis
Verlag GmbH, Kaufingerstraße 24, 80331 München, Germany